The Power of
An Introduction to Autodesk Inventor 2008

Elise Moss

ISBN: 978-1-58503-372-0

SDC
PUBLICATIONS

Schroff Development Corporation

www.schroff.com
www.schroff-europe.com

Schroff Development Corporation
P.O. Box 1334
Mission, KS 66222
(913) 262-2664
www.schroff.com

Trademarks
The following are registered trademarks of Autodesk, Inc.: AutoCAD, AutoCAD Mechanical Desktop, Inventor, Autodesk, AutoLISP, AutoCAD Design Center, Autodesk Device Interface, and HEIDI
Microsoft, Windows, Word, and Excel are either registered trademarks or trademarks of Microsoft Corporation.
All other trademarks are trademarks of their respective holders.

Copyright 2007 by Elise Moss
All rights reserved. No part of this book may be reproduced, stored in a retrieval system, or transcribed in any form or by any means – electronic, mechanical, photocopying, recording, or otherwise – without the prior written permission of Schroff Development Corporation.

Moss, Elise
    The Power of Design: An Introduction to Autodesk Inventor 2008
  Elise Moss
    ISBN: 978-1-58503-372-0

The author and publisher of this book have used their best efforts in preparing this book. These efforts include the development, research, and testing of material presented. The author and publisher shall not be held liable in any event for incidental or consequential damages with, or arising out of, the furnishing, performance, or use of the material herein.

Examination Copies:
Books received as examination copies are for review purposes only and may not be made available for student use. Resale of examination copies is prohibited.

Electronic Files:
Any electronic files associated with this book are licensed to the original user only. These files may not be transferred to any other party.

Printed and bound in the United States of America.

Preface

For users familiar with Inventor R10 or earlier releases, the 2008 release concentrates on refining and improving the software. No textbook can cover all the features of any software application. This book is meant for beginning users who want to gain a familiarity with the tools and interface of Inventor before they start exploring on their own. By the end of the text, the user should feel comfortable enough to create a standard model, and assemblies.

> The files used in this text are accessible from the Internet at
>
> **www.schroff.com/resources**
>
> They are free and available to students and teachers alike.

We value customer input. Please contact us with any comments, questions, or concerns about this text.

Elise Moss
elise_moss@mossdesigns.com

Acknowledgements from Elise Moss

This book would not have been possible without the support of some key Autodesk employees. The Inventor Development team has always been extremely supportive to my efforts. Thanks to Buzz Kross, Bill Bogan, Paul Clemens and Lance Grow.

Additional thanks to Inventor users who share my love for designing by computer. Their questions continue to challenge and astound me.

The effort and support of the editorial and production staff of Schroff Development Corporation is gratefully acknowledged. I especially thank **Stephen Schroff** and **Mary Schmidt** for helpful suggestions regarding the format of this text.

Finally, truly infinite thanks to Ari for his encouragement and his faith.

Elise Moss

Table of Contents

PREFACE

ACKNOWLEDGEMENTS

Lesson 1:
Parametric Modeling Fundamentals – Quick Start

Exercise 1-1: Creating an Extrusion	1-2
Review Questions	1-15

Lesson 2:
User Interface

Exercise 2-1: Controlling the Visibility of the Status Bar, Panel Bar, and Browser	2-2
Exercise 2-2: Using Toolbars instead of the Panel Bar	2-4
Exercise 2-3: Setting Up a Project	2-14
Exercise 2-4: Modifying a Project Definition	2-16
Exercise 2-5: Customizing your Toolbar	2-38
Review Questions	2-40

Lesson 3
2D Sketch Tools

Exercise 3-1: Mirror	3-4
Exercise 3-2: Auto Dimension	3-16
Exercise 3-3: Move and Copy	3-18
Exercise 3-4: Copy	3-19
Exercise 3-5: Scale	3-20

Exercise 3-6: Rotate	3-22
Exercise 3-7: Stretch	3-24
Exercise 3-8: Adding Horizontal and Vertical Constraints	3-27
Exercise 3-9: Deleting Constraints	3-30
Exercise 3-10: Inserting an AutoCAD file	3-33
Exercise 3–11: Creating Etched Text	3-36
Exercise 3-12: Insert Image	3-40
Exercise 3-13: Edit Coordinate System	3-42
Exercise 3-14: Import Points	3-45
Review Questions	3-46

Lesson 4
The Features Toolbar

Exercise 4-1: Performing a Revolve	4-3
Exercise 4-2: Adding a Hole using Linear	4-5
Exercise 4-3: Adding a Hole using Concentric	4-8
Exercise 4-4: Adding a Hole using Sketch	4-11
Exercise 4-5: Customizing the Thread Data	4-15
Exercise 4-6: Creating a Shell	4-16
Exercise 4-7: Creating a Rib	4-20
Exercise 4-8: Loft	4-23

Exercise 4-9: Loft with Rails	4-28
Exercise 4-10: Sweep	4-34
Exercise 4-11: Creating a Coil	4-38
Exercise 4-12: Adding a Thread	4-41
Exercise 4-13: Adding a Fillet	4-42
Exercise 4-14: Adding a Chamfer	4-44
Exercise 4-15: Applying a Face Draft	4-45
Exercise 4-16: Splitting a Part	4-46
Exercise 4-17: Emboss	4-48
Exercise 4-18: Decal	4-51
Exercise 4-19: Circular Pattern	4-53
Exercise 4-20: Bend Part	4-55
Exercise 4-21: Mirror Feature	4-57
Exercise 4-22: Sculpt	4-58
Review Questions	4-60

Lesson 5
Drawing Management

Exercise 5-1: Creating a Base View	5-2
Exercise 5-2: Create a Projected View	5-5
Exercise 5-3: Adding an Auxiliary View	5-8

Exercise 5-4: Adding a Section View	5-12
Exercise 5-5: Editing a Section View	5-15
Exercise 5-6: Creating a Detail View	5-17
Exercise 5-7: Creating a Broken View	5-19
Exercise 5-8: Creating an Overlay View	5-22
Exercise 5-9: Creating a Slice View	5-26
Exercise 5-10: Creating a Custom Title Block	5-28
Exercise 5-11: Adding Custom Properties	5-37
Exercise 5-12: Copy a Titleblock	5-40
Exercise 5-13: Define New Symbol	5-42
Exercise 5-14: Inserting a Symbol	5-44
Exercise 5-15: Creating a Symbol with Attributes	5-46
Exercise 5-16: Editing Symbols	5-47
Review Questions	5-50

Lesson 6
Drawing Annotation Toolbar

Exercise 6-1: Using the Style Manager Library	6-2
Exercise 6-2: Applying a General Dimension	6-10
Exercise 6-3: Applying a Baseline Dimension	6-12
Exercise 6-4: Applying Ordinate Dimensions	6-15

Exercise 6-5: Adding a Hole Note	6-17
Exercise 6-6: Adding Center Marks	6-19
Exercise 6-7: Adding Center Lines	6-20
Exercise 6-8: Adding a Centered Pattern	6-22
Exercise 6-9: Adding Balloons	6-24
Exercise 6-10: Adding a Parts List	6-28
Exercise 6-11: Adding a Hole Chart	6-35
Exercise 6-12 Adding Revision Tags and Tables	6-37
Review Questions	6-40

Lesson 7
Assembly Tools

Exercise 7-1: Place Component	7-4
Exercise 7-2: Place Component using Windows Explorer	7-6
Exercise 7-3: Create Component	7-8
Exercise 7-6: Place INSERT Constraint	7-14
Exercise 7-7: Place TANGENT Constraint	7-17
Exercise 7-8: Place MATE Constraint	7-18
Review Questions	7-20

Lesson 8
Presentations

    Exercise 8-1   8-2
    Creating an Exploded View

    Exercise 8-2:   8-7
    Creating an Animation

    Exercise 8-3:   8-14
    Recording an Animation

    Exercise 8-4:   8-15
    Changing Views in an Animation

    Review Questions   8-19

About the Author

Lesson 1
Parametric Modeling Fundamentals – Quick Start

Objectives

In Inventor, the parametric part modeling process involves the following steps:

1. Create a rough two-dimensional sketch
2. Apply geometric constraints as needed
3. Apply dimensions
4. Extrude/revolve/sweep
5. Add additional features, such as holes, fillets, chamfers, and shells
6. Analyze and refine the model
7. Create the drawing layout

This lesson will cover the process of going from sketch to feature to create a part.

The Power of Design: An Introduction to Autodesk Inventor

Exercise 1-1
Creating an Extrusion

Drawing Name: **(none, start from scratch)**
Estimated Time: 5 Minutes

When you have completed this lesson, you can:

- Create simple parametric models
- Understand the Basic Parametric Modeling Process
- Create Rough Sketches
- Understand the "Shape before Size" Approach
- Use the view commands
- Create and modify dimensions

1. Start Inventor by double-clicking on the Inventor icon on the desktop.

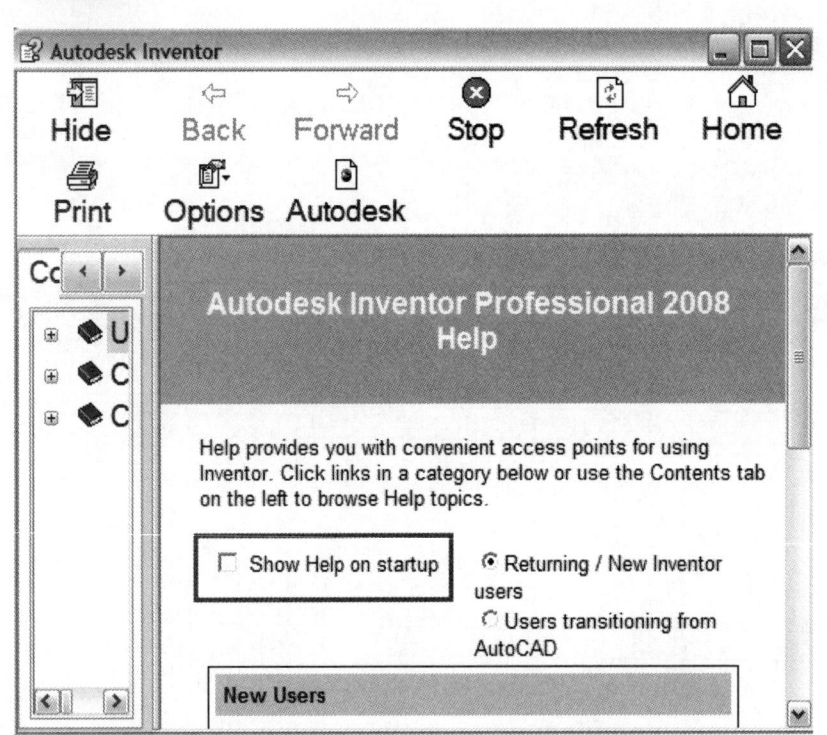

The Help will come up. This is the program default. Uncheck the Show Help on startup button. This will suppress the Help unless you need it.

2. Go to **File→New**.

1-2

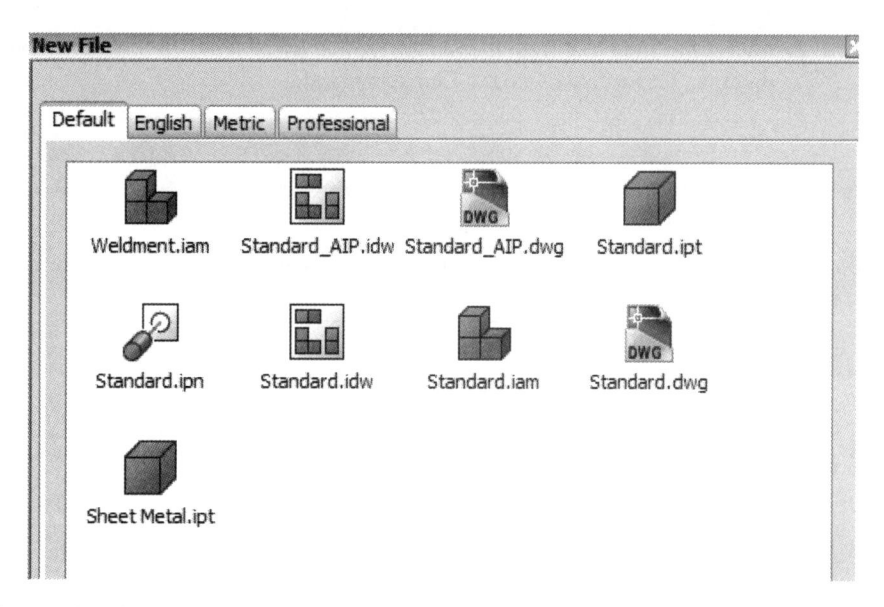

Parametric Modeling Fundamentals – Quick Start

The files displayed and used in the Start-Up dialog are templates. These templates contain default drafting standards. You can create your own templates using materials, dimension styles, and other set-up defaults.

Inventor uses different file extensions specific to Inventor. Refer to the table below to assist you in determining the type of file being created and accessed.

File Extension	File Type
*.ipt	Part File\Sheet Metal Part\Catalog (3D model)
*.iam	Assembly File (3D model)
*.idv	Design View File
*.ipj	Project File
*.idw	Drawing Layout (2D paper space)
*.ipn	Presentation (Scene\Rendering\Exploded Views)
*.ide	Design Element
*.dwg	AutoCAD drawing

3. Select the **Standard.ipt** icon and press **OK**.

We will start with a rough sketch. Keep the following guidelines in mind when creating a rough sketch.

- Create a sketch that is proportional to the desired shape.

- Keep sketches simple. Leave out fillets, rounds, and chamfers…those can be added later.

- Exaggerate the geometric features of the desired shape. Remember your sketch is parametric; you can adjust the size when we start adding dimensions. It's easier to go from big to small than vice versa.

- Draw the geometry so it does not overlap. Inventor looks for a closed polygonal shape. Overlapping lines can create confusion and errors.

The Power of Design: An Introduction to Autodesk Inventor

4. Line [L] Select the **Line** tool from the Sketch toolbar by clicking on it with the left-mouse button. This will activate the Line command.

TIP: You can also type 'L' at the keyboard to initiate the Line command. This similar to the command alias 'L' used in AutoCAD.

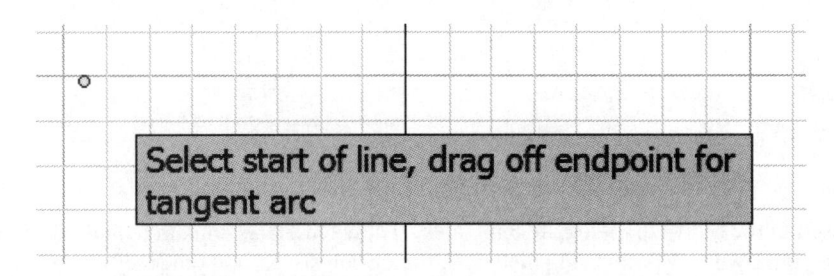

Select start of line, drag off endpoint for tangent arc

A prompt appears next to the cursor. Use the prompt as a guide. Inventor expects us to identify the starting location of a line. To switch to arc mode, merely hold down the left-mouse button and move the mouse to form an arc.

5. Move the mouse near the lower left corner of the Drawing Screen and create a freehand sketch as shown to the right. Create the sketch by starting at Point 1 and ending at Point 7. Do not be concerned with the actual size of the sketch. Do not worry about keeping the lines perfectly horizontal or vertical as Inventor automatically keeps lines orthogonal.

    *Note: There is no **CLOSE** command like there is in AutoCAD.*

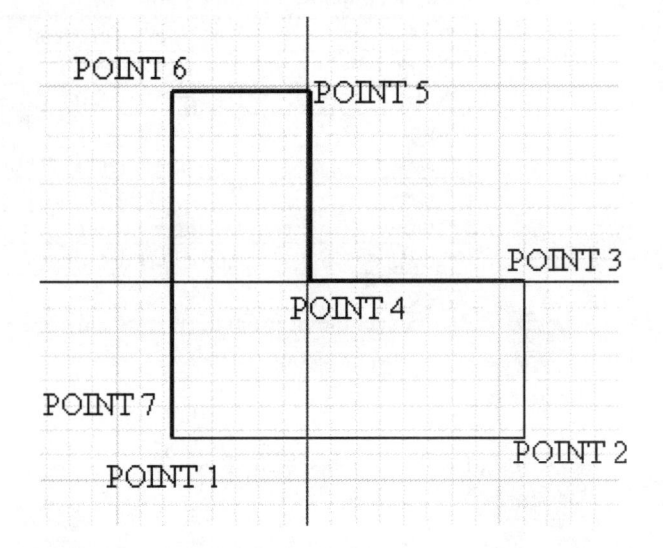

1-4

Parametric Modeling Fundamentals – Quick Start

6. 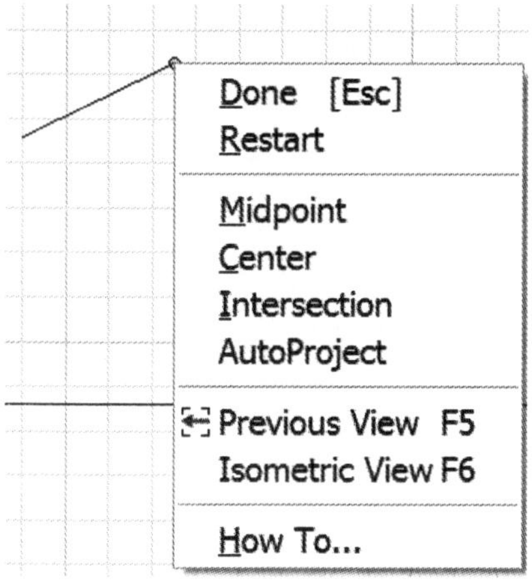 While in the **Line** command, we can right-click the mouse to get the pop-up shown.

   Done [Esc]
   Restart
   Midpoint
   Center
   Intersection
   AutoProject
   Previous View F5
   Isometric View F6
   How To...

   When we are done with the sketch, we right-click the mouse and select **Done**.

   Notice that we have the option to select Midpoint, Center, Intersection, or AutoProject to help us in our sketching.

7.  General Dimension — Select the **Dimension** tool by left-picking the Dimension icon with the mouse.

8. Select the line to dimension. Finally, move the mouse away from the line being dimensioned and select the dimension location.

   To modify the dimension, left pick on the dimension and an edit box appears.

9. Place your mouse inside the text box and enter the desired dimension.

   Select the green check mark when done editing.

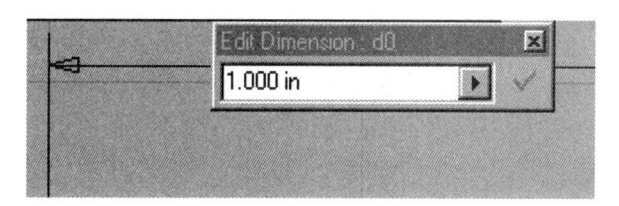

   Edit Dimension : d0
   1.000 in

1-5

10. Repeat using the Dimension tool applying the dimensions shown below.

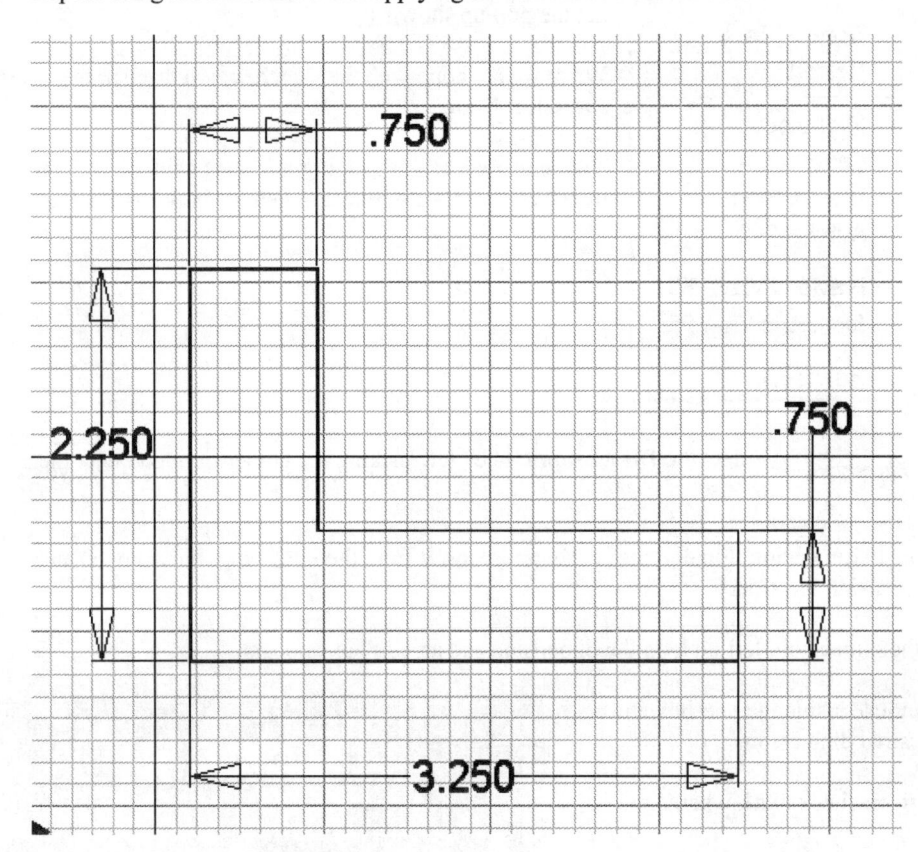

TIP: Typing the letter D on the keyboard will also initiate the DIMENSION command.

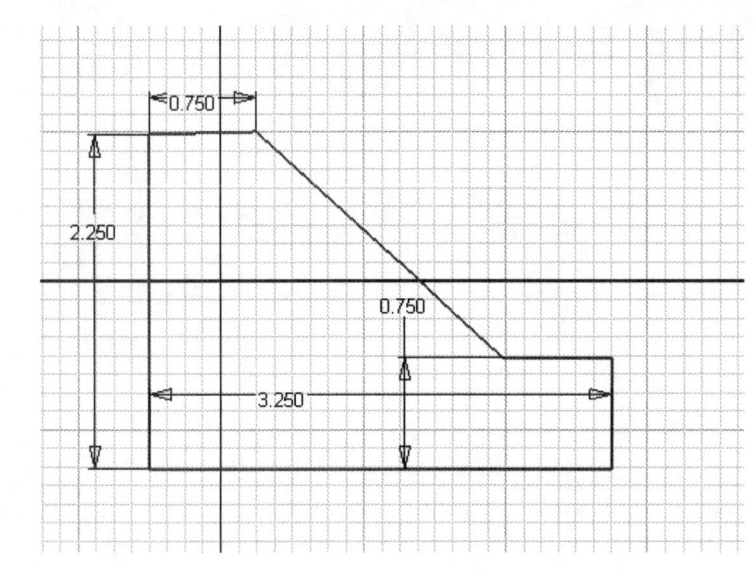

What if your sketch turns out like this?

Time to use geometric constraints.

Add a vertical constraint to the slanted line.

Add a horizontal constraint to the upper line.

If your sketch is different, apply the constraints as needed.

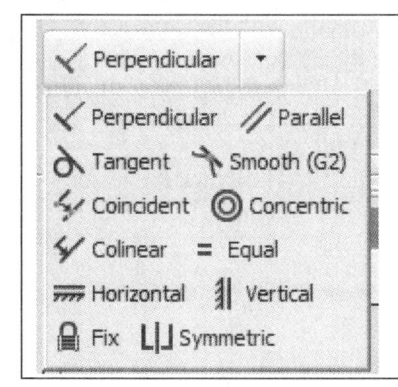

TIP: If you do not wish constraints to automatically be added with your geometry, hold down the CONTROL [Ctrl] key as you sketch.

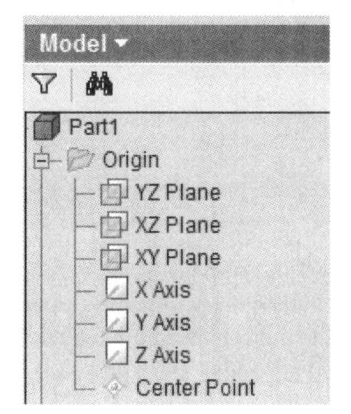

In the Browser, we see the base planes, axes, and a center point.

These can be used to help constrain a sketch in 3D space. Inventor does not automatically insert reference planes into your sketches. This allows the user to decide what reference geometry to use to constrain the sketch and keeps the file sizes smaller.

You need to copy the desired reference geometry into your current sketch in order to use it for a constraint.

11. To copy reference geometry into the current sketch, use **Project Geometry**.

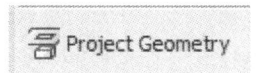

Select the **Project Geometry** tool.
Pick the **Center Point** in the Browser.
Right-click and select **Done**.

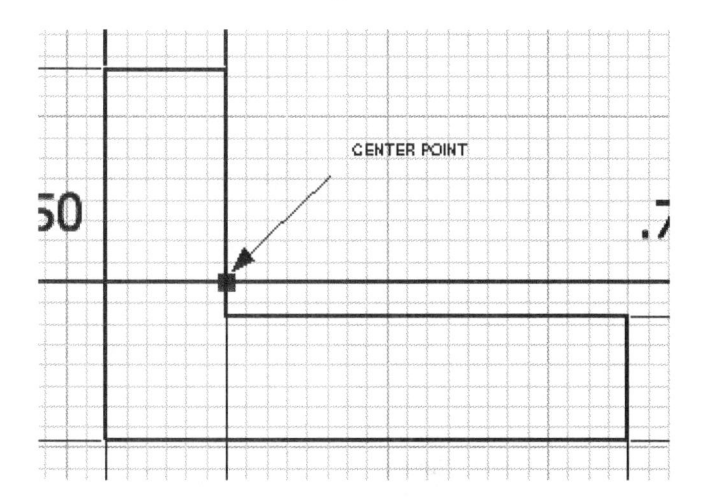

We see the center point copied in our sketch.

12. Select the **Coincident** constraint to shift and constrain your sketch. The Coincident constraint is located in the drop-down underneath the Perpendicular constraint.

1-7

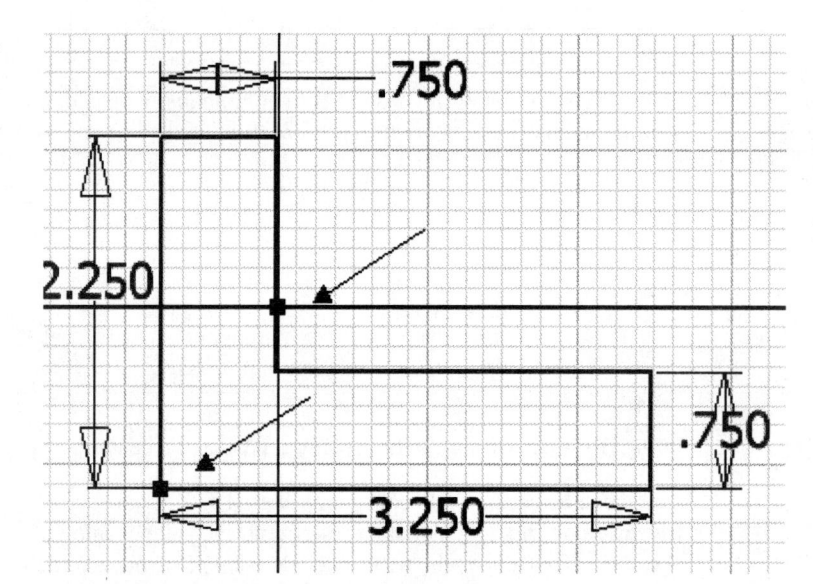

The Power of Design: An Introduction to Autodesk Inventor

13. With the Coincident constraint tool active, pick the lower left corner of the sketch.

   Then, pick the center point.

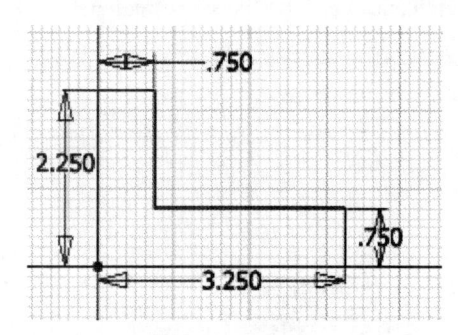

   Your sketch will automatically shift position.

   It will also change color to indicate it is now fully defined.

14. Right-click and select **Done** to indicate you are done adding Coincident constraints.

15. Right-click the mouse and Select **Finish Sketch** to indicate we are done defining our geometry.

TIP: Typing **S** on the keyboard will also initiate the **Finish Sketch** command.

1-8

We can see that we are no longer in sketch mode because our grid is no longer visible and the Browser updates so that the Sketch is no longer highlighted.

Completing the Base Solid Feature

Now that the 2D sketch is completed, we will proceed to the next step: create a 3D part from the 2D profile. Extruding a 2D sketch is the most common method used to create 3D parts. We can extrude planar faces along a path. We can also specify a height value, direction and a tapered angle.

16. Before we extrude, let's switch to an Isometric View. That will allow us to see what is going to happen more easily.

    To switch to an Isometric View, right-click the mouse anywhere in the Drawing Screen area and select **Isometric View** from the pop-up menu. We select by highlighting **Isometric View** and picking with the left mouse button.

TIP: You can also shift to an isometric view by hitting the **F6** key.

17. Once we selected **Finish Sketch**, Inventor automatically switched to Part Features mode.

    The Panel bar changes to our Features tools.

1-9

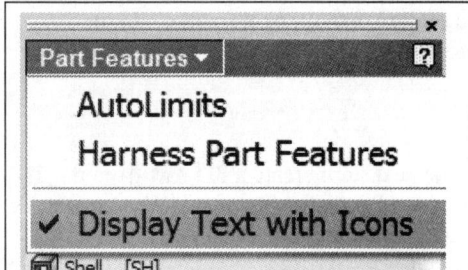

The Power of Design: An Introduction to Autodesk Inventor

> **TIP:** To hide the text next to the tool icons, disable the Display Text with Icons mode. Place your mouse over the panel title, left-click on the down arrow and select **Display Text with Icons**.
>
> *NOTE: The menu items available are different if you are using Inventor Standard vs. Inventor Professional.*

18.   Extrude [E]   We select the **Extrude** icon located in the Features toolbar. The Extrude icon is selected by picking it with the left mouse button.

Note that when we move the mouse over the Extrude icon, a message appears in the lower left corned of the display to indicate what the Extrude icon does.

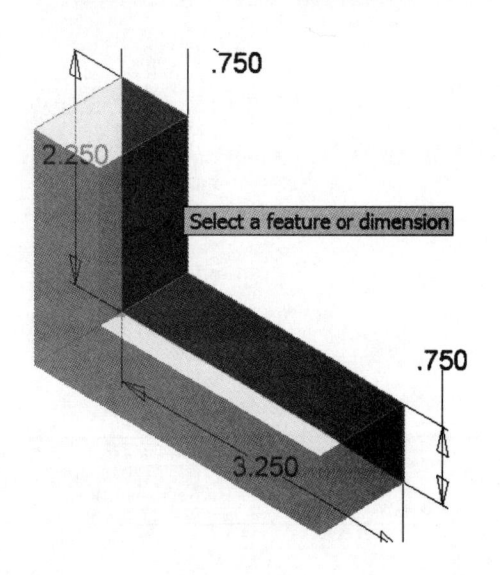

The sketch profile is automatically selected and a preview of the extrusion appears.

> **TIP:** The E next to the Extrude icon means that typing the letter **E** on the keyboard will also initiate the Extrude command.

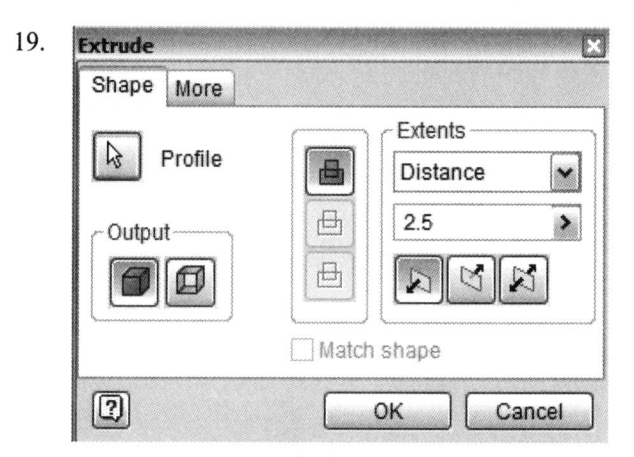

Parametric Modeling Fundamentals – Quick Start

19.   Select **Distance** under Extents and enter **2.5** as the value.

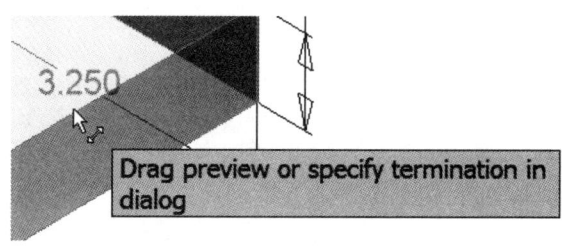

Note that we can modify the distance by using the mouse in the Drawing Screen area.

20. To do this, simply place the mouse near one of the outer corners and hold down the left mouse button. Now drag the mouse back and forth. The value in the Extrude dialog box will automatically update.

Using your mouse, set your distance to **3.25** units. Then click on the **OK** button.

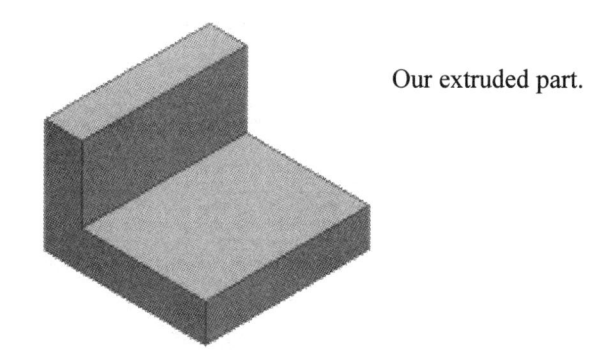

Our extruded part.

21.   Click on the **Zoom Realtime** icon in the Standard toolbar area.
Inside the Drawing Screen area, hold down the left mouse button and move the mouse upward.

Next, move the mouse downward. Note how the size of the object changes. We are not actually scaling or modifying the part size. We are changing our perspective view. Imagine holding a part at arms-length and then moving the part closer to our face. This is what Zoom Realtime does.

Right-click the mouse.

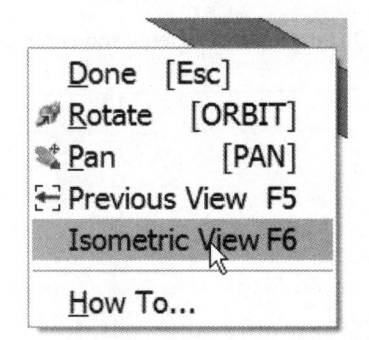

The Power of Design: An Introduction to Autodesk Inventor

22.   A pop-up menu appears that provides several View Options. Select the **Pan View** option.

```
Done        [Esc]
Rotate      [ORBIT]
Pan         [PAN]
Previous View  F5
Isometric View F6
How To...
```

Our mouse icon changes from an arrow to a small hand. Imagine a piece of paper on our desk and using our hand to move the paper around the tabletop. This is what 'pan' does. The size of our object does not change, but we can shift its position inside our Drawing Screen area.

23. Right-click the mouse again and select **Rotate**. The 'Rotate View' displays an arcball, which is a circle divided into four quadrants. This enables us to manipulate the view of 3D objects by clicking and dragging with the left mouse button.

Inside the arcball, press down the left mouse button and drag it up and down to rotate about the X-axis.

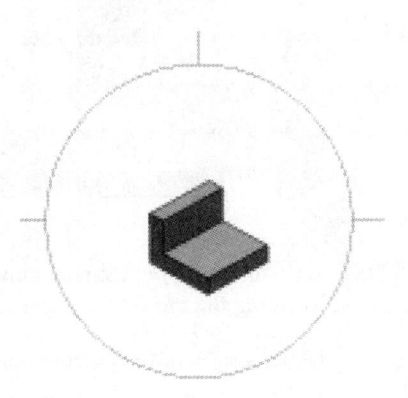

24. Move the cursor to different locations on the screen, such as outside the arcball or on one of the four small tickmarks, and experiment with the real-time dynamic rotation feature.

NOTE: The graphics card installed on the computer system as well as the amount of memory installed will affect how well the dynamic rotation feature works.

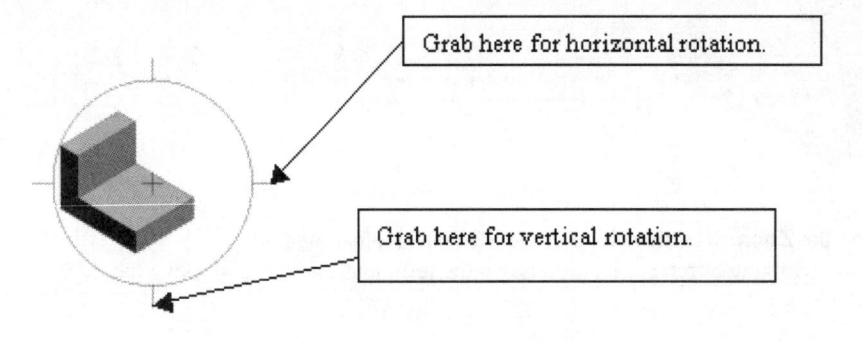

Grab here for horizontal rotation.

Grab here for vertical rotation.

TIP: To rotate the model in the horizontal direction, grab one of the horizontal bars. To rotate the model in the vertical direction, grab one of the vertical bars.

Use the 3D Rotate button on the Standard toolbar to:

- Rotate a part or assembly in the graphics window
- Display standard isometric and orthographic views of a part or assembly.

Rotation can be around the center mark, free in all directions, or around the X or Y axes in the 3D Rotate symbol view. You can rotate the view while other tools are active.

TIP: The 3D Rotate tool remembers the last mode used when you exit the command. When the command is active, press the spacebar to switch modes.

25.

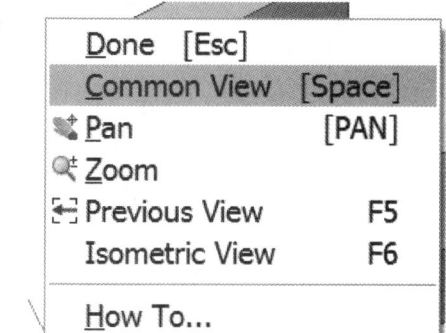

    | Done | [Esc] |
    | Common View | [Space] |
    | Pan | [PAN] |
    | Zoom | |
    | Previous View | F5 |
    | Isometric View | F6 |
    | How To... | |

    To switch to Common View mode, right-click and select **Common View** from the menu, or press the spacebar.

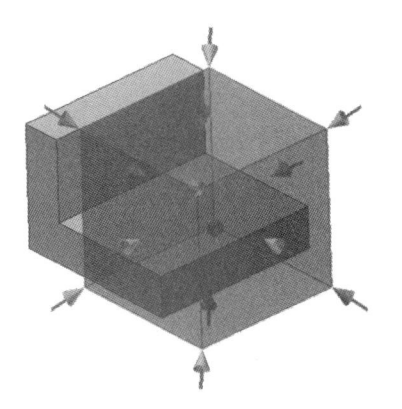

26. Common View mode allows the user to quickly switch view orientations. Simply pick on the arrow to select the desired view.

    You can also select an edge to rotate the view when the model is positioned to show a face.

27. Right click and select **Done** to exit the Orbit/Common View mode.

The Power of Design: An Introduction to Autodesk Inventor

Inventor comes with three modes for model appearance: Wire frame, Hidden Edge, and Shaded. To change the mode, use the left mouse button to select the desired mode.

Shaded mode uses the least amount of your system's resources.

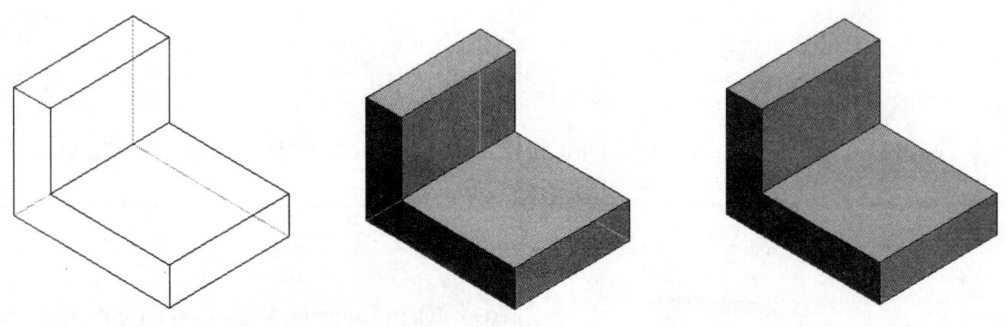

Our model in all three modes: Wire frame, Hidden Edge, and Shaded.

28. Save the file as *ex1-1.ipt*.

You may get an error message that the file is being saved in a folder not in your active project. Press **OK**. We will be covering Projects in the next lesson.

Review Questions

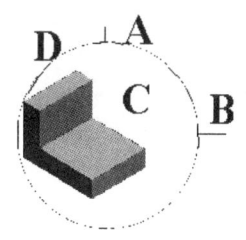

1. Select the area to rotate the model in a horizontal direction.

2. The shortcut key to place a General Dimension is:

    A. I
    B. G
    C. D
    D. GD

3. To edit or modify a dimension:

    A. Left pick on a dimension and type in the edit box.
    B. Right pick on a dimension and select Edit from the menu.
    C. Select the Edit Dimension tool.
    D. Select Edit Dimension from the Menu.

4. To quickly switch to an Isometric View:

    A. Right click anywhere in the drawing window and select 'Isometric View' from the menu.
    B. Select the Isometric View tool from the View toolbar.
    C. Select the 3D Orbit tool, right click and select 'Isometric View.'
    D. Select Isometric View from the View menu.

5. To draw a line, you should be in this mode:

    A. Sketch
    B. Features
    C. Solids
    D. Assembly

6. The three modes of appearance for a model are:

    A. Wire Frame, Hidden, Shaded with Edges On
    B. Wire Frame, Shaded, Colored
    C. Wire Frame, Hidden, Rendered
    D. Wire Frame, Hidden, Shaded

7. To prevent constraints from automatically being added to a sketch:

    A. Turn off constraint mode under Options.
    B. Hold down the CONTROL key while sketching.
    C. Hold down the ALT key while sketching.
    D. You can not prevent constraints from being added.

8. The file extension for an Inventor part file is:

    A. ipt
    B. dwg
    C. iam
    D. ipn

9. The files displayed in the start up dialog are:

    A. Dummy files
    B. Blank files
    C. Templates with default settings
    D. Transparent files – they don't exist, they are just icons used for starting a new file

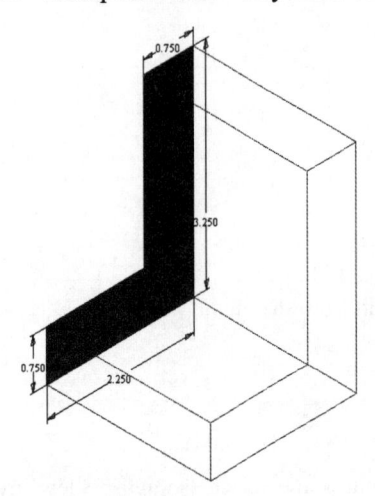

10. To change the extrusion distance:

    A. Modify the value in the distance edit box of the Extrude dialog box
    B. Use the mouse to drag the extrusion into position
    C. Right click on the extrusion and enter a value.
    D. A & B but not C.

ANSWERS: 1) B; 2) C; 3) A; 4) A; 5) A; 6) D; 7) B; 8) A; 9) C; 10) D

Lesson 2
User Interface

Toolbars and Menus

Autodesk Inventor adheres to Microsoft Windows standards and features user interface elements common to Windows-based applications.

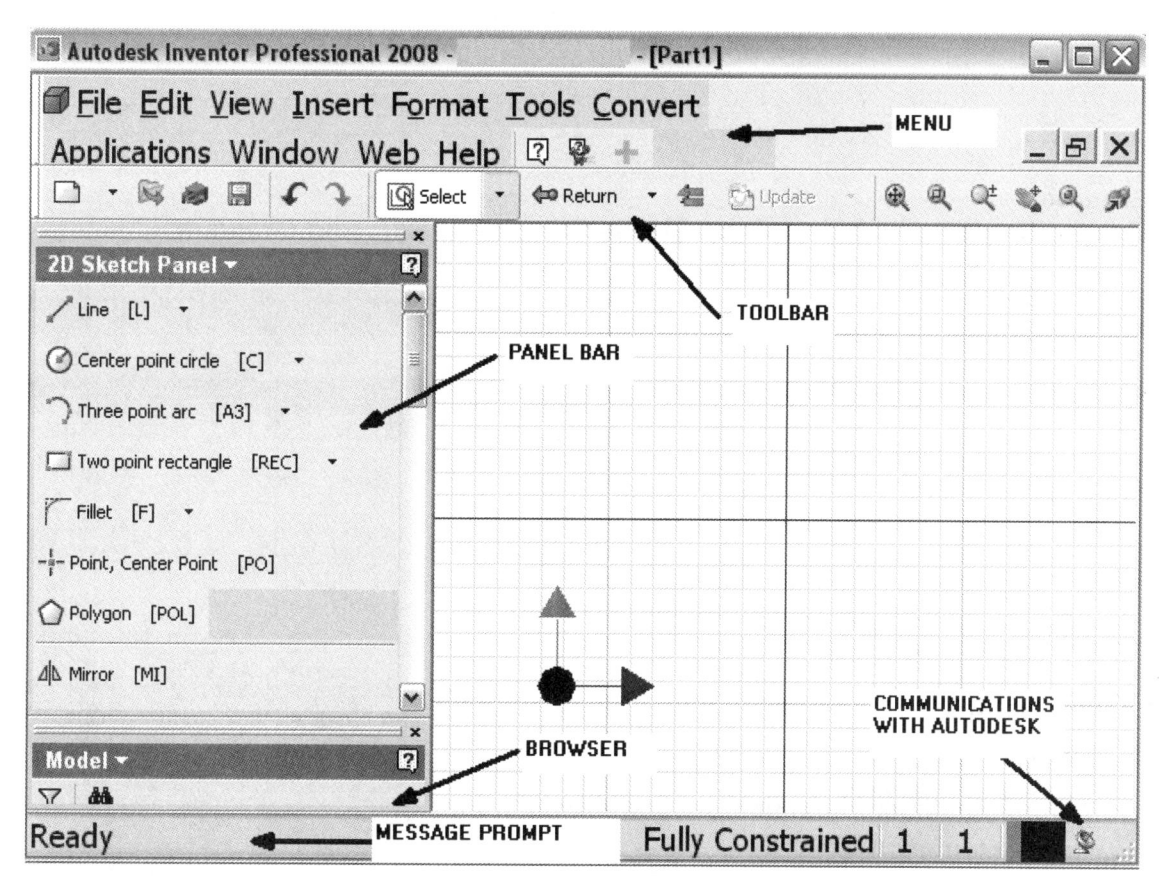

Status bar

2-1

The Power of Design: An Introduction to Autodesk Inventor

The user interface within Inventor is completely customizable. Experiment by pulling and dragging the toolbars and placing them in different areas of the screen. We can also customize toolbars, write macros, and use a Visual Basic Editor to create our own custom programs.

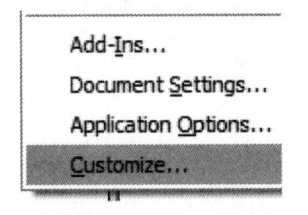

You can access the various toolbars through the menu using **Tools→Customize**.

You can also access the Customize dialog by right-clicking on the gray area of any toolbar.

There are six standard toolbars: Inventor Standard, Menu Bar, Part Features, 2D Sketch Panel, Inventor Collaboration, and Part Standard.

Exercise 2-1:
Controlling the Visibility of the Status bar, Panel Bar, and Browser

Drawing Name: New
Estimated Time: 10 minutes

When you have completed this lesson, you will:

- Be able to turn on/off the Status Bar
- Be able to turn on/off the Panel Bar
- Be able to turn on/off the Browser

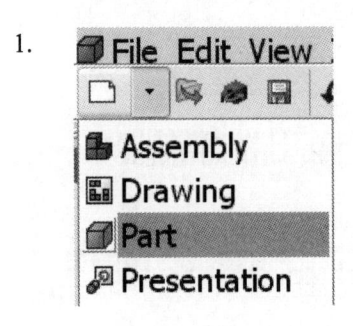

1. Select the down arrow next to the **New** tool.

   Select **Part** from the drop-down list.

   This will automatically open a new part file using the default template.

   The default template to be used is based on the default units selected when you installed Inventor.

2. Go to **View→Status Bar**.
   Left-click on the check mark to turn off the visibility of the Status Bar.

   This turns off the bar at the bottom of your screen that shows helpful messages.

3. Go to **View→Toolbar→Panel Bar**.

   Uncheck the Panel Bar to turn off the visibility of the Panel Bar.

2-2

User Interface

4. Right-click on the grey area of the Standard toolbar.

   You see the same list as under the View menu.

   Left-click to restore the Panel Bar visibility.

   - Panel Bar
   - ✓ Browser Bar
   - ✓ Inventor Standard
   - 2D Sketch Panel

5. Go to **View→Toolbar→Browser Bar**.

   Uncheck the Browser Bar to turn off the visibility.

6. Restore the visibility of the Browser Bar using the right-click option on the toolbar.

7. Close the drawing without saving.

The Power of Design: An Introduction to Autodesk Inventor

Exercise 2-2:
Using Toolbars instead of the Panel Bar

Drawing Name: New
Estimated Time: 10 minutes

When you have completed this lesson, you will:

- Be able to turn on/off the Panel Bar
- Be able to turn on/off desired toolbars

Once you get comfortable with Inventor, you may want to turn off the Panel Bar and use toolbars instead. This will give you more screen space, which means a greater work area.

1. Select the down arrow next to the **New** tool.

   Select **Part** from the drop-down list.

   This will automatically open a new part file using the default template.

   The default template to be used is based on the default units selected when you installed Inventor.

2. Right-click and select **Finish Sketch** to exit the sketch.

3. Go to **View→Toolbar→Panel Bar**.

   Uncheck the Panel Bar to turn off the visibility of the Panel Bar.

4. Right-click on the grey area of the Standard toolbar.

   Select **Customize**.

5. Select the **Toolbars** tab.

2-4

User Interface

Highlight **2D Sketch Panel** and press **Show**.

Highlight **Part Features** and press **Show**.

Close the dialog.

6.

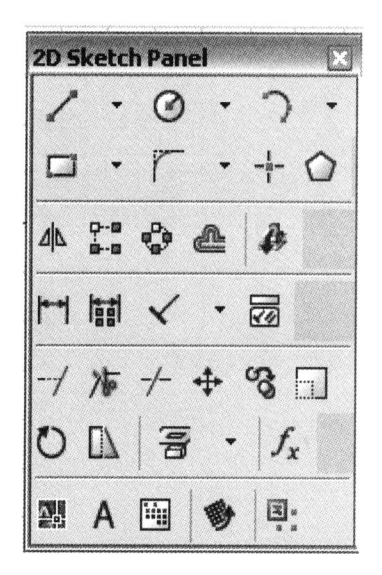

The toolbars appear on-screen. Like AutoCAD, you can have floating toolbars or docked toolbars. You can change the shape of the toolbars by using stretch grips just like in AutoCAD.

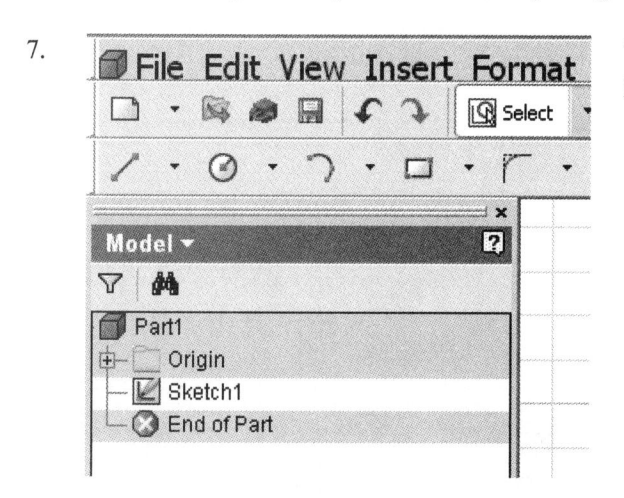

7. Try docking the toolbars by dragging them into a position above and along the Browser Bar.

Note how the two toolbars take up less space than the Panel Bar. The Panel Bar has more information.

2-5

The Power of Design: An Introduction to Autodesk Inventor

TIP: Use the Panel Bar while you are learning. Once you gain familiarity with the various icons, disable the Panel Bar and enable the toolbars to conserve space on your desktop.

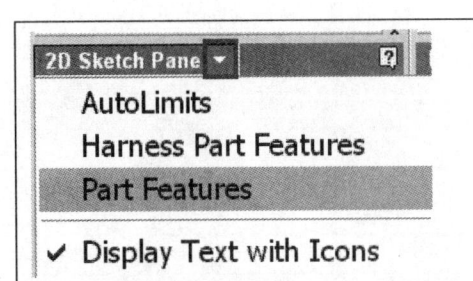

TIP: A left mouse pick on the arrow at the top of the panel bar allows the user quickly to switch between Panel Bars.

The iProperties Dialog Box

Selecting the iProperties item from the File menu allows the user to set up file statistics as well as material properties for sheet metal parts.

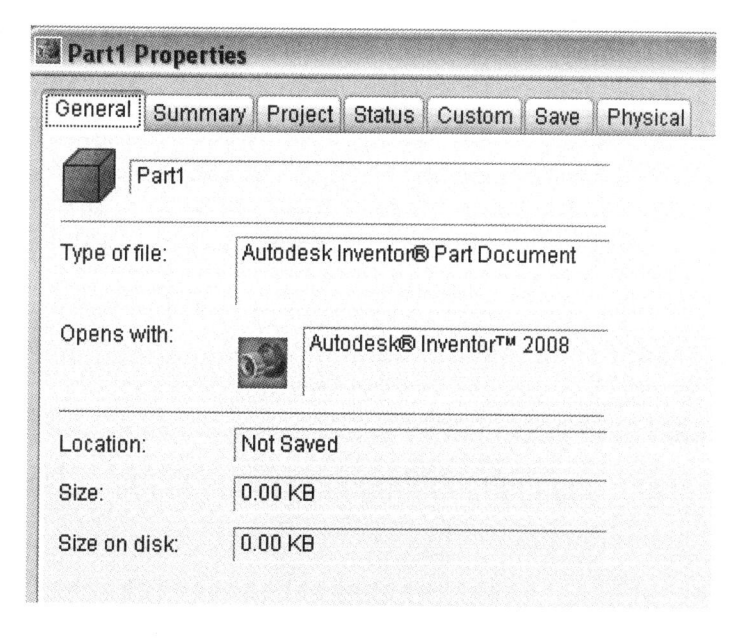

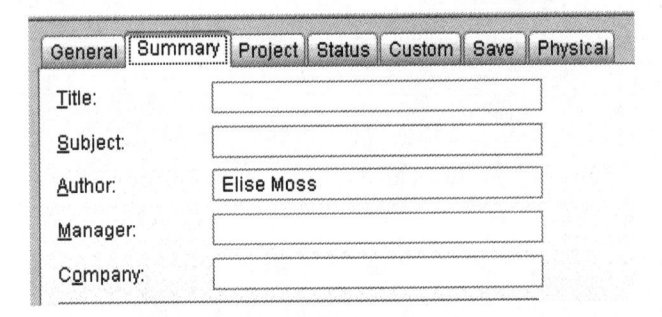

Summary Tab

The Summary window defines summary properties for the selected part, assembly, drawing, or template file. You can use summary properties to categorize and manage your Autodesk Inventor files, search for files, and create reports. The fields are automatically linked to title blocks and parts lists in drawings, and bills of materials in assemblies.

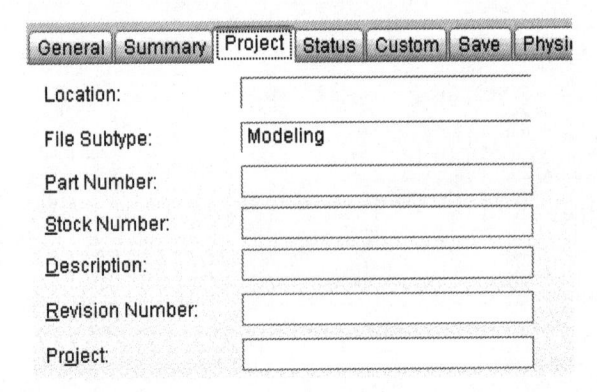

Project Tab

The Project window allows the user to embed additional data into the file. This information can be extracted for use in a document control system or used when performing a search for a specific file.

Location	Displays the location of the selected file.
File Subtype	Displays the Autodesk Inventor file type for the selected file.
Part Number	Specifies the part number. If you do not enter a part number, the file name is automatically assigned as the part number.
Stock Number	Specifies a stock or tracking number.
Description	Adds a description for a part or assembly file. This can be used in a bill of materials.
Revision Number	Specifies the revision number of the file. This can be used in a bill of materials.
Project	Specifies a project name. This can be used by Design Assistant to help manage files.
Designer	Specifies the name of the drafter/designer.
Engineer	Specifies the name of the cognizant engineer.
Authority	Specifies the name of the project/team leader.
Cost Center	Specifies a cost center.
Estimated Cost	Assigns a cost to the file. Enter a real number.
Creation Date	Shows the date that Autodesk Inventor created the file. To change the date, click the arrow and select a new date.
Vendor	Manufacturer or supplier name for components obtained from a third party.
WEB Link	Displays a website address.

User Interface

[Screenshot of Status Tab showing fields: Part Number, Stock Number, Status, Design State (WorkInProgress), Checked By]

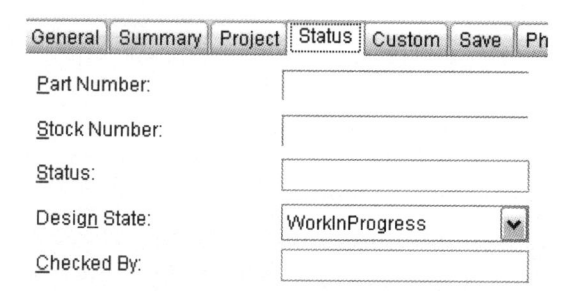

Status Tab

The Status window can be incorporated into an Engineering Control process. Files could be attached to an ECO and signed off in this window by appropriate personnel.

Part Number	Displays the Part Number that is set on the Project tab.
Status	Sets the status for the file. You can enter any status as it is simply text.
Design State	Design State allows the user to set where the file is in process. Options are: Work In Progress, Pending, or Released.
Checked By	Names the person who checked the file.
Checked Date	Shows the date that the file was checked. To change the date, click the arrow and select a date.
Eng Approved By	Names the person who approved the file for Engineering.
Eng Approved Date	Shows the date that the file was approved in Engineering. To change the date, pick the arrow and select a date.
Mfg Approved By	Names the person who approved the file for Manufacturing.
Mfg Approved Date	Shows the date that the file was approved for Manufacturing. To change the date, pick the arrow and select a date.
File Status	Shows the checked-out status of the file for collaborative projects. To use the Checked Out option, you must have Vault installed and set up.

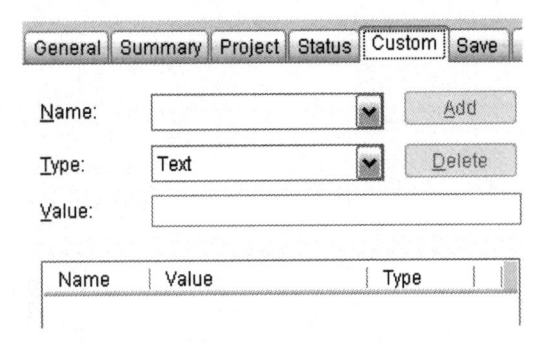

Custom Tab

Adds custom properties to the selected part, assembly, drawing, or template file. You can use custom properties to classify and manage your Autodesk Inventor files, search for files, create reports, and automatically update title blocks and parts lists in drawings and bills of materials in assemblies.

Name	Enter a name for the custom attribute/property field.
Type	Sets the data type for the field. Pick the arrow and select from the drop-down. Options are: Text, Date, Number, and Yes or No.
Value	Provides a default value for the property field. The value must conform to the Type selected.
Properties	Lists the custom properties currently defined in the active window.
Add/Modify	Allows you to add additional property fields or modify existing ones. The Modify button does not work.
Delete	To delete, highlight the property in the list window and select the Delete button.

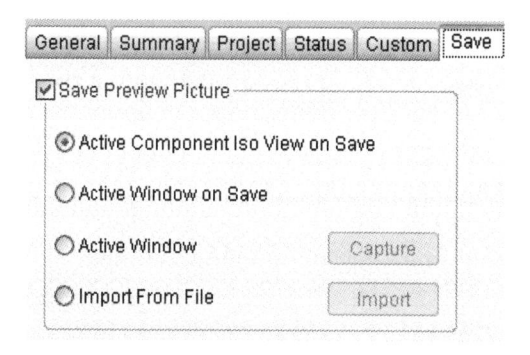

Save Tab

You can use the Save tab to capture a thumbnail image of the model to be displayed as a preview in the File Open dialog. This dialog specifies the origin of the preview image of the model.

Save Preview Picture	Saves a thumbnail image of the model; the default is Enabled. In order to select any remaining options, this option must be enabled.
Active Window on Save	Sets thumbnail image to the view in the active graphics window when the file is saved.
Active Window	If you want your preview to always be a specific view, i.e., isometric, set your model in the desired view and select the Capture button.
Import From File	Imports a BMP file. Select 'Import' to locate the image file. The image must be 120 x 120 pixels.

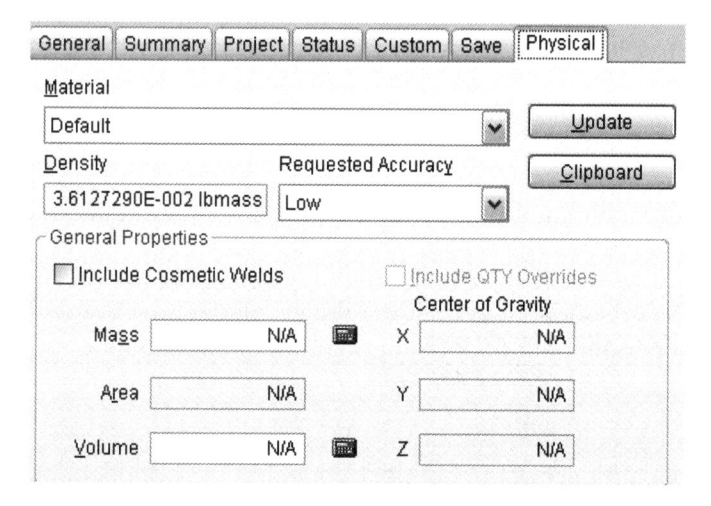

Physical Tab

The Physical Tab allows the user to assign physical properties to a model for use in later analysis. This tab calculates physical and inertial properties for a part or assembly to demonstrate how differences in materials, analysis tolerances, and other values affect the model.

If you only need to calculate mass, surface area, and volume, you do not need to click Update. They are automatically calculated.

Physical properties are affected whenever you add, delete, or modify a feature on a part, or add or delete a part from an assembly. Any time a change has occurred, you must click Update to recalculate. If required components are not loaded into memory, a message asks if you want to load them. When physical properties are up-to-date, the Update button is dimmed.

The clipboard is used to copy the physical properties to a Notepad or text file.	
Material	The user must assign a material to the model in order for Inventor to calculate the physical properties. Inventor comes with a library of materials which can be modified and expanded.
Density	Lists the density of the assigned material.
Accuracy	Sets the number of decimal places to use for the purposes of calculating the material properties. The greater the accuracy, the more time and memory the calculations may require.
General Properties calculates mass, surface area, volume, and center of gravity of the selected part or assembly. The units are determined by the template used when starting the model or the Units set using the Document Settings.	
Mass	Mass in default units.
Area	Surface area of the selected part or assembly.
Volume	Volume of the selected part or assembly.
Center of Gravity	Lists the x, y, z coordinates of the center of gravity of the selected component, relative to the assembly/part origin. This may not be the origin of the assembly/part model if the model/assembly is not constrained to the origin.
Principal or XYZ are two different methods for calculating inertial properties. Principal allows you to select a component. XYZ allows you to select a coordinate.	
Principal Moments	Calculates principal moments of inertia.
Rotation to principal	Calculates rotation to principal relative to the reference frame.
Mass Moments	Calculates mass moments relative to the reference frame.

The icon next to Mass and Volume indicates how the values were established.

Mass N/A [calculator icon]	A calculator indicates that the values were calculated.
Mass 0.600 lbmass [hand icon]	A hand symbol indicates that the user entered in a value in the field to override the calculated value.
[hand with calculator icon]	A hand holding a calculator indicates that the value was calculated using data typed in by the user.

TIP: You can search on materials to locate parts and assemblies that meet specified criteria. Searches are performed using the Design Assistant.

User Interface

Projects

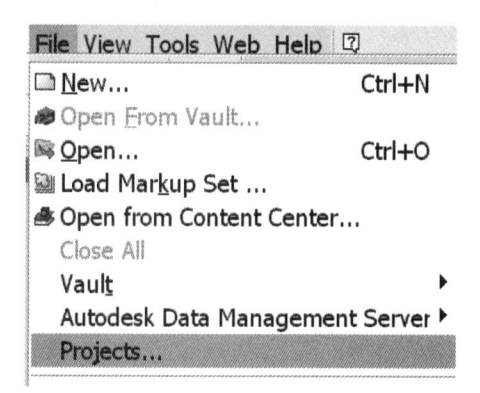

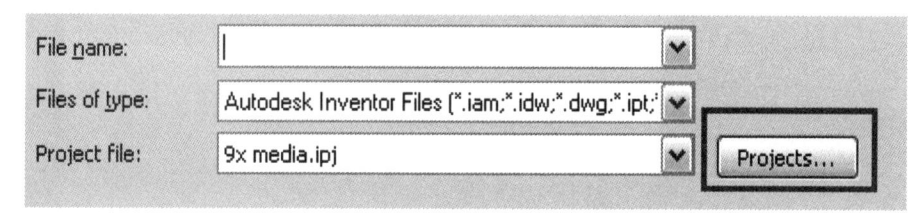

Projects are set up under **File→Projects** or in the 'New/Open' dialog.

Projects are really just shortcuts that allow the user to quickly select the subdirectory where files are located.

Tips & Tricks
- All files must be closed in order to define a new project.
- For better performance, restrict your paths to three levels of sub-folders or less. This is especially important if you use a network to store your files.
- For better performance, limit the number of library and support paths.
- For better performance, keep your project files on your local drive, not on the network.
- For better performance, keep filenames to less than 16 characters.
- You can only modify settings on projects in the Project Editor when Inventor is closed.
- You can only change project settings on inactive projects.
- You can launch the Project Editor inside Windows Explorer by opening the *.ipj file.

The Power of Design: An Introduction to Autodesk Inventor

Exercise 2-3:
Setting Up a Project

Drawing Name: None
Estimated Time: 10 minutes

When you have completed this lesson, you will:

♦ Be able to create a project definition

1   Close any open files.

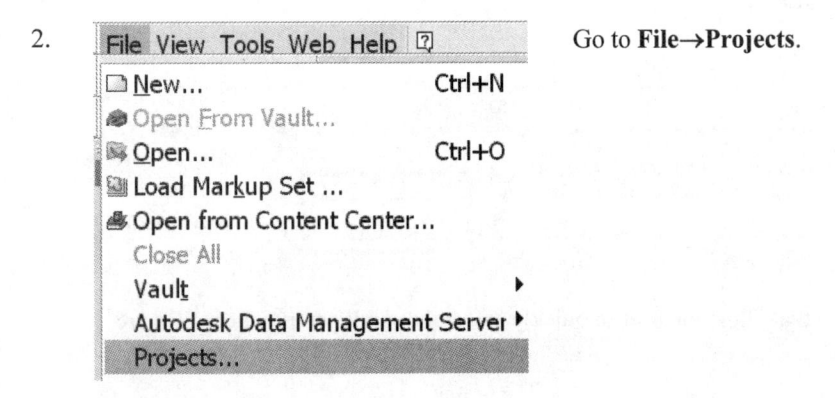

2. Go to **File→Projects**.

3. Select the **New** button.

You can create two types of project: **Vault** or **Single User**.

A **Vault** project is selected if you are working as a member of a team and you will be checking files in and out.

When you check a file out, a copy will be saved to a folder you designate as your workspace. When you check a file in, the file will be moved from your local workspace onto the server. Your local copy will be deleted.

4. Enable **New Single User Project**.
    Press **Next**.

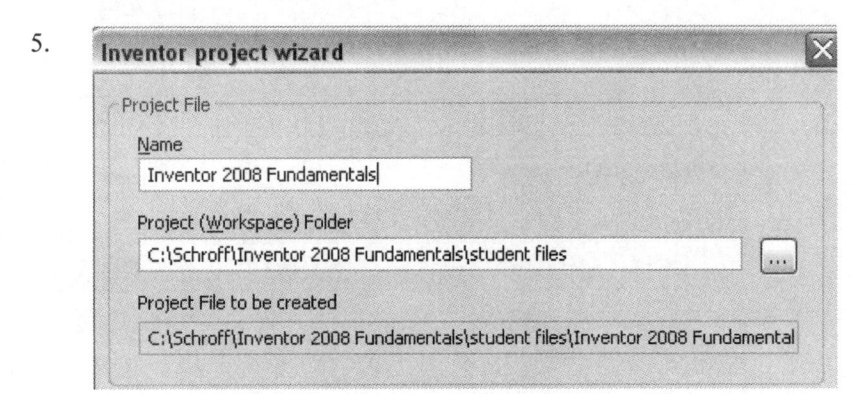

5. Use the **...** button to select a location to store your files. Select a location on your local drive, if possible (this will speed up performance).

    If the workspace path does not exist, it will be created for you automatically.

2-14

User Interface

6. Enter a name for your project.
   Press **Next**.

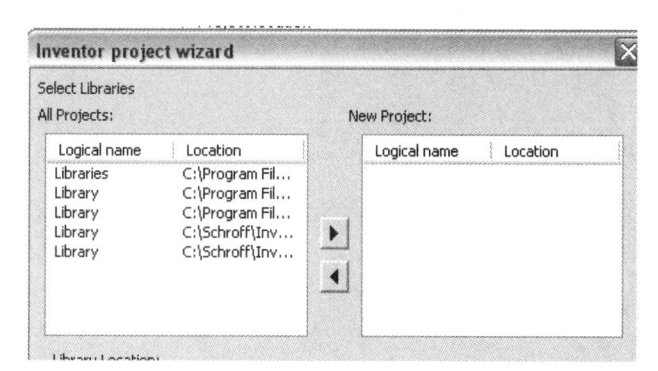

The next dialog allows you to link to content libraries. These are libraries where fasteners and other common parts are stored. If you work on a network, your CAD Manager may have created common parts that are to be used by the team members.

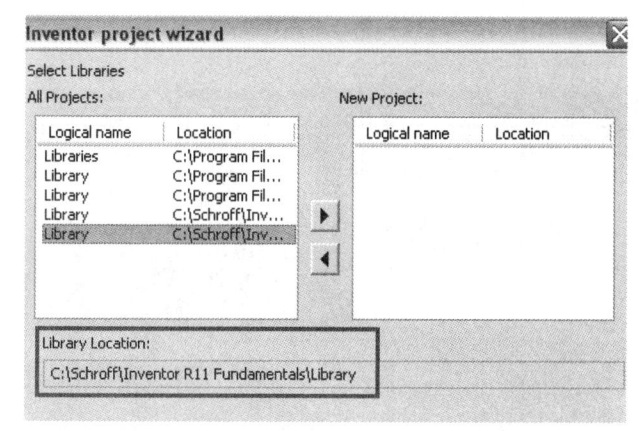

If you highlight a library, you can see the full path under the Library Location to determine it is the library and location you wish to use.

7. Press **Finish**.

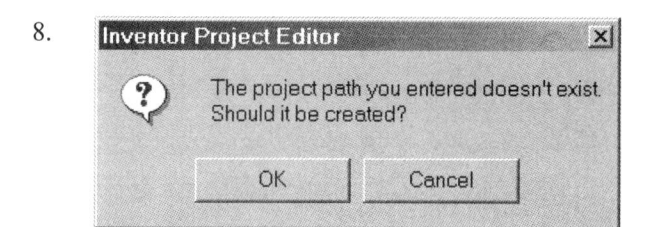

8. If you are creating a new path, you will see this dialog advising you that the path does not exist.

   Press **OK**.

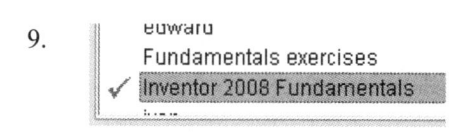

9. Left-click on the project you just created to set it active. You will see a check mark next to the name when it is activated.

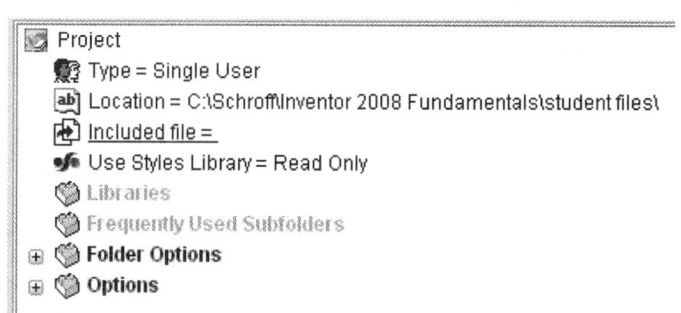

   Your project settings will be shown in the lower panel of the Project Dialog.

10. Close Inventor to work on the next exercise.

2-15

The Power of Design: An Introduction to Autodesk Inventor

Exercise 2-4:
Modifying a Project Definition

Drawing Name: None
Estimated Time: 10 minutes

When you have completed this lesson, you will:

- Be able to edit a project definition
- Add libraries to a project
- Change a project from Single-User to Vault and vice-versa
- Change the number of versions to be saved
- Change the path where files are stored

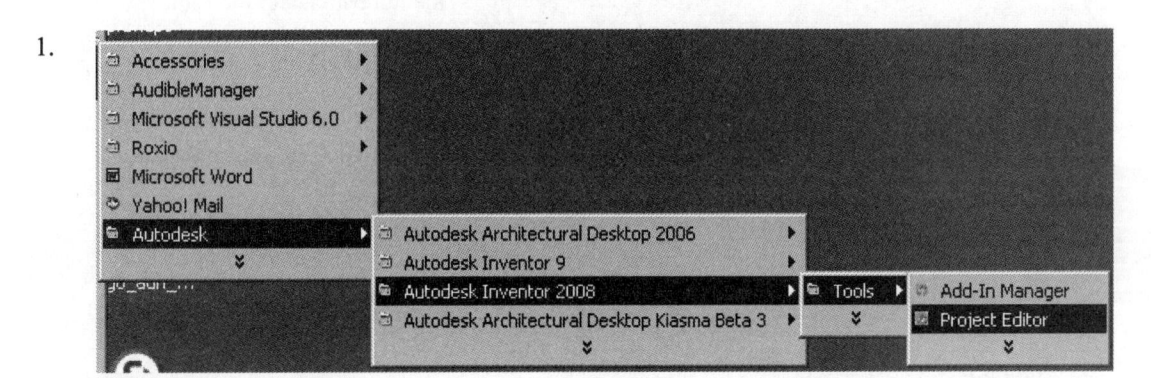

1. Go to **Start → Programs →Autodesk→Inventor 2008→Tools→Project Editor**.

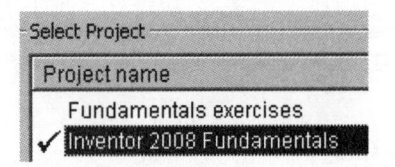

    You should see the active project listed.

2. Locate the **More** button located in the lower right of the dialog and select.

    You see a Workspace path and a Workgroup path. The Workgroup path is grayed out because you opted for a single user project.

    You cannot edit your project unless it is not active.

2-16

User Interface

3.
Project name	Project location
shaft	C:\Program Files\Autodesk\Inventor 2008\Tutorial Files\
tutorial_files	C:\Program Files\Autodesk\Inventor 11\Tutorial Files\
tutorial_files	C:\Program Files\Autodesk\Inventor 9\Tutorial Files\
tutorials	C:\ttutorials\
✓ samples	C:\Program Files\Autodesk\Inventor 2008\Samples\

Set the samples project active by left-clicking on it.

Now you can make changes to your project.

4. Fundamentals exercises   C:\Schroff\Inventor R11 Fundamentals\
   Inventor 2008 Fundamentals   C:\Schroff\Inventor 2008 Fundamentals\student files

Locate your project file and highlight it to select it.

Workspace
   Workspace -

You can edit the location where you want to store your files by highlighting the Workspace folder and selecting the **Edit** tool on the right.

Edit Project
Project
Type = Single User
Location   Single User   es\A
Included   Vault
Use Styles Library = Read Onl

You can switch projects from Single User to Vault or vice versa by selecting the Project Type and right-clicking.

5. **Libraries**   Highlight the **Libraries** folder.

6. **+**   Select the **+** button to add a library.

7. Select the folder button to browse for a folder.

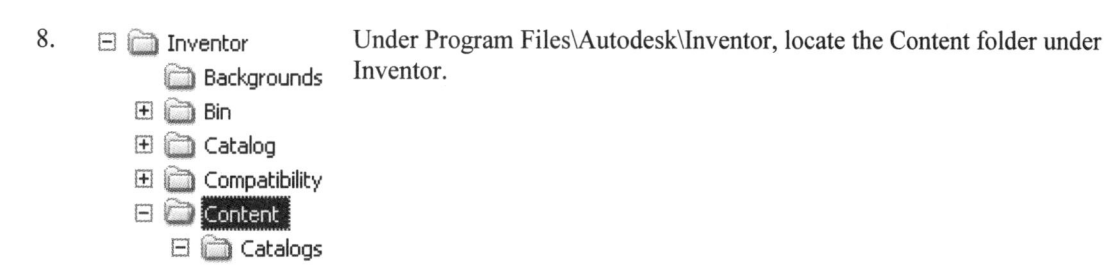

8. Inventor
   Backgrounds
   Bin
   Catalog
   Compatibility
   Content
   Catalogs

Under Program Files\Autodesk\Inventor, locate the Content folder under Inventor.

9. e Styles Library = Read Only
   raries
   Inventor Content Center
   quently Used Subfolders

Left-click on the word **Library** and change the library name to **Inventor Content Center**.

This is useful if you want to set up several different libraries for company parts, and you want to be able to identify each one for easy searching.

2-17

The Power of Design: An Introduction to Autodesk Inventor

Expand the **Options** category.

Note you can set how many old versions of each file to save.

Click on this line.

10. Enter the number **4** in the edit box.
You will now save up to four versions of each file in the project.

Note: The more file versions you elect to save the larger your file size can become as Inventor keeps track of each previous version in your file.

11. Press **Save** and **Close**.

You can modify items by highlighting and then right-click to access a context-sensitive menu.

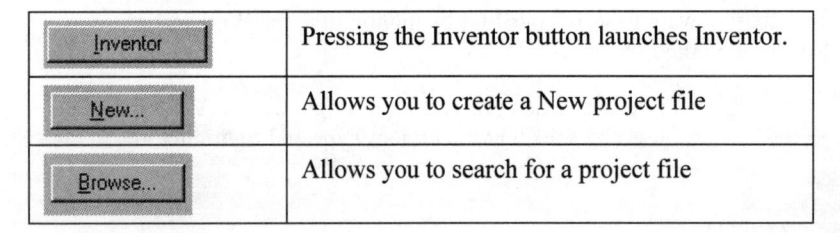

Inventor	Pressing the Inventor button launches Inventor.
New...	Allows you to create a New project file
Browse...	Allows you to search for a project file

Options Dialog

Users can set system options using the **Options** dialog box. It is accessed under the **Tools** menu.

Go to **Tools→Application Options**.

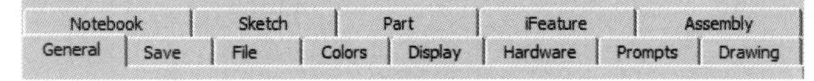

The tabs are:

- General
- Save
- File
- Colors
- Display
- Hardware
- Prompts
- Drawing
- Notebook
- Sketch
- Part
- IFeature
- Assembly

The **Options** dialog controls the color and display of your Autodesk Inventor work environment, the behavior and settings or files, the default file locations, and a variety multiple-user functions.

2-18

General Options

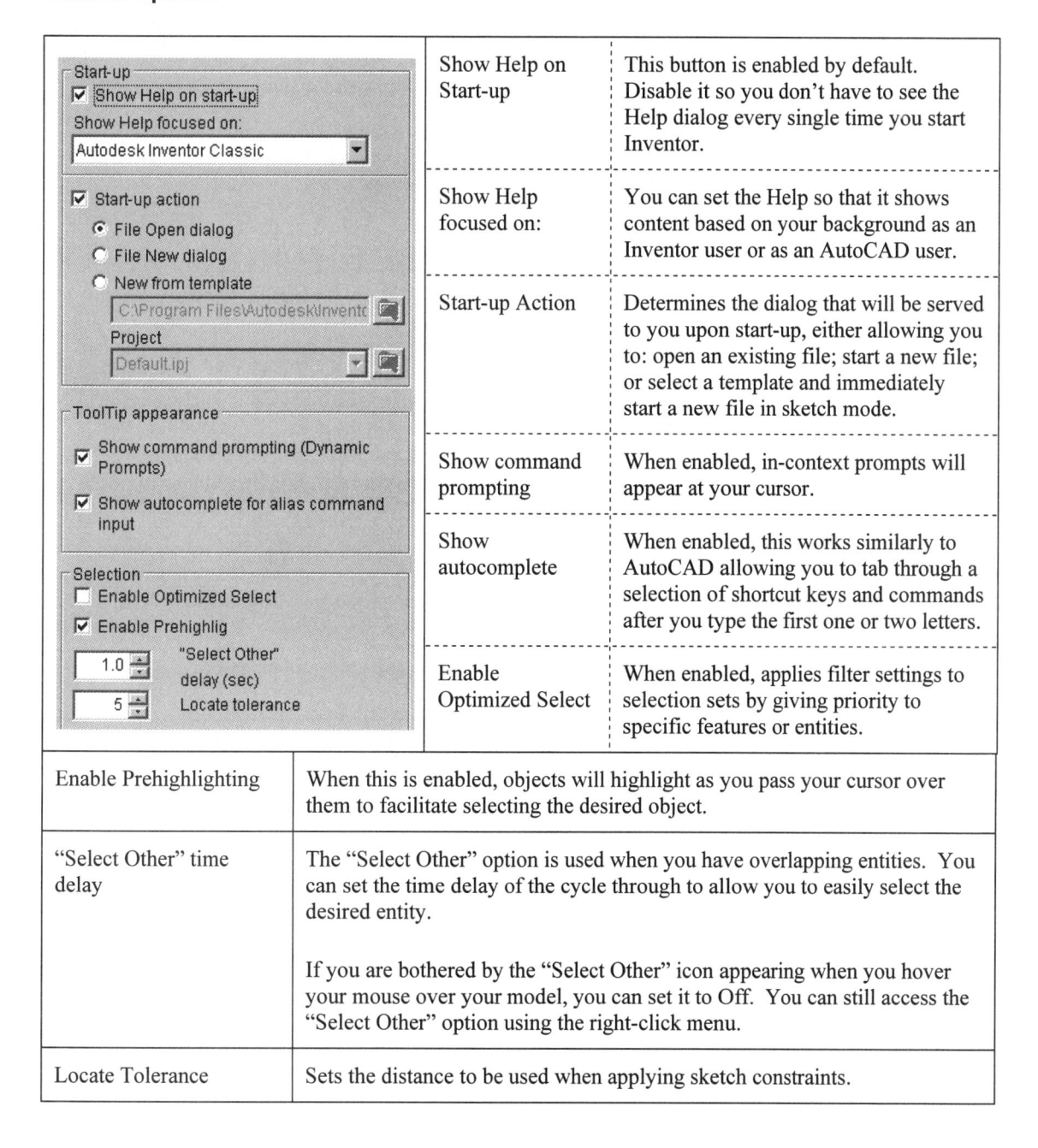

	Show Help on Start-up	This button is enabled by default. Disable it so you don't have to see the Help dialog every single time you start Inventor.
	Show Help focused on:	You can set the Help so that it shows content based on your background as an Inventor user or as an AutoCAD user.
	Start-up Action	Determines the dialog that will be served to you upon start-up, either allowing you to: open an existing file; start a new file; or select a template and immediately start a new file in sketch mode.
	Show command prompting	When enabled, in-context prompts will appear at your cursor.
	Show autocomplete	When enabled, this works similarly to AutoCAD allowing you to tab through a selection of shortcut keys and commands after you type the first one or two letters.
	Enable Optimized Select	When enabled, applies filter settings to selection sets by giving priority to specific features or entities.
Enable Prehighlighting	When this is enabled, objects will highlight as you pass your cursor over them to facilitate selecting the desired object.	
"Select Other" time delay	The "Select Other" option is used when you have overlapping entities. You can set the time delay of the cycle through to allow you to easily select the desired entity. If you are bothered by the "Select Other" icon appearing when you hover your mouse over your model, you can set it to Off. You can still access the "Select Other" option using the right-click menu.	
Locate Tolerance	Sets the distance to be used when applying sketch constraints.	

The Power of Design: An Introduction to Autodesk Inventor

General Options

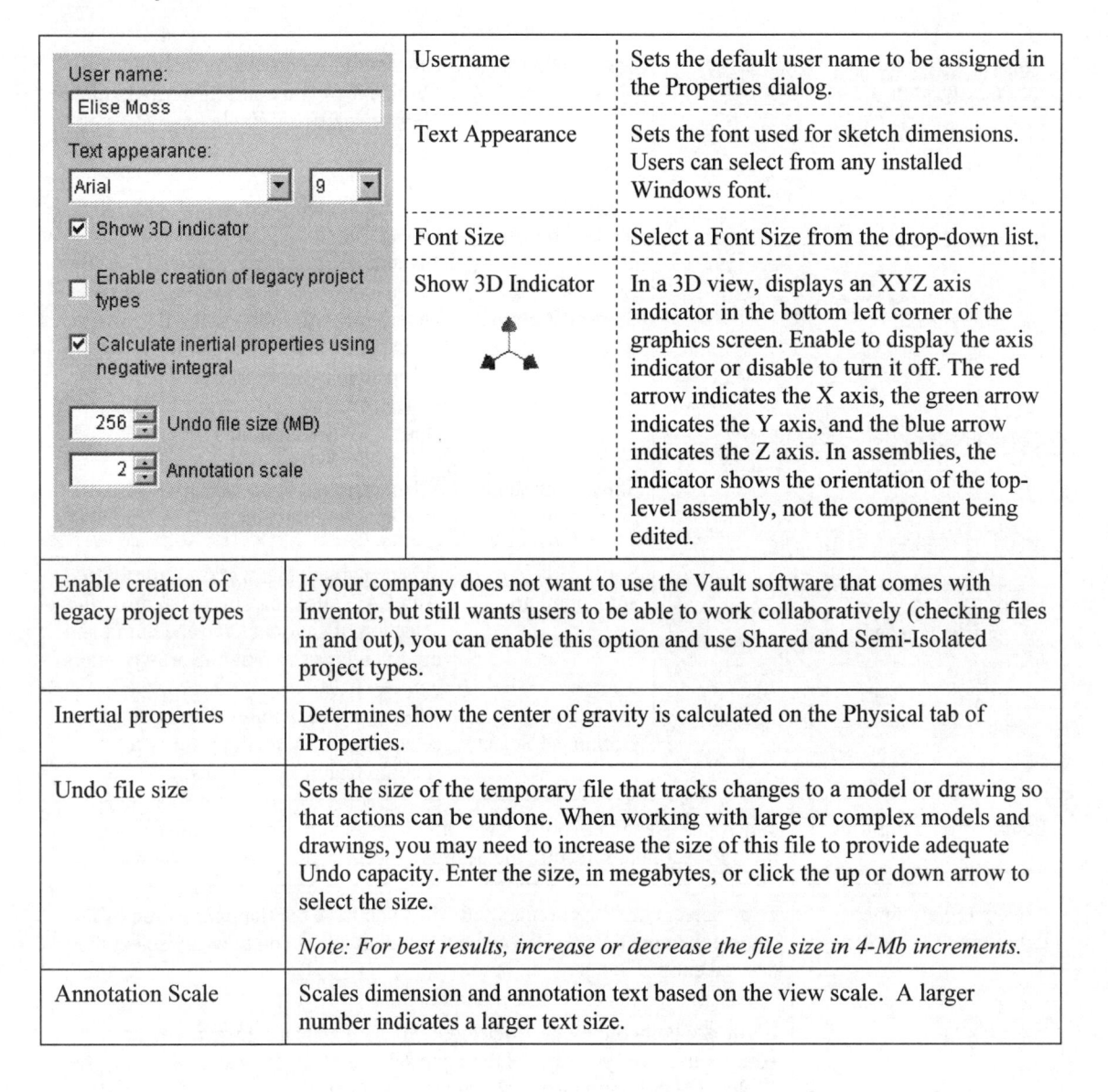

User name: Elise Moss / Text appearance: Arial 9 / ☑ Show 3D indicator / ☐ Enable creation of legacy project types / ☑ Calculate inertial properties using negative integral / 256 Undo file size (MB) / 2 Annotation scale	Username	Sets the default user name to be assigned in the Properties dialog.
	Text Appearance	Sets the font used for sketch dimensions. Users can select from any installed Windows font.
	Font Size	Select a Font Size from the drop-down list.
	Show 3D Indicator	In a 3D view, displays an XYZ axis indicator in the bottom left corner of the graphics screen. Enable to display the axis indicator or disable to turn it off. The red arrow indicates the X axis, the green arrow indicates the Y axis, and the blue arrow indicates the Z axis. In assemblies, the indicator shows the orientation of the top-level assembly, not the component being edited.
Enable creation of legacy project types	If your company does not want to use the Vault software that comes with Inventor, but still wants users to be able to work collaboratively (checking files in and out), you can enable this option and use Shared and Semi-Isolated project types.	
Inertial properties	Determines how the center of gravity is calculated on the Physical tab of iProperties.	
Undo file size	Sets the size of the temporary file that tracks changes to a model or drawing so that actions can be undone. When working with large or complex models and drawings, you may need to increase the size of this file to provide adequate Undo capacity. Enter the size, in megabytes, or click the up or down arrow to select the size. *Note: For best results, increase or decrease the file size in 4-Mb increments.*	
Annotation Scale	Scales dimension and annotation text based on the view scale. A larger number indicates a larger text size.	

User Interface

Save Options

You can eliminate prompting during Save operations on this Save options tab. However, if you frequently save over files unintentionally when you meant to use the 'Save As Copy' feature, it would be best to retain the prompts.

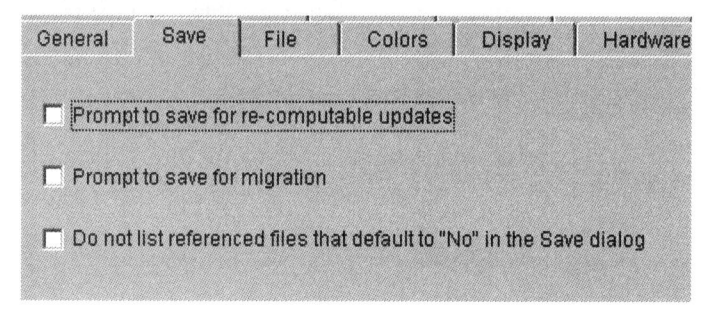

File Options

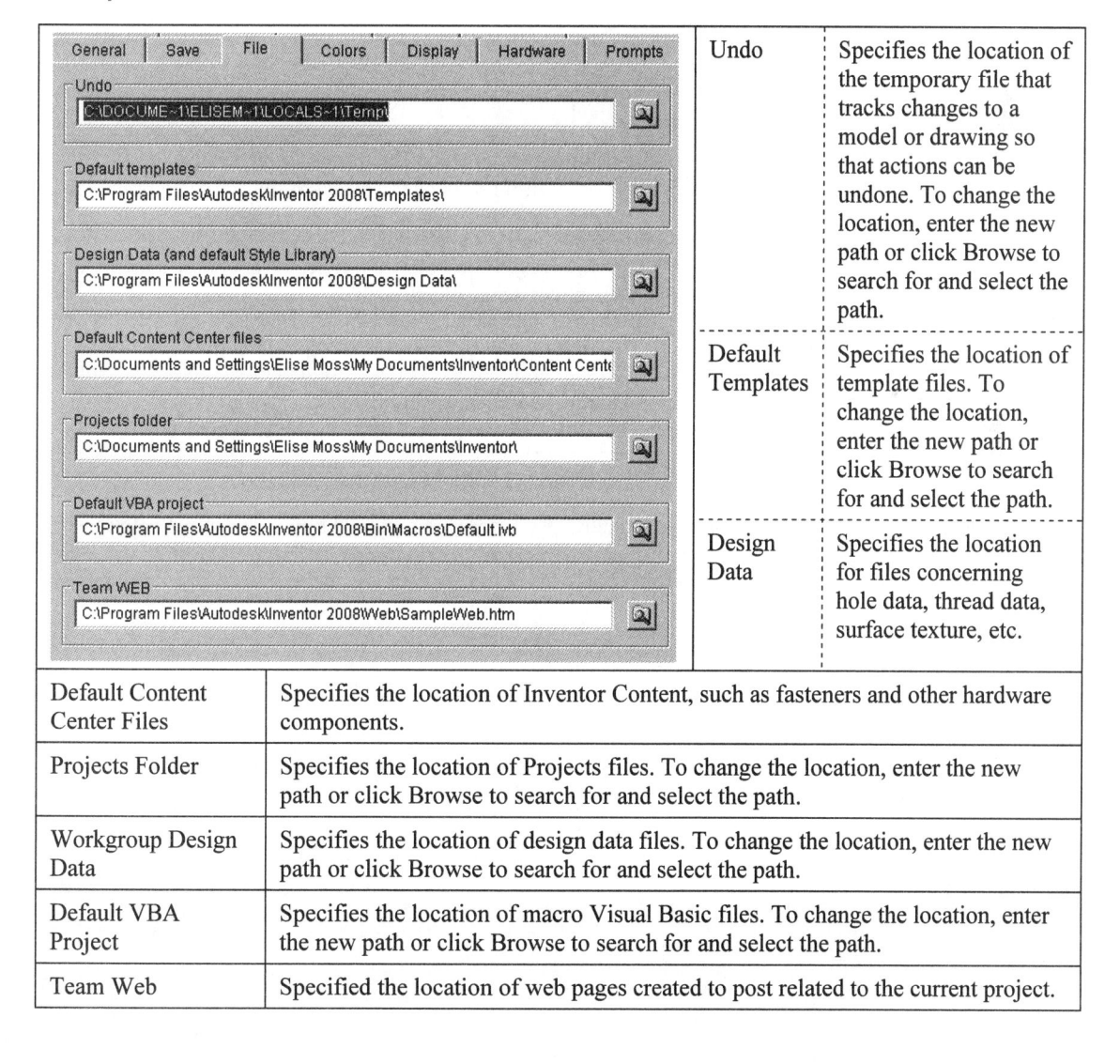

Undo	Specifies the location of the temporary file that tracks changes to a model or drawing so that actions can be undone. To change the location, enter the new path or click Browse to search for and select the path.
Default Templates	Specifies the location of template files. To change the location, enter the new path or click Browse to search for and select the path.
Design Data	Specifies the location for files concerning hole data, thread data, surface texture, etc.

Default Content Center Files	Specifies the location of Inventor Content, such as fasteners and other hardware components.
Projects Folder	Specifies the location of Projects files. To change the location, enter the new path or click Browse to search for and select the path.
Workgroup Design Data	Specifies the location of design data files. To change the location, enter the new path or click Browse to search for and select the path.
Default VBA Project	Specifies the location of macro Visual Basic files. To change the location, enter the new path or click Browse to search for and select the path.
Team Web	Specified the location of web pages created to post related to the current project.

2-21

The Power of Design: An Introduction to Autodesk Inventor

Colors

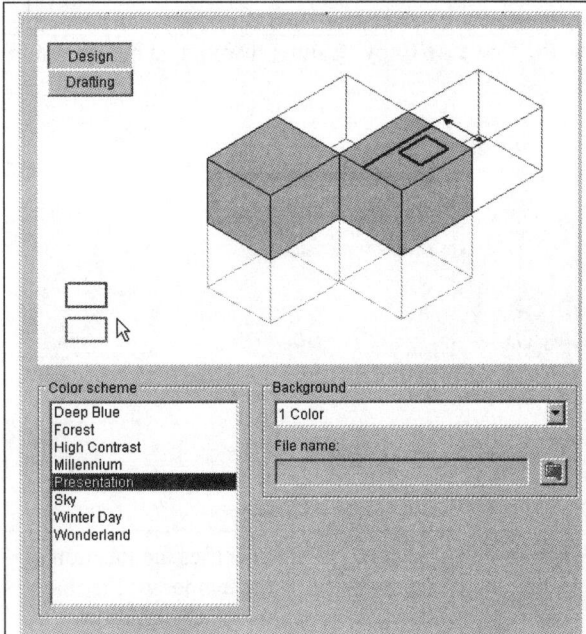

Select the **Color** tab.

Highlight the various colors and note how the display changes. When you find a color you like, select **Apply**.

The two rectangles in the lower left corner indicate the highlighting that will be used for that color scheme.

Select **OK**.

Design/Drafting	Displays the effect of the color choice in either the design or the drafting environment. Click Design or Drafting to view the effect of the active color choice in the view box.
Color Scheme	Lists the available color schemes. Highlight to select from the available schemes. The view box displays the result of your selection.
Background	**1 Color** – Applies a single color. **Gradient** – Applies a saturation gradient to the background color. Enable to apply a gradient; disable to apply a solid background color. **Background Image** – Allows you to apply a bmp file as the background.
File Name	Browse for the bmp file to be used as your background.
Show Reflections	Adds a reflected image to parts when they are assigned highly reflective color and lighting styles.

TIP: To preview your selections in the graphics window before closing the dialog box, click Apply.

Display Options

	Sets the wire frame and shaded display of models and assemblies.
	Depth Dimming — Sets a dimming effect to better convey the depth of a model. Enable to turn on depth dimming; disable to turn it off.
	Active — Sets the preferences for wireframe display of a part or of the active components in an assembly. **Silhouettes** enable the display of silhouettes. Enable to display silhouettes; disable to suppress the display. **Hidden Edges** dims the display of edges hidden behind other geometry. Enable to dim the display of hidden edges; disable to display the hidden edges in full intensity.
Enabled	Sets the preferences for wireframe display of a typical enabled component in an assembly. **Silhouettes** enable the display of silhouettes. **Hidden Edges** dims the display of edges hidden behind other geometry.
Background	Sets the preferences for wire frame display of parts that are not enabled in an assembly. Silhouettes enable the display of silhouettes.
Display Quality	Sets the resolution for the display of the model. Generally, the smoother the resolution, the longer it takes to redisplay the model when changes are made. When working with a very large or complex model, you may want to lower the quality of the display to speed up operation. For example, the Rough setting temporarily simplifies detail on large parts but updates faster, while the Smooth setting temporarily simplifies fewer details but updates slower.

The Power of Design: An Introduction to Autodesk Inventor

Shaded display modes ☐ Depth dimming ◉ Blending transparency ○ Screen door transparency **Active** ☐ Silhouettes ☑ Edges ☐ ▇ **Enabled** ☐ Silhouettes ☐ Edges ☐ ▇ ☑ Shaded [25] % opaque **Background** ☐ Silhouettes ☑ Shaded [15] % opaque		Sets the preferences for shaded display of the model.

Depth Dimming	Sets a dimming effect to better convey the depth of a model.
Transparency	Sets the quality of the transparency display. If you do not have a 3D graphics board, the screen door display option speeds up operation. **Blending** specifies a high quality transparency display that is achieved by averaging the colors of overlapping objects. **Screen Door** specifies a lower quality transparency display that is achieved by using a pattern that allows the color of the hidden object to show through.
Active	Sets the preferences for shaded display of a part or of the active components in an assembly. **Silhouettes** enable the display of silhouettes. **Edge Display** sets the display of edges. If Edge Display is selected, you can change edge color by clicking on the color pad and choosing a color from the Color dialog box.
Enabled	Sets the preferences for shaded display of a typical enabled component in an assembly. **Silhouettes** enable the display of silhouettes. **Edge Display** sets the display of edges when shaded display is selected for enabled components. If Edge Display is selected, you can change edge color by clicking on the color pad and choosing a color from the Color dialog box. **Shaded** enables contrasting shading for enabled components except the active component when a single component is activated in an assembly. **% Opaque** – if Shaded is selected, you can set the opacity for the shading. Enter the percent opaque, or use the up or down arrow to select the value.
Background	Sets the preferences for shaded display of components that are not enabled in an assembly. **Silhouettes** enable the display of silhouettes. **Shaded** enables contrasting shading, rather than outline presentation, for background components. **% Opaque** – if Shaded is selected, you can set the opacity for the shading. Enter the percent opaque or click the up or down arrow to select the value.

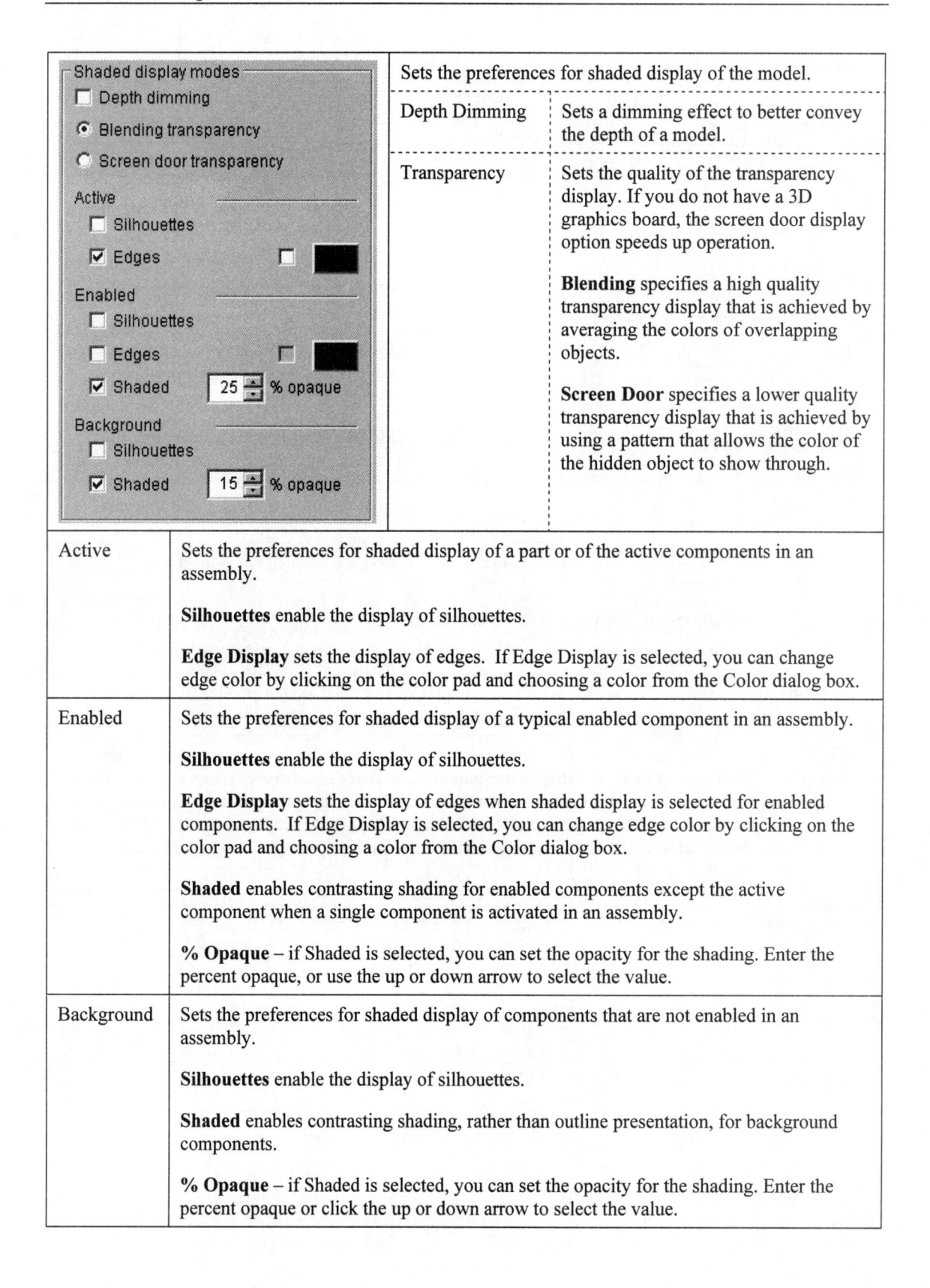

User Interface

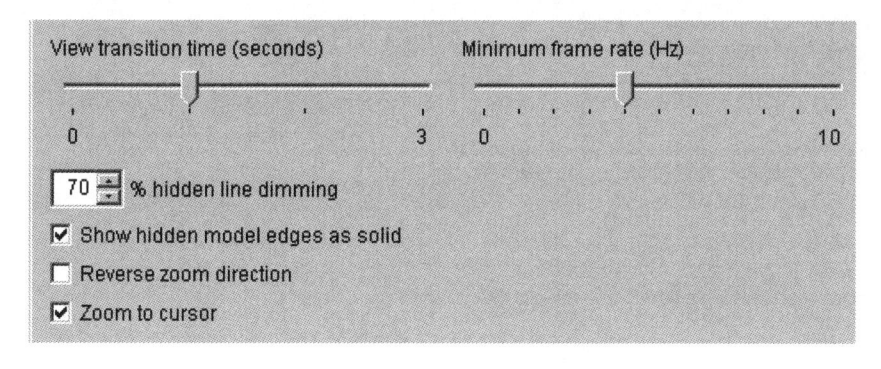

View transition time	Controls the time required to smoothly transition between views when using viewing tools (such as Isometric View, Zoom All, Zoom Area, Look At, and so on). Zero transition time causes transition to be abrupt, which might make it difficult to understand changes in position and orientation. Three sets the greatest amount of time to transition between views.
Minimum Frame Rate	With complex views (such as very large assemblies), use this setting to specify how slowly you are willing to update the display during interactive viewing operations (like Rotate, Pan, and Zoom). Autodesk Inventor tries to maintain the frame rate you set, but to do so, may need to simplify or discard parts of the view. All parts are restored to the view when movement ends. Set zero to always draw everything in the view, regardless of the time required. Set one to have Autodesk Inventor try to draw the view at least one frame per second; set five to draw at least five frames per second. *Note: Usually, this setting has no effect on views because they update more quickly than this rate.*
% Hidden Line Dimming	Sets the percent of dimming for hidden edges when one or more of the Hidden Edges check boxes is enabled. Enter the percentage of dimming to apply, or click the up or down arrow to select the value.
Show hidden model edges as solid	When enabled, the model edges in the background are displayed as solid, not hidden, edges.
Reverse zoom direction	When enabled, the left mouse button is reversed during dynamic zoom.
Zoom to cursor	When enabled, the display will zoom to the cursor upon a left mouse click.

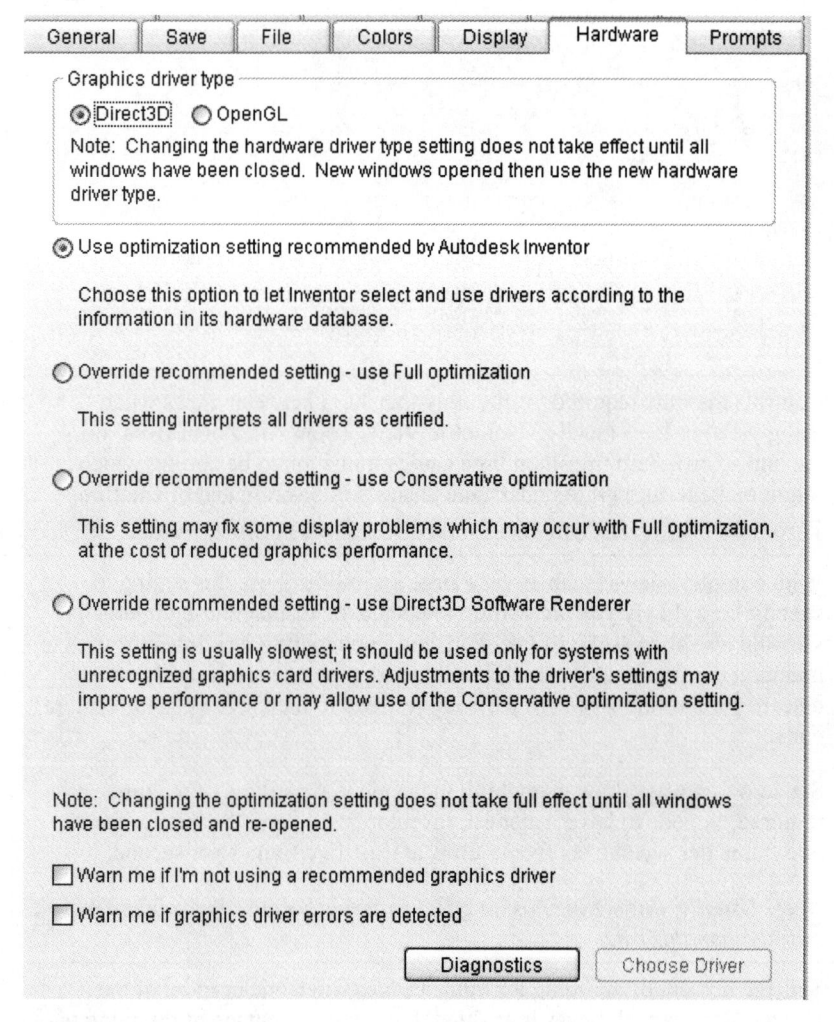

The Power of Design: An Introduction to Autodesk Inventor

[Screenshot of Hardware tab in Autodesk Inventor options dialog showing Graphics driver type settings with Direct3D selected, optimization options, and Diagnostics/Choose Driver buttons.]

Hardware

This tab was added because so many users complained about what a resource hog Inventor is. If you are running under the Windows Vista operating system, note that Vista prefers to run under Direct3D mode while many CAD graphics cards prefer OpenGL. You will observe better performance if you run in Direct3D mode, but verify that your graphics card also uses Direct3D.

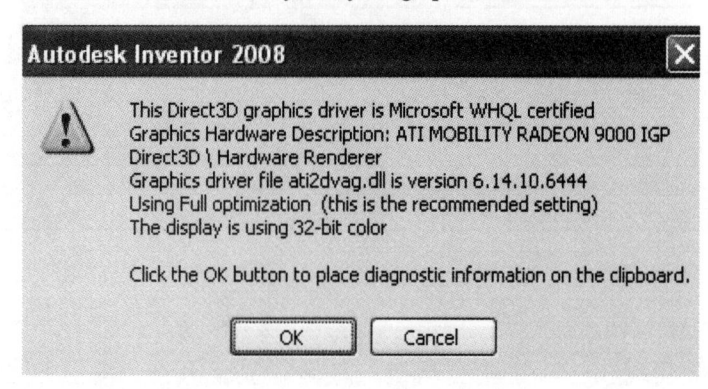

[Screenshot of Autodesk Inventor 2008 diagnostic dialog showing: This Direct3D graphics driver is Microsoft WHQL certified; Graphics Hardware Description: ATI MOBILITY RADEON 9000 IGP; Direct3D \ Hardware Renderer; Graphics driver file ati2dvag.dll is version 6.14.10.6444; Using Full optimization (this is the recommended setting); The display is using 32-bit color. Click the OK button to place diagnostic information on the clipboard. OK and Cancel buttons.]

Pressing the Diagnostics button brings up this window with all your system information. You can then copy and paste this into an email to be sent to Autodesk's Technical Support if you are having problems with your graphics display.

2-26

User Interface

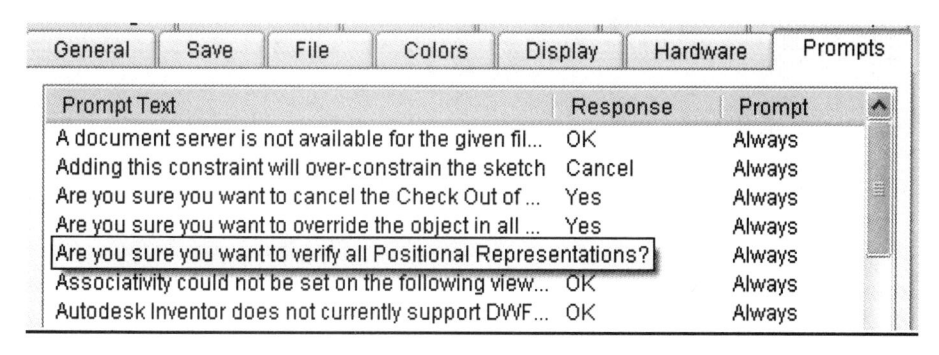

Prompt Text	Response	Prompt
A document server is not available for the given fil...	OK	Always
Adding this constraint will over-constrain the sketch	Cancel	Always
Are you sure you want to cancel the Check Out of...	Yes	Always
Are you sure you want to override the object in all...	Yes	Always
Are you sure you want to verify all Positional Representations?		Always
Associativity could not be set on the following view...	OK	Always
Autodesk Inventor does not currently support DWF...	OK	Always

Prompts

This tab allows the user to set how they will be prompted as they proceed through their design process. You can set the default response and how you will be prompted.

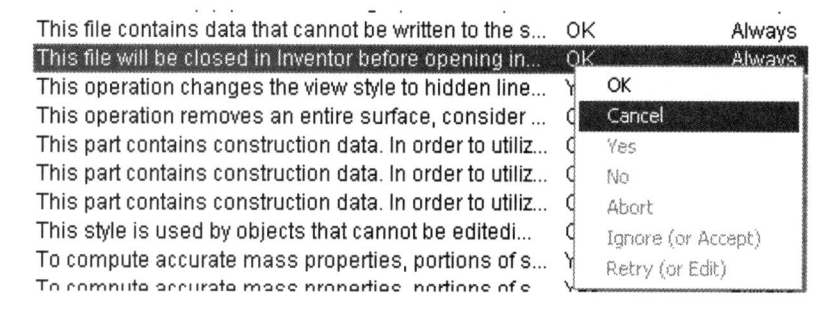

To change how you are prompted, right-click in the Response field and change the Response to 'Yes'.

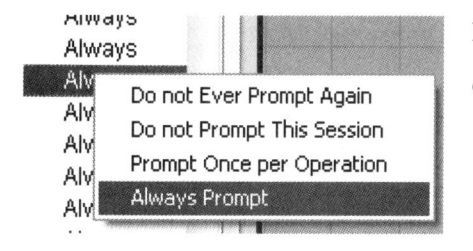

Right-click in the **Prompt** column.

Change the prompt frequency to **Prompt Once Per Operation**.

TIP: Take a minute to set your prompt options. This will save you time in the long run as it will reduce the number of dialogs you have to click through to get your work done.

2-27

The Power of Design: An Introduction to Autodesk Inventor

Drawing | Notebook | Sketch | Part

Defaults
- [ ] Retrieve all model dimensions on view placement
- [x] Center dimension text on creation

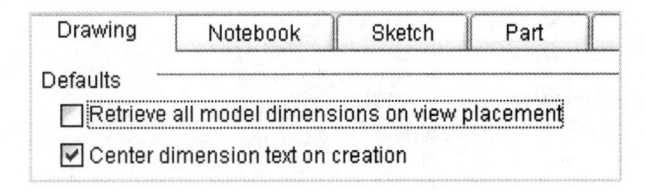

Drawing Options

Sets the drawing options in the idw format.

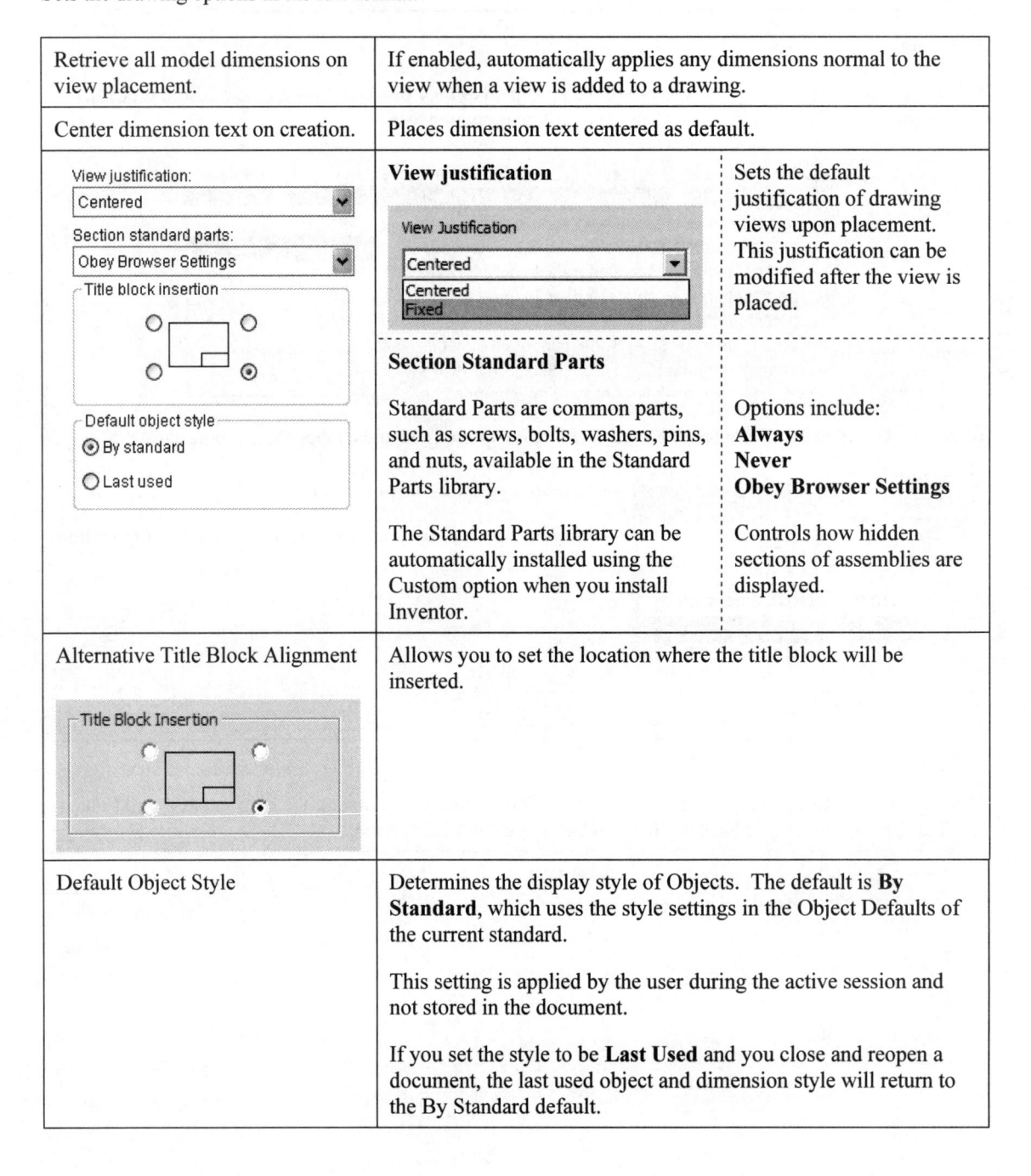

Retrieve all model dimensions on view placement.	If enabled, automatically applies any dimensions normal to the view when a view is added to a drawing.	
Center dimension text on creation.	Places dimension text centered as default.	
View justification: Centered Section standard parts: Obey Browser Settings Title block insertion Default object style: By standard / Last used	**View justification** View Justification: Centered / Fixed	Sets the default justification of drawing views upon placement. This justification can be modified after the view is placed.
	Section Standard Parts Standard Parts are common parts, such as screws, bolts, washers, pins, and nuts, available in the Standard Parts library. The Standard Parts library can be automatically installed using the Custom option when you install Inventor.	Options include: **Always** **Never** **Obey Browser Settings** Controls how hidden sections of assemblies are displayed.
Alternative Title Block Alignment Title Block Insertion	Allows you to set the location where the title block will be inserted.	
Default Object Style	Determines the display style of Objects. The default is **By Standard**, which uses the style settings in the Object Defaults of the current standard. This setting is applied by the user during the active session and not stored in the document. If you set the style to be **Last Used** and you close and reopen a document, the last used object and dimension style will return to the By Standard default.	

2-28

User Interface

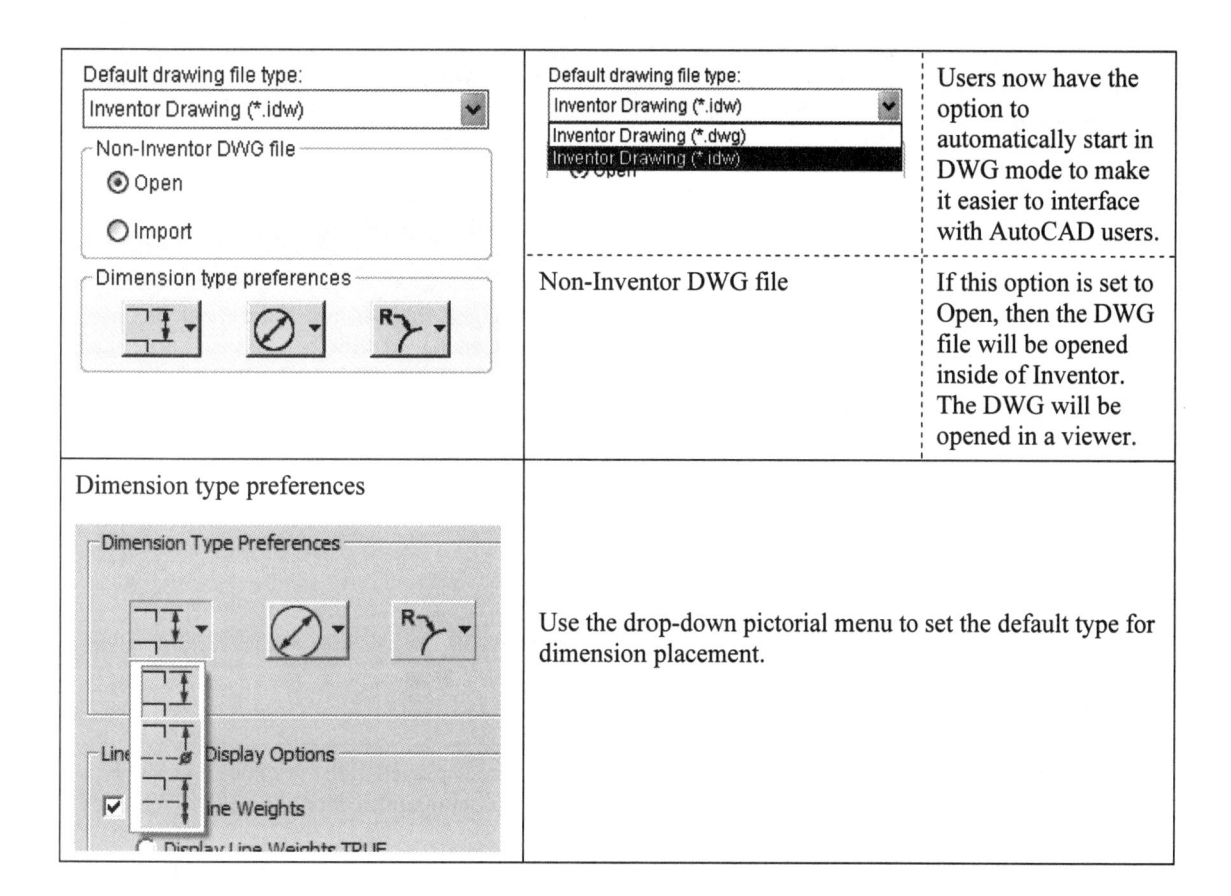

![Default drawing file type: Inventor Drawing (*.idw), Non-Inventor DWG file: Open/Import, Dimension type preferences]	![Default drawing file type dropdown showing Inventor Drawing (*.dwg) and Inventor Drawing (*.idw)]	Users now have the option to automatically start in DWG mode to make it easier to interface with AutoCAD users.
	Non-Inventor DWG file	If this option is set to Open, then the DWG file will be opened inside of Inventor. The DWG will be opened in a viewer.
Dimension type preferences ![Dimension Type Preferences panel with dropdown]		Use the drop-down pictorial menu to set the default type for dimension placement.

2-29

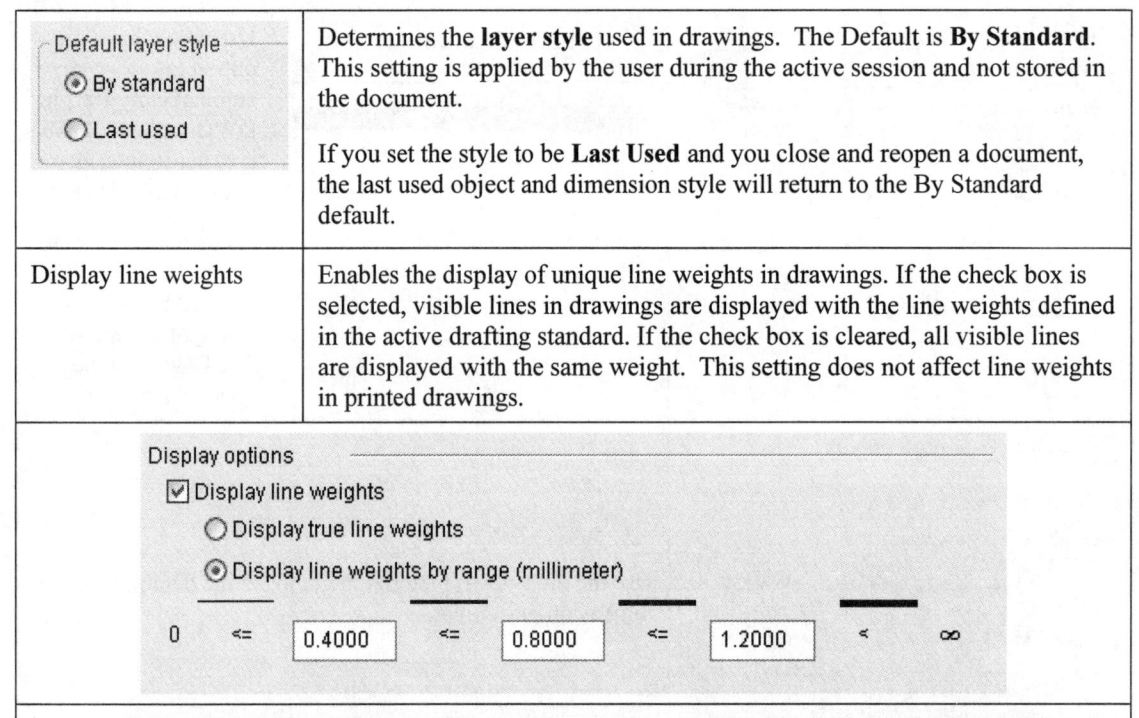

Default layer style ⊙ By standard ○ Last used	Determines the **layer style** used in drawings. The Default is **By Standard**. This setting is applied by the user during the active session and not stored in the document. If you set the style to be **Last Used** and you close and reopen a document, the last used object and dimension style will return to the By Standard default.
Display line weights	Enables the display of unique line weights in drawings. If the check box is selected, visible lines in drawings are displayed with the line weights defined in the active drafting standard. If the check box is cleared, all visible lines are displayed with the same weight. This setting does not affect line weights in printed drawings.

Display options
☑ Display line weights
  ○ Display true line weights
  ⊙ Display line weights by range (millimeter)
  0 <= 0.4000 <= 0.8000 <= 1.2000 < ∞	

• If **Display Line Weights** is ENABLED, line weights are displayed in the graphics window as they would appear plotted. For example, regardless of zoom magnification, a line 0.5-inch thick is the same as the height of 0.5-inch text. • If **Display Line Weights by Range** (millimeter) is ENABLED, line weights are displayed in the graphics window according to values you enter. Values range from smallest (left) to largest (right).	

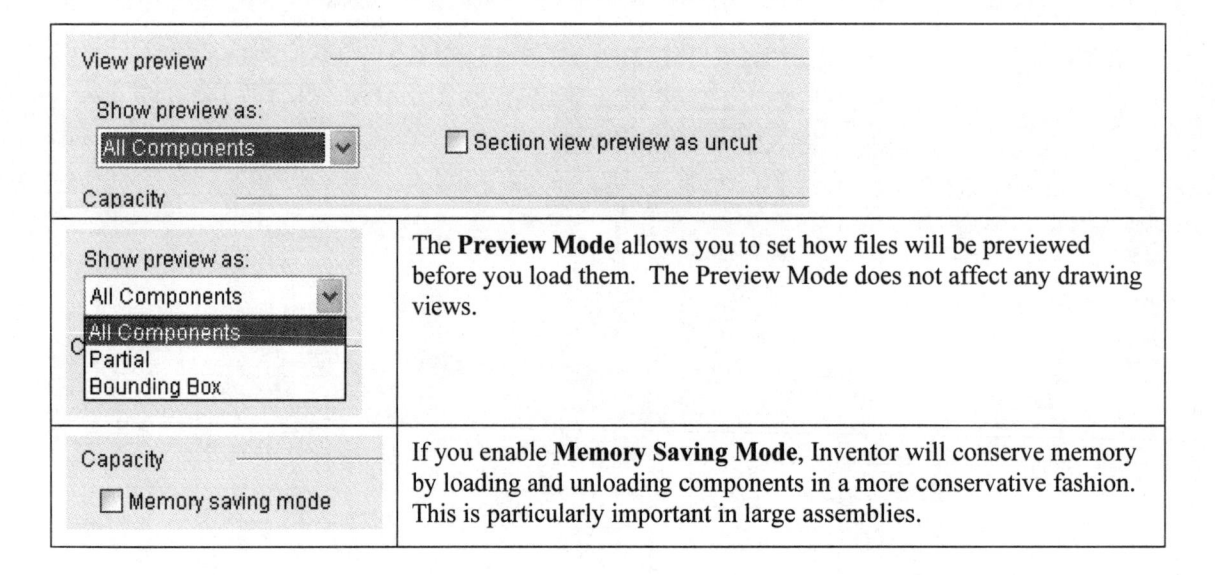

View preview Show preview as: [All Components ▼] □ Section view preview as uncut Capacity	
Show preview as: [All Components ▼] All Components Partial Bounding Box	The **Preview Mode** allows you to set how files will be previewed before you load them. The Preview Mode does not affect any drawing views.
Capacity □ Memory saving mode	If you enable **Memory Saving Mode**, Inventor will conserve memory by loading and unloading components in a more conservative fashion. This is particularly important in large assemblies.

TIP: To display full precision for an imprecise view, right-click the view and choose Make View Precise from the menu. A precise view cannot be changed to imprecise

Enable 'Get model dimensions' to automatically have dimensions appear when placing drawing views.

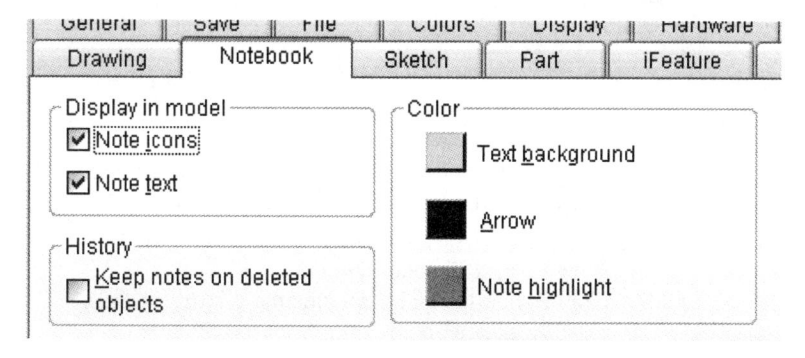

Notebook Options

Controls the display of design notes in the Engineer's Notebook.

Display in Model	Sets the display of note indicators in the model.	
	Note icons	Displays note icons in the model.
	Note text	Displays note text in pop-up windows in the model.
History	Sets archival options for design notes.	
	Keep notes on deleted objects	Keeps notes attached to deleted geometry.
Color	Sets the colors of elements in design notes. The color pad next to each item shows the current color setting. To change the color for an item, click the color pad to open the color dialog box and select the color.	
	Text background	Sets the background color for the comment boxes in design notes.
	Arrow	Sets the color for arrows in design notes.
	Note highlight	Sets the color for the highlighted component in note views.

TIP: If you attach multiple notes to a single item, only the first note displays an icon.

Changing the user name affects only those notes created after the change is made. You cannot change a name on an existing note.

Sketch Options

Sets the sketch options.

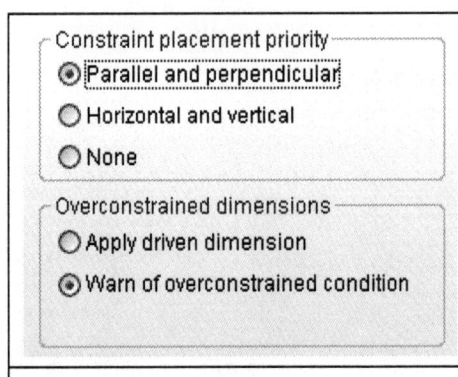

Constraint placement priority ◉ Parallel and perpendicular ○ Horizontal and vertical ○ None Overconstrained dimensions ○ Apply driven dimension ◉ Warn of overconstrained condition	**Constraint placement priority** sets the preferred constraint type for automatic placement of constraints. **Parallel and Perpendicular** Applies parallel and perpendicular constraints using existing sketch geometry. **Horizontal and Vertical** Applies horizontal and vertical constraints using the active grid as the guide.

Overconstrained Dimensions sets the preferred behavior for dimensions on overconstrained sketches.

Apply Driven Dimension
Applies a non-parametric dimension enclosed in parentheses. Dimension updates when the sketch changes, but does not change the sketch geometry.

Warn of Overconstrained Condition
Displays warning message when a dimension will overconstrain a sketch. You must then click OK to place the dimension or Cancel to prevent creating the dimension.

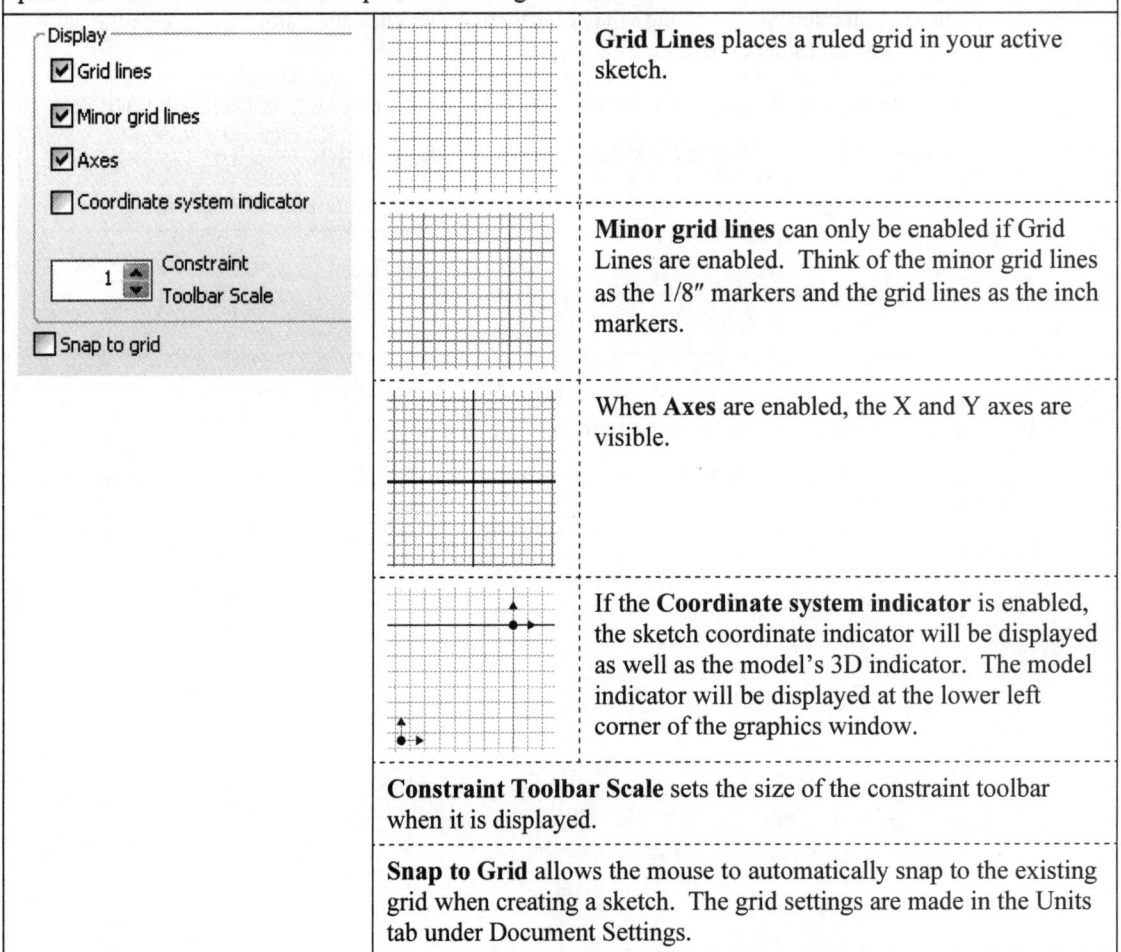

Display ☑ Grid lines ☑ Minor grid lines ☑ Axes ☐ Coordinate system indicator [1] Constraint Toolbar Scale ☐ Snap to grid	(ruled grid)	**Grid Lines** places a ruled grid in your active sketch.
	(grid with minor lines)	**Minor grid lines** can only be enabled if Grid Lines are enabled. Think of the minor grid lines as the 1/8″ markers and the grid lines as the inch markers.
	(grid with axes)	When **Axes** are enabled, the X and Y axes are visible.
	(grid with coordinate indicator)	If the **Coordinate system indicator** is enabled, the sketch coordinate indicator will be displayed as well as the model's 3D indicator. The model indicator will be displayed at the lower left corner of the graphics window.
		Constraint Toolbar Scale sets the size of the constraint toolbar when it is displayed.
		Snap to Grid allows the mouse to automatically snap to the existing grid when creating a sketch. The grid settings are made in the Units tab under Document Settings.

User Interface

☑ Edit dimension when created ☐ Autoproject edges during curve creation ☐ Autoproject edges for sketch creation and edit ☐ Parallel view on sketch creation ☐ Autoproject part origin on sketch create	**Edit dimension when created** presents the dimension edit box when a dimension is created. The default is enabled. **Autoproject edges during curve creation** automatically projects edges onto the current sketch plane. If **Autoproject edges for sketch creation and edit** is enabled, whenever you create a new sketch the edges of the plane selected will be projected as reference geometry to be used as part of the new sketch. **Parallel view on sketch creation** switches the display to a normal view when the user initiates sketch mode **Autoproject part origin on sketch create** automatically projects the part origin into a new sketch. You can then use this point to further constrain the sketch.
3D Sketch ☐ Auto-bend with 3D line creation	Automatically places tangent corner bends on 3D lines as you sketch them. Select the check box to automatically place corner bends; clear the check box to suppress automatic creation of corner bends.

Part Options

Sets the defaults for creating new parts.

Sketch on new part creation ○ No new sketch ⦿ Sketch on x-y plane ○ Sketch on y-z plane ○ Sketch on x-z plane	Sets the preference for creating the sketch when a new part file is created. Select one of the options: • **No new sketch** disables automatic sketch creation when creating a new part. • **Sketch on x-y plane** sets X-Y as the sketch plane when creating a new part. • **Sketch on y-z plane** sets Y-Z as the sketch plane when creating a new part. • **Sketch on x-z plane** sets X-Z as the sketch plane when creating a new part.
☑ Auto-hide in-line work features ☑ Auto-consume work features and surface features	When this is enabled, the user will not see the edges of in-line features. **Auto-consume work features and surface features** automatically deletes work features and surface features used to create dependent features. Once consumed, you can not roll-back to the consumed/deleted features.

2-33

Construction ☐ Opaque surfaces	Sets the visibility of construction geometry as see-through when enabled.	
3D grips ☑ Enable 3D grips ☑ Display grips on selection	If 3D grips are enabled, then you can use 3D grips to modify features on a part. If **Display grips on selection** is enabled, then when you select a face or edge, the 3D grip control will appear.	
Dimensional constraints ○ Never relax ● Relax if no equation ○ Always relax ○ Prompt	Determines the method to use when feature changes initiated by 3D Grip editing conflict with existing constraints.	
	Never Relax	Halts a feature from being grip-edited in a direction that has a dimensional constraint applied.
	Relax If No Equation	Features with a dimensional equation applied can not be grip-edited.
	Always Relax	A feature can be grip-edited regardless of any dimensions applied.
	Prompt	Provides a warning prompt prior to grip-editing a feature where a dimension is currently applied.
Geometric constraints ● Never break ○ Always break ○ Prompt	Determines how to cope when conflicts occur between 3D Grip-editing and geometric constraints.	
	Never Break	If a constraint exists, 3D grip-editing is ignored.
	Always Break	Deletes existing geometric constraints to allow 3D grip-editing.
	Prompt	Prompts the user before deleting any existing geometric constraint.

User Interface

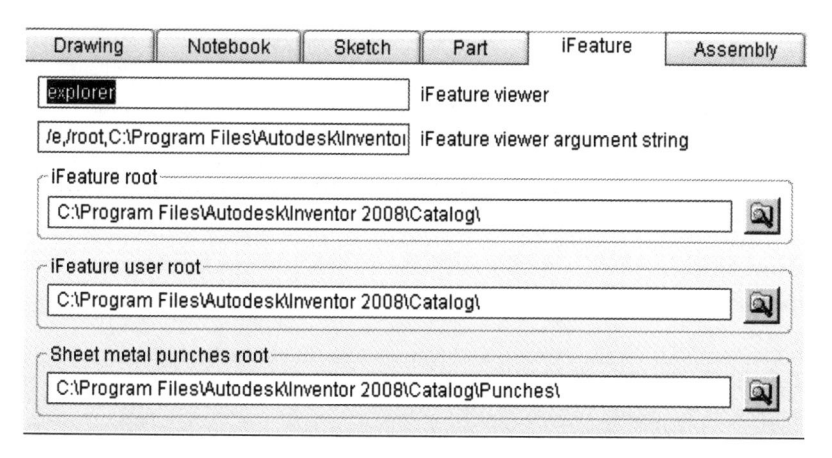

	iFeature viewer
explorer	
/e,/root,C:\Program Files\Autodesk\Inventor	iFeature viewer argument string

iFeature root
C:\Program Files\Autodesk\Inventor 2008\Catalog\

iFeature user root
C:\Program Files\Autodesk\Inventor 2008\Catalog\

Sheet metal punches root
C:\Program Files\Autodesk\Inventor 2008\Catalog\Punches\

iFeature Options

iFeature Viewer	Specifies the viewer application used to manage the design element files. Enter the name of the executable file for the viewer application in the box. The default is Windows Explorer.
iFeature Viewer Argument String	Sets the viewer command line arguments for run-time options. The default is /n. Windows Explorer opens with the folder specified in the Design Elements Root box. *Note: To find out if your viewer application supports command line arguments, refer to the Help for your viewer.*
iFeature Root	Specifies the location of design element files used by the View Catalog dialog box. The location can be on your local computer or on a network drive accessed by other users. Enter the path in the box or click Browse to search for and select the path. The default path is the path to the Catalog folder installed with Autodesk Inventor.
iFeature User Root	Specifies the location of Design Element files used by both the Create Design Element and Insert Design Element dialog boxes. The location can be on your local computer or on a network drive accessed by other users. Enter the path in the box or click Browse to search for and select the path. The default path is the path to the Catalog folder installed with Autodesk Inventor. If desired, you can define a Design Elements Root location on a network drive that can be accessed by others in your company and a Design Elements User Root location on your computer hard drive. *Note: Use Windows shortcuts to quickly access other folders. For example, you can place a shortcut to a shared folder on a network in your Design Elements User Root folder.*
Sheet Metal Punches Root	Specifies the location of iFeature files used by the sheet metal Punch Tool dialog box. The location can be on your local computer or on a network drive accessed by other users. Enter the path in the box or click Browse to search for and select the path. The default path is the path to the Catalog folder installed with Autodesk Inventor.

Use the Browse button to locate the design element files.

The Power of Design: An Introduction to Autodesk Inventor

> **TIP:** It is a good idea to locate any custom files, such as iFeatures, away from the standard Inventor directories. Otherwise, if you have to re-install Inventor or when you perform an upgrade, you may lose your work.

iFeatures were called Design Elements in previous releases of Inventor.

Assembly Options

Sets the preferences for creating assemblies.

☐ Defer update ☐ Delete component pattern source(s) ☐ Enable constraint redundancy analysis ☐ Enable related constraint failure analysis ☐ Features are initially adaptive ☐ Section all parts ☑ Constraint audio notification	**Defer Update** sets preference for updating assemblies when you edit components. Enable to defer updates of an assembly until you click the Update button for the assembly file. Clear the check box to update an assembly automatically after you edit a component. **Delete Component Pattern Sources** sets the default behavior when deleting pattern elements. Enable to delete the source component when deleting a pattern. Disable to retain the source component instance(s) when deleting a pattern.
Enable Constraint Redundancy Analysis	Used to control constraint checking when you use adaptive parts. If enabled, Inventor will alert you to redundant constraints that may affect adaptivity.
Enable Related Constraint Failure Analysis	Default is OFF. If it is turned on, the Constraint Doctor will determine why applied constraints conflict or fail.
Section All Parts	Controls whether standard parts can be sectioned in an assembly. Standard parts are placed from the Standard Parts library. These are primarily fasteners, washers, bolts, and pins. The default is NOT to section these small components.
Constraint Audio Notification	Enables/disables the audio sound when a constraint is applied.

┌─ In-place features ──────────────────────────────┐ │ From/to extents (when possible): │ ☐ Mate plane │ ☐ Adapt feature │ │ Cross part geometry projection │ ☑ Enable associative edge/loop geometry projection during in-place modeling └──┘		
In-Place From/To Feature Extent	When creating a part in place in an assembly, you can set options to control feature termination. **Mate plane** and **Adapt feature**, when possible, construct an adaptive relationship automatically. When the plane on which the feature is constructed changes size or position, the in-place feature adapts.	
Cross part geometry projection	When you create an in-place part or feature, the edges of the selected face will automatically be projected into the base sketch if this is enabled.	
┌─ Component opacity ─┐ │ ○ All │ ⦿ Active only └─────────────────────┘	Controls how parts in an assembly are viewed when editing a single component within an assembly. The default is to dim all components except the one being modified.	
┌─ Zoom target for place component with iMate ─┐ │ ○ None │ │ ○ Placed component │ │ ○ All └──┘	Sets the View when using an iMate. **None** means the view will not change. **Placed Component** zooms into the component you are inserting into an assembly. **All** zooms to fit the entire assembly	
Default level of detail: ┌──────────────────┬─┐ │ Last Active │▼│ ├──────────────────┴─┤ │ Master │ │ All Components Suppressed │ │ All Parts Suppressed │ │ Last Active │ └────────────────────┘	Sets the default level of detail to be used when saving an assembly.	

The Standard Toolbar

The Inventor Standard toolbar looks similar to the Microsoft Windows standard toolbar. It has New File, Open File, Undo and Redo tools.

Many users will miss some of the standard Windows tools: Cut, Copy, Paste, and Print.

These tools can be added back using **Tools→Customize**.

Exercise 2-5:
Customizing your Toolbar

Drawing Name: New
Estimated Time: 10 minutes

When you have completed this lesson, you can:

- Customize Toolbars

1. Start a new **Part** file.

   A quick way to do this is to left-click on the down arrow next to the **New** button and select **Part**.

2. Right-click in the grey area of the toolbar.

   Select **Customize**.

3. Select the **Commands** tab.

   Highlight **Management**.

4. Locate the **Cut** tool in the right hand pane.

5. Drag and drop the **Cut** tool into position on the Standard toolbar.

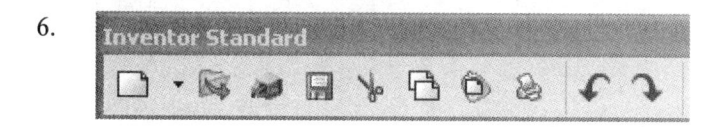

6. Repeat to place the **Paste**, **Copy**, and **Print** tools on the toolbar.

7. To remove a tool, simply reverse the process.
   Hold down the left mouse button and drag the tool back into the Commands window.

8. Close the Customize dialog to finish.

Review Questions

| General | Save | File | Colors | Display | Hardware | Prompts | Drawing |
| Notebook | | Sketch | | Part | | iFeature | Assembly |

1. To turn off the 3D indicator in the drawing window:
    A. Go to Tools→Options→Display
    B. Go to Tools→Options→Part
    C. Go to Tools→Options→General
    D. Go to Tools→Options→iFeature

2. In the 3D Indicator, the red arrow is:
    A. the X axis
    B. the Y axis
    C. the Z axis
    D. the active sketch plane

3. The maximum number of versions of a file that can be saved is:
    A. 1
    B. 2
    C. 5
    D. 10

4. To change the units used in a file:
    A. Go to Tools→Document Settings
    B. Go to File→Properties→Units
    C. Go to Tools→Options→General
    D. Go to File→Properties→Physical

5. In assembly mode, the 3D indicator indicates the orientation of:

    A. The active part
    B. The active assembly
    C. The base part of the assembly
    D. The active sketch plane

| General | Save | File | Colors | Display | Hardware | Prompts | Drawing |
| Notebook | | Sketch | | Part | | iFeature | Assembly |

6. The Options tab to select to set the color of the drawing window:

    A. General
    B. Colors
    C. Display
    D. Drawing

User Interface

[Tabs shown: General | Save | File | Colors | Display | Hardware | Prompts | Drawing | Notebook | Sketch | Part | iFeature | Assembly]

7. The tab to select the set the sketch plane to be used when creating a new part:

    A. General
    B. Sketch
    C. Part
    D. iFeature

8. The tab to select to determine how constraints are applied.

    A. General
    B. Sketch
    C. Part
    D. iFeature

[ ] Retrieve all model dimensions on view placement

9. You place a check mark next to 'Retrieve all model dimensions on view placement'. This accomplishes the following:

    A. Model dimensions will automatically update when placing a view.
    B. Model dimensions will become visible when placing a drawing view.
    C. Model dimensions will become visible when rotating a model.
    D. Model dimensions will automatically be erased when placing a drawing view.

10. The information entered in the Properties dialog box may be used in:

    A. Parts lists
    B. Title blocks
    C. Mass Analysis
    D. All of the above

11. In order to define a new project, you must have all files closed.

    A. True
    B. False

12. The active project file is automatically set to Read-Only.

    A. True
    B. False

The Power of Design: An Introduction to Autodesk Inventor

13. The Old Versions folder is used to store previous file versions of parts, drawings, and assemblies.

    A. True
    B. False

14. The new Section Standard Parts setting determines:

    A. Whether parts inserted from the Standard Parts library are sectioned in a section view of a an assembly
    B. Whether first-level components in an assembly are sectioned
    C. Whether all the parts in an assembly are sectioned
    D. None of the above

15. The audio constraint notification can be turned off using the Assembly tab on the Application Options dialog.

    A. True
    B. False

ANSWERS: 1) C; 2) A; 3) D; 4) A; 5) C; 6) B; 7) C; 8) B; 9) B; 10) D; 11) A; 12) A; 13) A; 14) A; 15) A

Lesson 3
2D Sketch Panel Tools

Inventor's Sketch toolbar contains tools for creating the basic geometry to create features and parts.

On the surface, the Geometry tools look fairly standard: Line, Circle, Arc, Rectangle, Fillet/Chamfer, Point and Polygon.

Line/Spline

Let's start with the **Line** tool. Its drop-down has two options: **Line** or **Spline**. Run the mouse over the button and look in the lower left-hand section of the screen; a help description will appear describing the tool function. In this case, the tool creates lines and tangent arcs. This means filleted corners can be created without having to exit the line mode and performing a fillet command.

When you are in Sketch mode, these tools will appear on the Standard toolbar.

⊥	To create a Construction line instead of an object line, toggle the **Construction Line** option on the Standard toolbar.
⊕	To create a Center/Axis line, toggle the **Center Line** option tool.
·	To create an **Object Line**, this tool is toggled.
H×H	To add an **Associative/Driven Dimension**, toggle this tool.

3-1

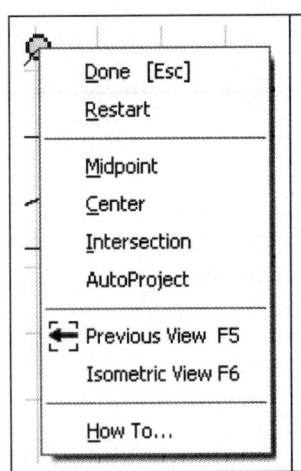

Right-click the mouse while in 'Line' mode; this will bring up a submenu to assist in the construction of your sketch.

To create a tangent arc while in 'Line' mode, select an endpoint and hold down the left mouse button. When the arc is located properly, release the left mouse button and Inventor will automatically return to 'Line' mode.

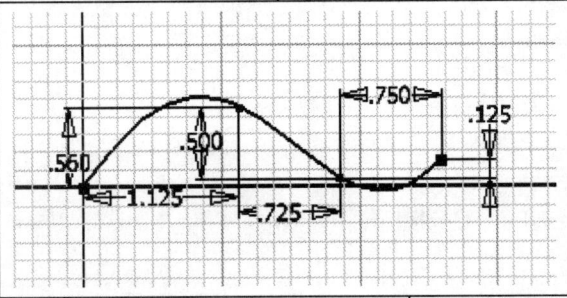

The **Spline** tool allows you to add and remove points, change the fit, and adjust the shape.

You can add dimensions to control the size of the spline and fully constrain it.

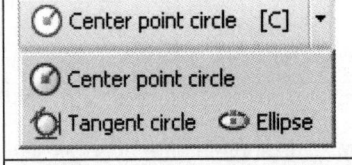

Circle/Ellipse

The default **Circle** tool creates a circle using **Center point** and **Radius**. Access the drop down toolbar to see that there are two other circle options: **Ellipse** and **Tangent, Tangent, Tangent**.

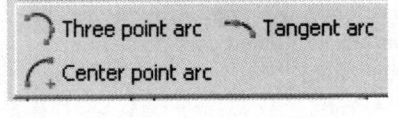

Arc

The default **Arc** tool creates an arc using three points. The drop down toolbar provides two additional options: **Center, Start, End** and **Start, Tangent**. All three methods will draw arcs either clockwise or counter-clockwise.

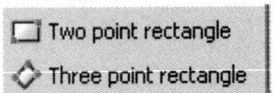

Rectangle

The **Rectangle** tool provides two options:
- The default is to select the two opposite corners of the rectangle.
- The second option has the user define the length of one side and then the length of the adjacent side.

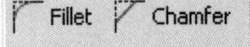

Fillet/Chamfer

The **Fillet** tool is actually a flyout that includes **Fillet** and **Chamfer**. Inventor recommends that it is better to add fillets and chamfers as placed features. The reason is that the user can then suppress fillets for faster regens and to conserve memory. It also makes it easier to modify values. However, there are instances where it is preferable to include the fillet or chamfer in the sketch.

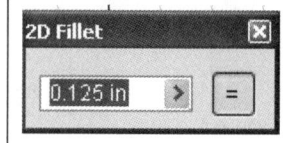

Sketch Tools

[2D Fillet dialog: 0.125 in]	The **Fillet** tool prompts the user to select the edges of the sketch to be modified and brings up a dialog box where the user can modify the radius value. To modify the value of a fillet you've already placed, just double-click using the left mouse button and a dialog box will pop up allowing you to edit the value. Pressing the equal button allows the user to select an existing fillet and apply that fillet's value to the fillet being defined.
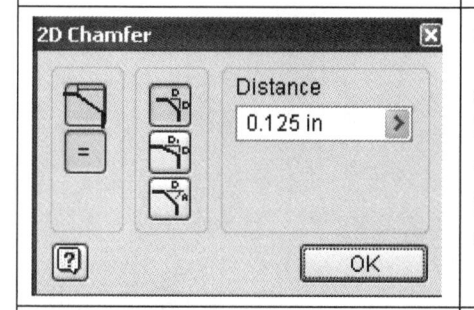 [2D Chamfer dialog: Distance 0.125 in]	**Chamfer** Chamfers can be defined in three ways: **Equal Distance**, **2 Distance**, and **Distance-Angle**. The user also has the option of selecting an existing chamfer in the sketch and applying that value to the chamfer being defined.
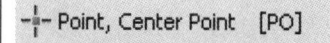 Point, Center Point [PO]	**Point, Hole Center** The **Point** tool is used to determine the location of holes as well as points. To create a **Sketch Point** (used to constrain geometry), select the Sketch Point under Style.
Polygon [Polygon dialog: 6, Done]	[Inscribed icon] **Inscribed** uses the vertex between two edges to determine the size and orientation of the polygon.
	[Circumscribed icon] **Circumscribed** uses the midpoint of an edge segment to determine the size and orientation of the polygon.
	 [6 dropdown] The **number dropdown** specifies the number of edges used to create the polygon shape. The maximum number allowed is 120.

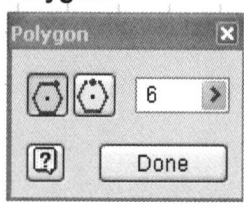

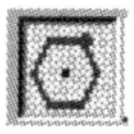

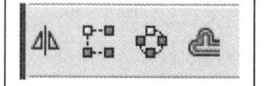

[Pattern toolbar icons]	The next section of the Sketch toolbar contains **Pattern** tools: **Mirror**, **Rectangular Pattern**, **Circular Pattern** and **Offset**.

[Mirror icon]	**Mirror** Use the **Mirror** tool on the Sketch toolbar to mirror sketch geometry across a centerline. Equal constraints are automatically applied to the mirrored geometry and source geometry. You can delete or edit segments after you mirror them and the remaining segments will retain their symmetry and constraints.

TIP: You must keep the mirror axis selection separate from the geometry selection. If you accidentally include the mirror axis in the selection of geometry to be mirrored, you will get an error message.

Use of Mirror when creating symmetric parts will use up less system resources.

Exercise 3-1:
Mirror

File: New (Standard using Inches)
Estimated Time: 30 minutes

This lesson reinforces the following skills:

- Rectangle
- Project Geometry
- Dimension
- Extrude
- Redefine Sketch
- New Sketch
- Mirror
- Close Loop
- Show Dimensions
- Update

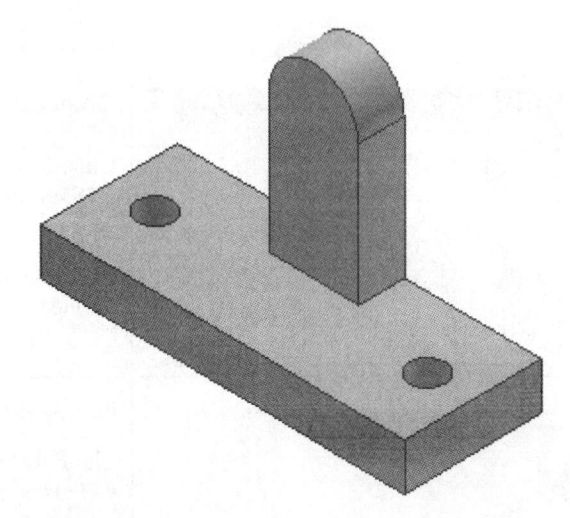

1. Start a new file using Standard units.

   Assembly
   Drawing
   Part
   Presentation

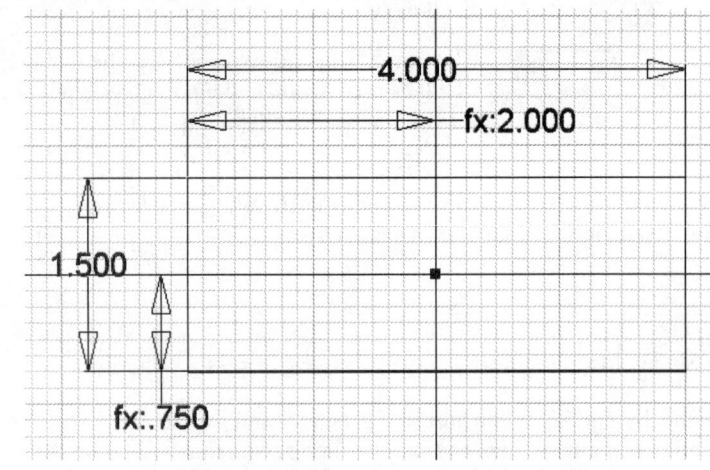

2. Draw a 4 x 1.5 in. rectangle and center it using a projected center point.

   First draw the rectangle using the rectangle tool.

   To project the center point, select **Project Geometry** then select the **Origin** point in the Browser.

3. Draw a **Rectangle** in the graphics window.

4. Select the **Project Geometry** tool.

3-4

5. 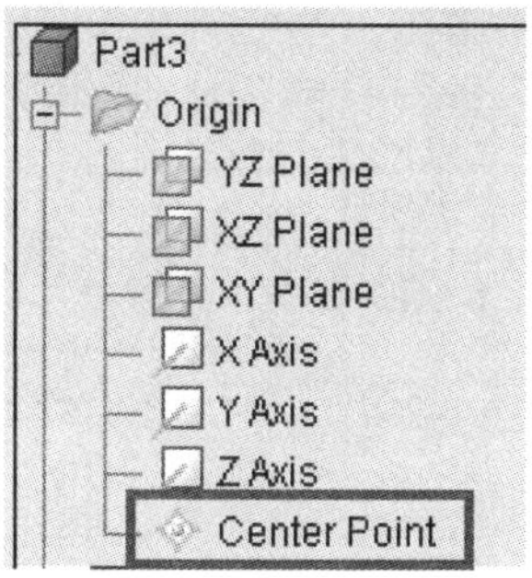 Left pick on the **Center Point** located in the Panel Browser.

This will project or add the center point to the active sketch.

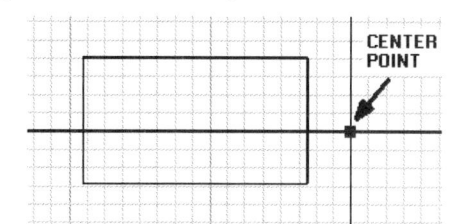

6.  Select the **Dimension** tool.

7. 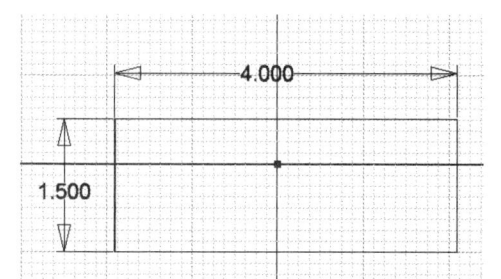 Add the **4.00** and **1.500** dimensions to the rectangle.

The size of the rectangle will automatically adjust to the new dimensions.

8.  Select the **Dimension** tool.

9. 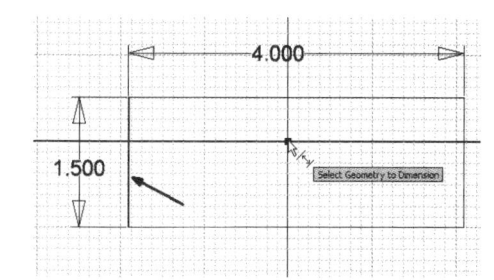 Select the left vertical line and the center point and place a horizontal dimension.

10. 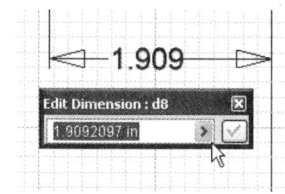 Erase the dimension shown by highlighting and pressing the backspace or delete button.

11. 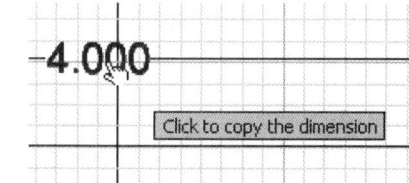 Left-click on the **4.00** dimension and that value will be copied into the edit box.

12. 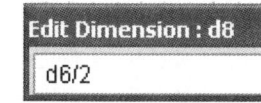 The dimension name will appear in the **Edit Dimension** box. (This is the name assigned by Inventor to keep track of dimensions.)
Add a 'divided by' symbol (/) and the number **2**.
This will automatically center the rectangle in the horizontal direction.
Press the green check symbol on the dimension edit box to end.

13. *NOTE: The fx: displayed on the dimension indicates that a formula was used to define the dimension.*

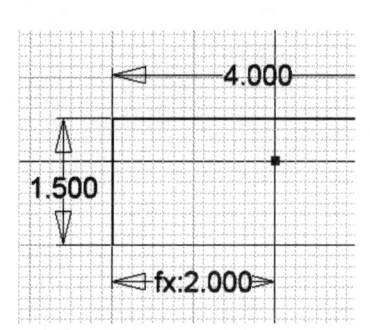

14. Repeat for the horizontal direction to center the rectangle on the origin.

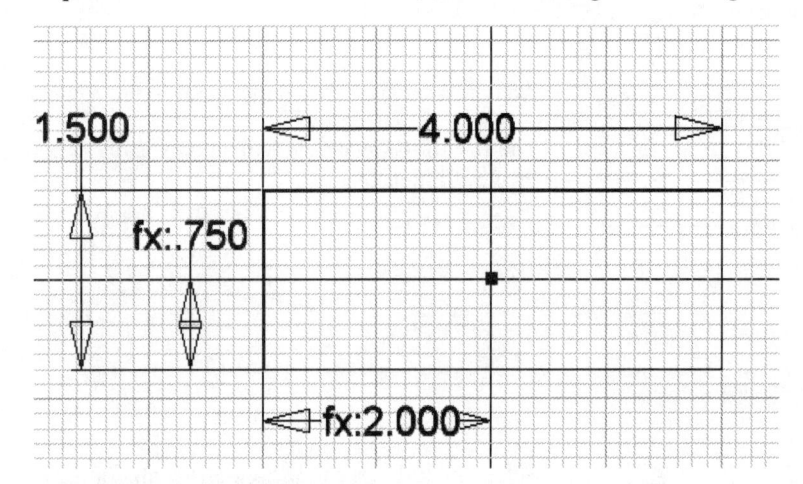

15. Right-click and select **Finish Sketch**.

16. Press **F6** to switch to an Isometric View or right-click in the graphics window and select **Isometric View**.

17. Select the **Extrude** tool.

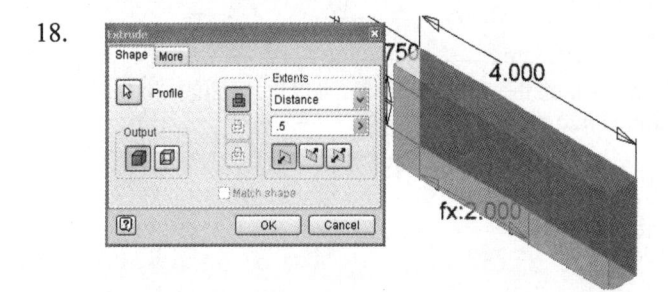

18. Extrude **0.5** inches.

    Set the Extents to **Distance.**
    Set the Distance value to **.5**.

    Press **OK**.

We would like to reorient our isometric view so the block is lying flat.
We can do this two ways:
- We can redefine the isometric view, or
- We can redefine the sketch to a different workplane.

In this exercise, we will redefine the sketch.

Sketch Tools

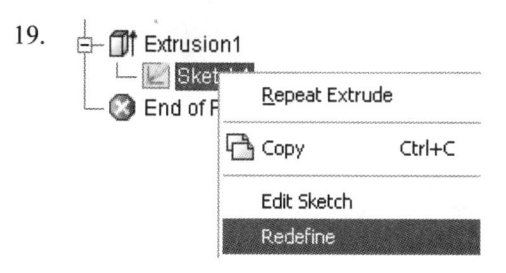

19. Highlight the sketch in the Browser.
Right-click and select **Redefine**.

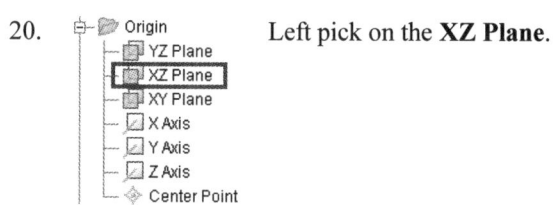

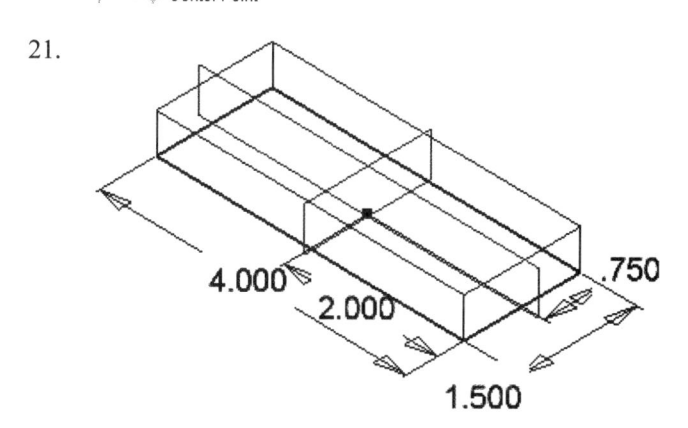

20. Left pick on the **XZ Plane**.

21. The box flips to the correct orientation.

Note how it remains centered on the center point.

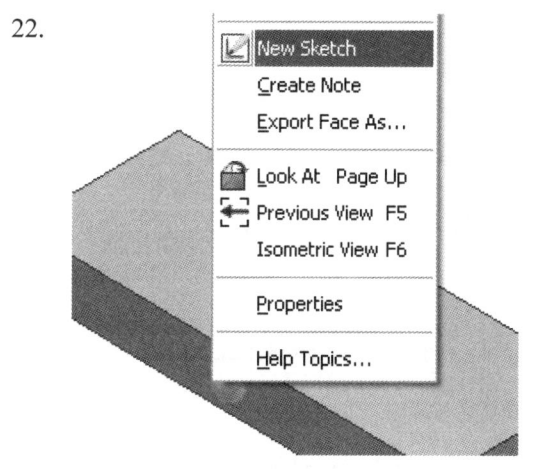

22. Highlight the front face.
Right-click and select **New Sketch**.

23. Select **Project Geometry**.

3-7

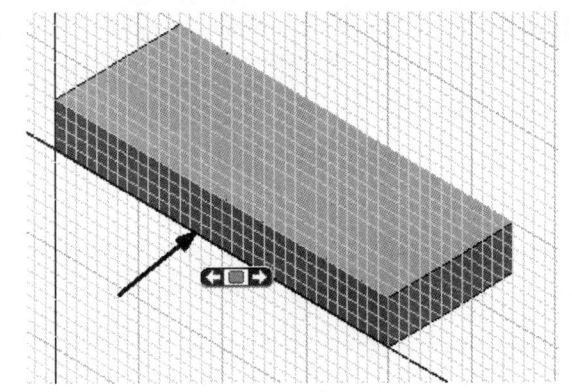

24. Select the bottom edge of the block.

    This copies a line using the edge into your active sketch.

    In order to locate the midpoint of the bottom edge, you need to project the bottom edge.

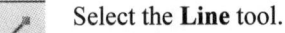

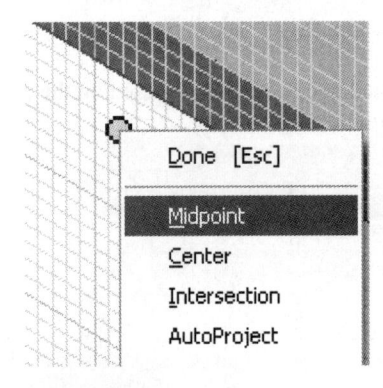

25. Select the **Line** tool.

26. Draw a vertical line at the midpoint of the front side. Right-click to select **Midpoint** to have your mouse locate the midpoint. A green point will appear at the end of the cursor to indicate that the midpoint has been selected.

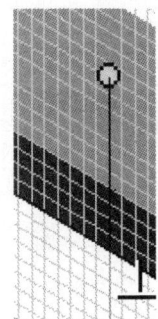

27. A Perpendicular constraint symbol will appear when the line is straight.

    Select a point above the block to end the line.

28. Highlight the line and toggle **Centerline** on the Standard toolbar.

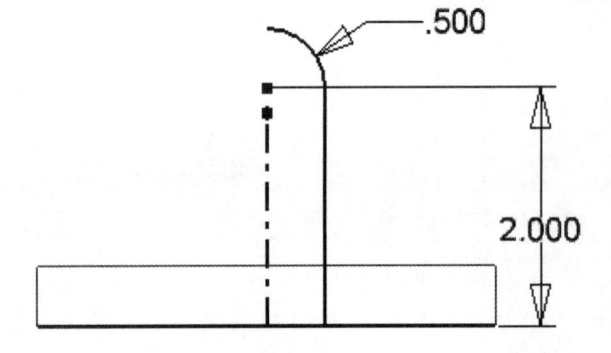

29. Create the sketch shown.

    Draw a vertical line and an arc.

    Add a **Coincident** constraint between the arc center and the centerline.

    Add a **Vertical** constraint between the open arc end and the centerline.

3-8

30. Select the **Mirror** tool.

31. Select the line and arc. You can do this using a window or by picking each object.

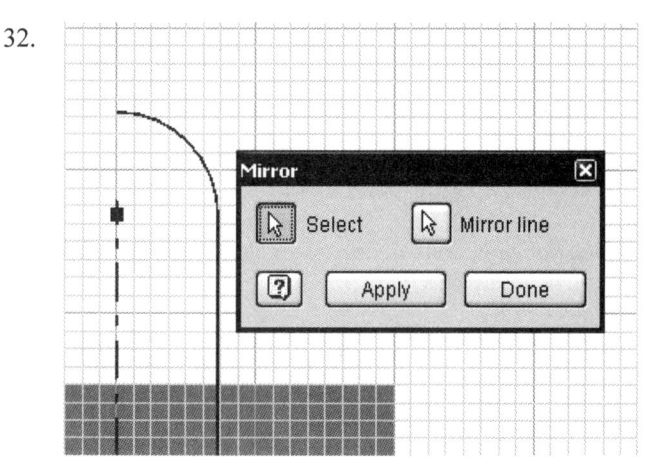

32. You may need to deselect the centerline to specify it as the **Mirror line**.

    To deselect, press down the Control key then pick the centerline.

    Select the **Mirror line** select button.

    Then select the center line.

    Once all the selections are done, press **Apply** and then **Done**.

33. Draw a horizontal line between the two vertical lines to create a closed profile.

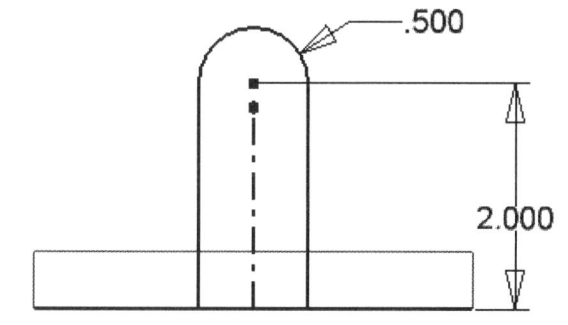

34. Use the **Dimension** tool to add dimensions.

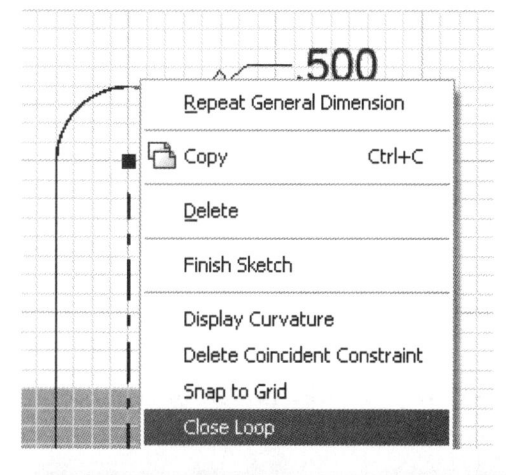

35. Our mirrored geometry does not define a closed loop.

    Select the right arc of the sketch.
    Right-click and select **Close Loop**.
    Then select the remainder of the sketch.

36. Press **OK**.

Select the vertical line, horizontal line, left vertical line, and left arc, in order.

37. *[Dialog: Autodesk Inventor 2008 — A Coincident constraint is required between the highlighted sketch curve and sketch point. Would you like to add a Coincident constraint? Yes / No / Cancel]*

As you select geometry, you will see messages to add Coincident constraints or to close gaps. When those messages appear, press **Yes**.

38. *[Dialog: Close Loop — The loop has been successfully closed. OK]*

    When you have fully defined your closed polygon, this message will appear.

    NOTE: If you continue to get errors, select the bottom horizontal line and then the remaining geometry to form the closed polygon.

    Press **OK**.

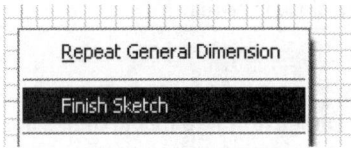

39. Right-click and select **Finish Sketch**.

40. Press **F6** to switch to an Isometric View, or right-click in the graphics window and select **Isometric View**.

41. Select the **Extrude** tool.

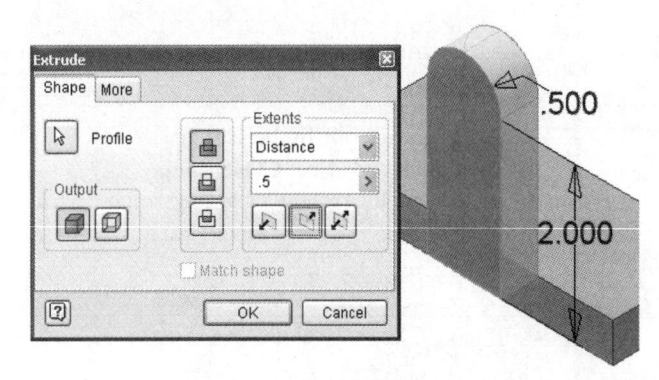

42. Extrude the geometry into the block a depth of **0.5** units.

    If the sketch is not selected, press the select **Profile** button, then select the sketch.

    Use the **Direction** buttons to determine the direction of the extrusion.

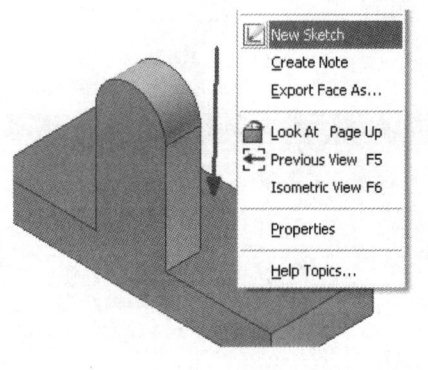

43. Select the top face, right-click, and select **New Sketch**.

44. Draw a circle on the right side and dimension as shown.

    NOTE: It is generally not considered a good idea to extrude a circle to create a hole. The preferred method is to use the Hole tool. The reason is that, when you create your 2D layout, you will be able to use the Hole Note tool to designate the hole rather than a basic dimension.

45. Select the **Project Geometry** tool.

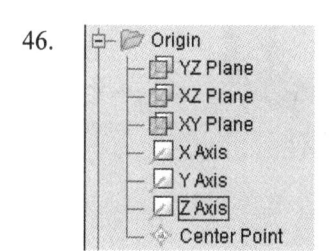

46. Select the **Z-axis** to add it to the sketch.

47. Select the projected axis.

48. Change the style to **Centerline** using the Centerline toggle.

49. Select the **Mirror** tool.

    Select the circle to be mirrored and the axis as the **Mirror line**.

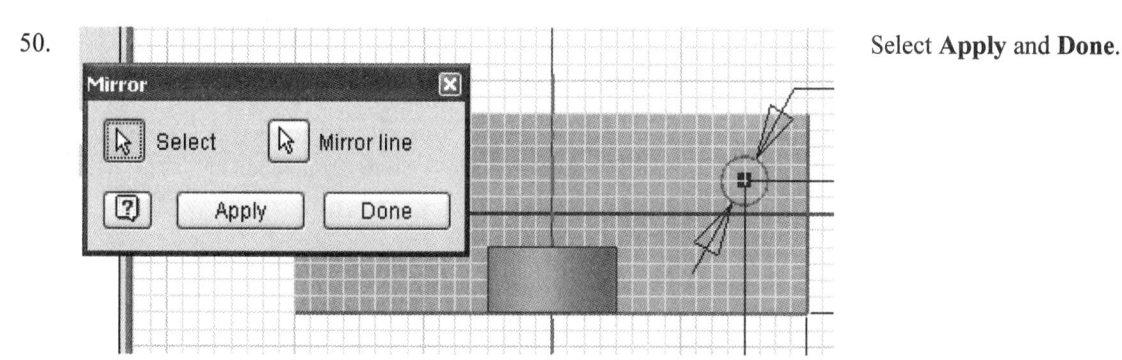

50. Select **Apply** and **Done**.

    NOTE: One advantage to mirroring the hole is that you only need to change the dimensions on one hole and both holes will automatically update.

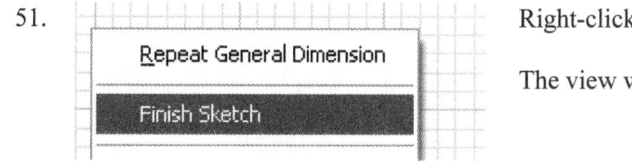

51. Right-click and select **Finish Sketch**.

    The view will automatically switch to an isometric view.

52. Select the **Extrude** tool.

3-11

53. Select both holes.
    Set to **Cut**.
    Set Extents to **All**.
    Extrude the holes as a cut through all.

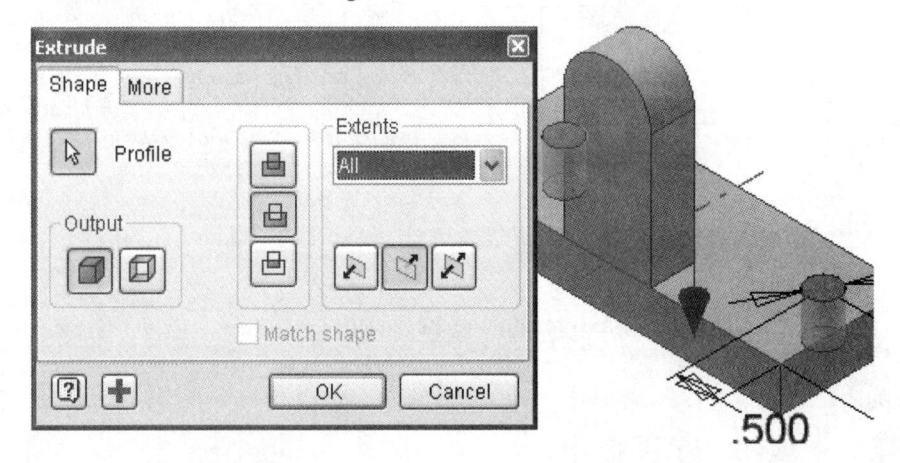

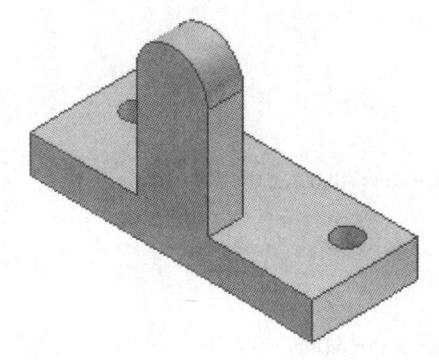

    Our model so far.

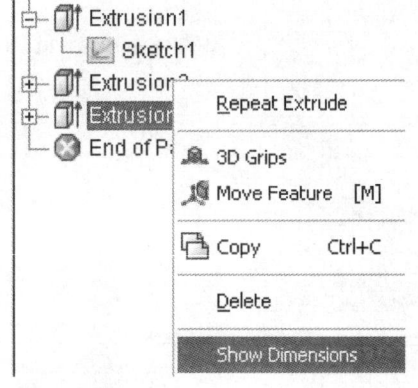

54. Select the **hole extrusion** in the Browser.

    The holes will highlight in the graphics window when selected.

    Right-click and select **Show Dimensions**.

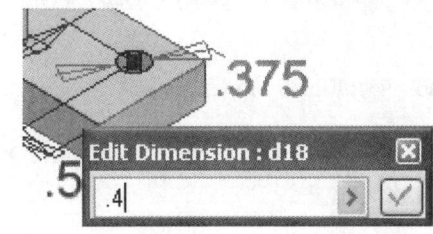

55. Change the **hole diameter** to **0.400**.

Sketch Tools

56. Change the **1.00 dimension** to **0.75**.

57. Press **Update**.

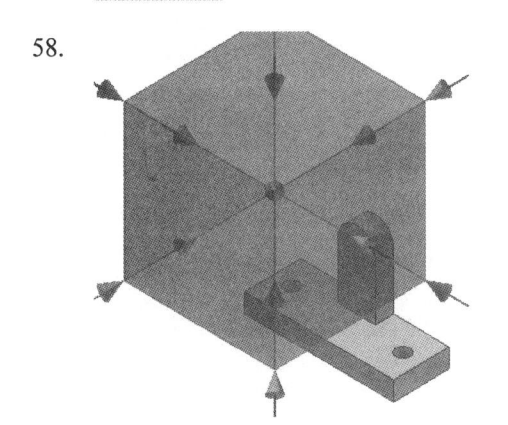

58. Use **Common View** to change the view orientation of the part.

    Select the **3D Orbit** tool thEn press the **SPACE** bar to switch to Common View mode.

    Select the green arrows to change the display.

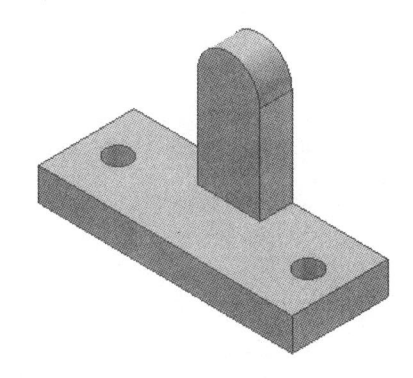

    Note that both holes shift and change size. This is because the mirror definition causes them to be linked.

59. Save as *ex3-1.ipt*.
    Close the file.

3-13

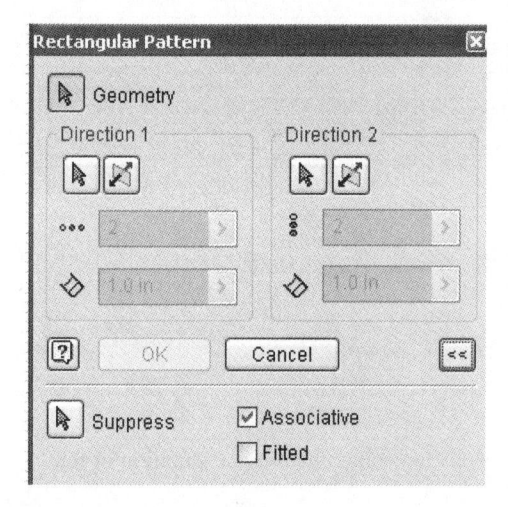

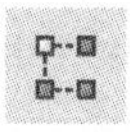

Rectangular Pattern

This **Rectangular Pattern** tool only works in Sketch mode.

Select the geometry to be patterned.
Select the edge to be used as the axis for the pattern.

To suppress an instance, pick the Suppress button and then select the instance to suppress. A dashed line designates suppressed instances.

TIP: When you edit pattern dimensions, you can use parametric equations to drive the position of your sketch patterns.

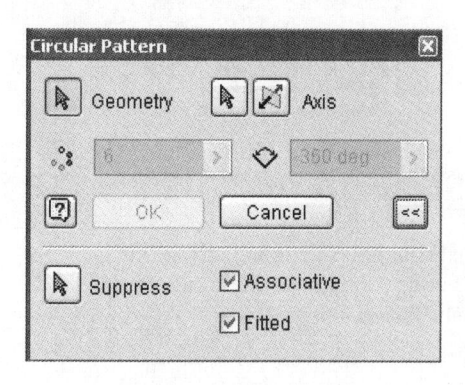

Circular Pattern

Select the geometry to pattern.

For the axis, select an arc or circle.

To suppress an instance, select the Suppress button and then the sketch geometry to suppress. A suppressed instance is designated by a dashed line.

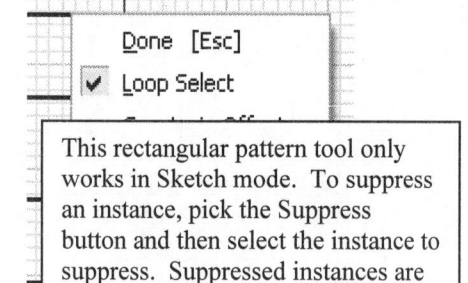

Offset

The **Offset** tool prompts the user to select the object to offset and the user then uses the mouse to drag and drop the offset copy to the approximate location. To constrain the offset object, the user can add dimensions using the **Dimension** tool.

A right mouse click brings up a submenu where the user can determine the constraints used for the offset or change views to facilitate editing.

This rectangular pattern tool only works in Sketch mode. To suppress an instance, pick the Suppress button and then select the instance to suppress. Suppressed instances are designated by a dashed line.

More than one object can be selected at a time for **Offset**. The selected objects will highlight in green. When we have completed our selections, right-click the mouse and select **Continue** in the submenu. Then drag the offset to the approximate location desired.

The default setting automatically selects loops (curves joined at the endpoints) and constrains the offset curve to be equidistant from the original curve. To offset one or more individual curves or omit the Equal constraint, right-click and clear the checkmarks on Loop Select and Constrain Offset in the submenu.

Sketch Tools

General Dimension

General Dimension

The first icon, which resembles a paintbrush roller, is used for General Dimensioning. Inventor automatically knows whether the object being dimensioned is a line or an arc.

If an arc is being dimensioned, you can right-click the mouse and bring up a sub-menu. This submenu allows you to switch from Radius mode to Diameter mode simply by selecting that option.

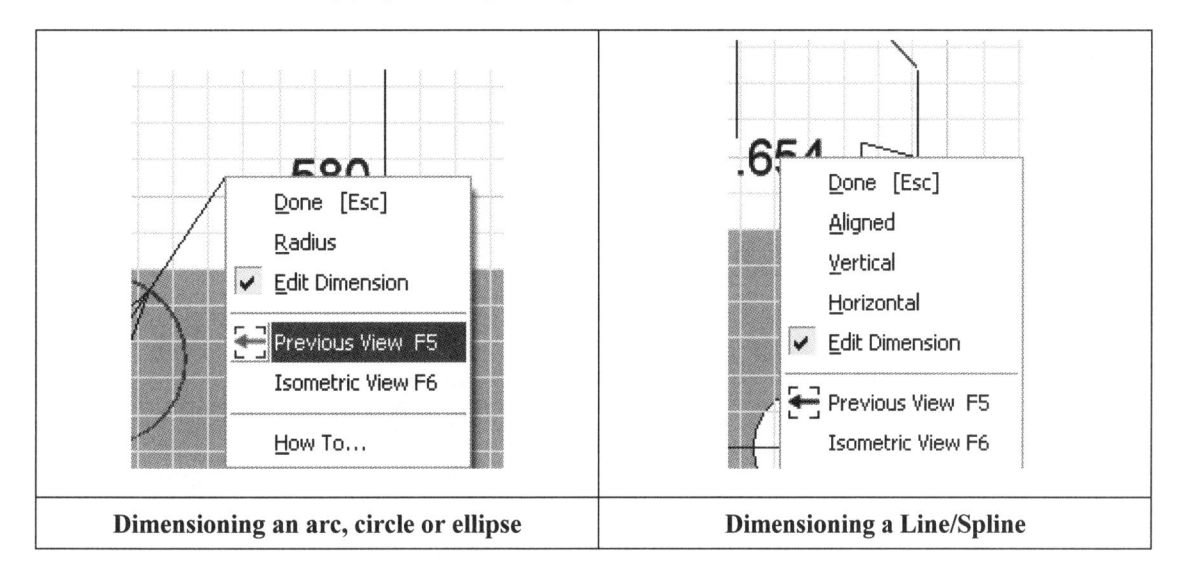

Dimensioning an arc, circle or ellipse	Dimensioning a Line/Spline

When dimensioning a line, right-clicking the mouse will bring up a submenu with the options for **Aligned**, **Vertical** or **Horizontal** linear dimensions.

Simply selecting a dimension and then editing the value in the dialog box that appears will modify any dimension.

TIP: When **Edit Dimension** is enabled, as displayed in the dimension shortcuts, this means that the Edit Dimension dialog will automatically display whenever you place a dimension. You can toggle this user option off or on from the shortcut menu.

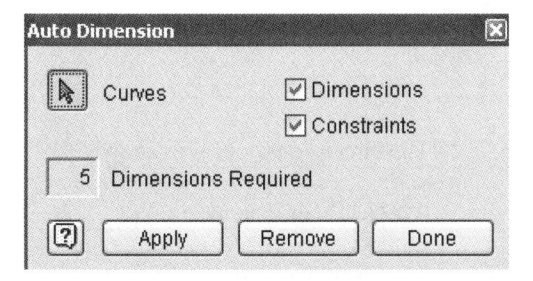

Auto Dimension

Auto Dimension tells the user how many dimensions are required to fully define a sketch and applies constraints as needed.

3-15

The Power of Design: An Introduction to Autodesk Inventor

Exercise 3-2:
Auto Dimension

File: New (Standard using Inches)
Estimated Time: 30 minutes

This exercise reinforces the following skills:

- Sketch
- Auto Dimension
- Sketch Constraints

1. Start a new file using Standard units.

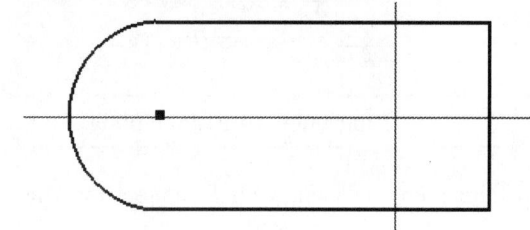

2. Draw the sketch shown.

   Do not add any dimensions or constraints.

3. Select **Auto Dimension**.

4. A dialog appears indicating how many dimensions are required to fully constrain the sketch.

   Press the **Apply** button.

5. The dimensions appear as shown.

   NOTE: Your dimensions will probably be different depending on how you drew your sketch.

   Press **Done**.

   You can now select dimensions and edit them as needed.

3-16

Sketch Tools

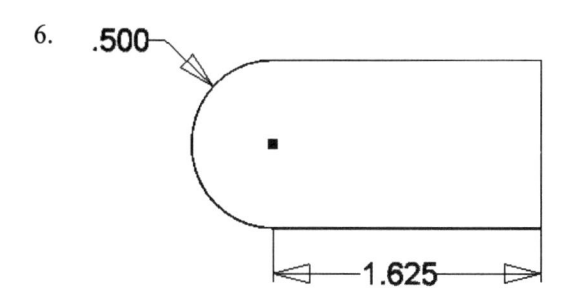

6. [sketch showing dimensions .500 and 1.625]

Modify the dimensions as shown. You will have to delete two of the redundant dimensions in order to make the changes. Right-click and select **Finish Sketch**.

7. Save the file as *ex3-2.ipt*.
    Close the file.

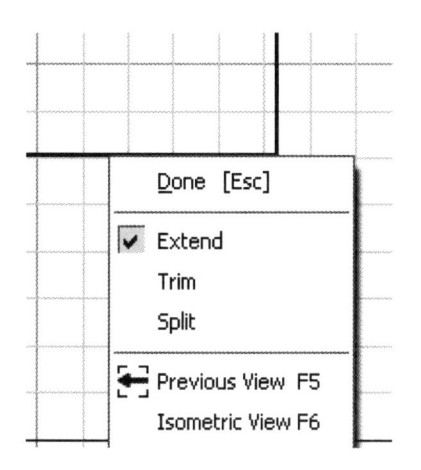

[context menu showing: Done [Esc], ✓ Extend, Trim, Split, Previous View F5, Isometric View F6]

Extend

The **Extend** tool works differently than in AutoCAD. The user is prompted for the object to extend. The object then highlights in red and the user moves the mouse to indicate how far to extend the object. Inventor previews the object as modified and the user left-clicks the mouse to accept the modification.

Right-clicking the mouse while in 'Extend' mode will bring up a submenu giving the option to switch to 'Trim' mode or change views.

Trim

The **Trim** tool prompts the user to select the object to trim and automatically uses any intersecting edges as the cutting tool. Inventor previews the modification in red for the user and the user accepts by left-clicking the mouse. A right mouse click brings up the same submenu as the Extend right mouse click, only with the check mark appearing next to the Trim option. Thus, the user can easily switch from 'Trim' mode to 'Extend' mode.

Split

The **Split** tool works similarly to the Break @ function in AutoCAD. Select the tool, select the line, and select a break point. The line will then be broken into two separate segments.

> **TIP:** Press and hold SHIFT to temporarily enable Trim when in Extend mode, or to enable Extend when in Trim mode.

> **TIP:** You can double left-click on a sketch name to activate Edit Sketch mode.

The Power of Design: An Introduction to Autodesk Inventor

Exercise 3-3:
Move and Copy

File: Ex3-2.ipt
Estimated Time: 15 minutes

This exercise reinforces the following skills:

- Using the Move Sketch tool

1. Open *ex3-2.ipt*.

2. 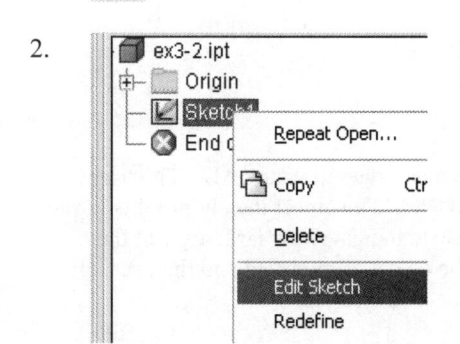 Highlight the **Sketch** in the Browser. Right-click and select **Edit Sketch**.

3. Select the **Move** tool from the 2D Sketch toolbar.

4.

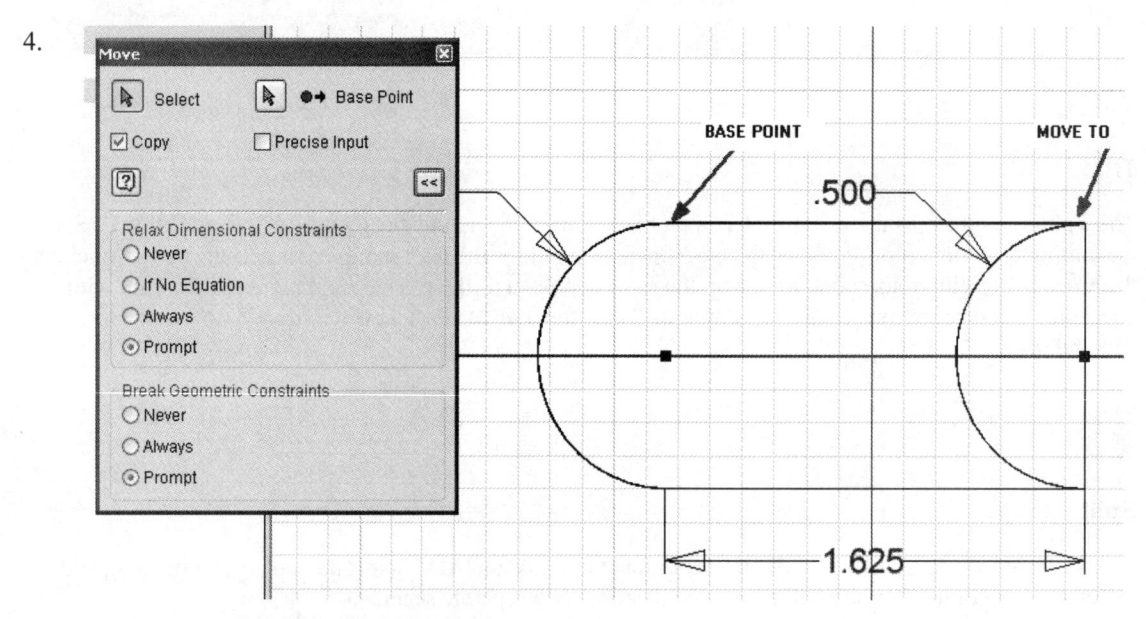

   Press the **Select** button and select the arc.
   Enable the **Copy** button.
   The arc will highlight to indicate it has been selected.
   Press the **Base Point** button and select the top arc endpoint.
   Select the top endpoint of the rectangle as the destination point.

5. 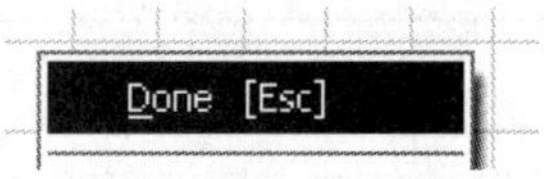 Right-click and select **Done**.

   The arc is now copied to the new position.

6. Exit Sketch mode.

7. Save the file as *ex3-3.ipt*.
   Close the file.

Exercise 3-4:
Copy

File: copy.ipt
Estimated Time: 5 minutes

This exercise reinforces the following skills:

- Using the Copy Sketch tool

1. Open *copy.ipt*.
   This file must be downloaded from the publisher's website at *www.schroff.com/resources*.

2. Highlight **Sketch** in the Browser.
   Right-click and select **Edit Sketch**.

3. Select the **Copy** tool.

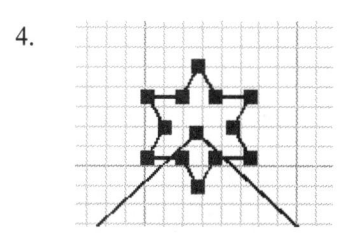

4. Window around the star to select it.

5. Select the **Base Point** button.

3-19

6. Select the tip of the tree as the base point.

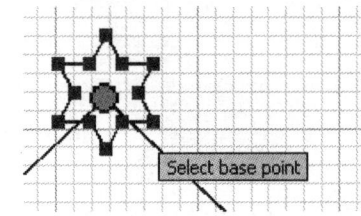

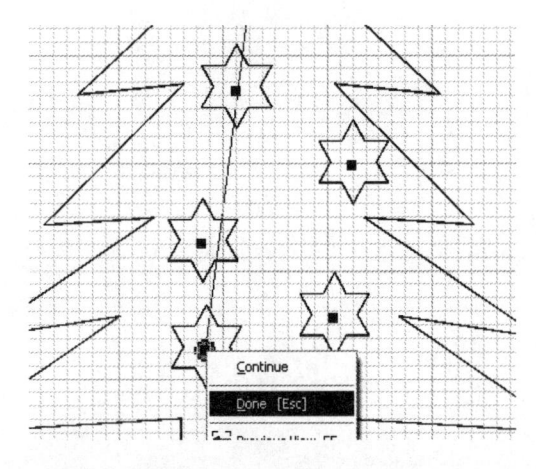

7. Place several stars on the tree.

   Right-click and select **Done**.

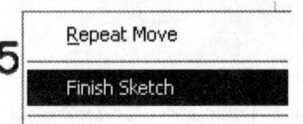

8. Exit Sketch mode.

9. Close the file without saving.

Exercise 3-5:
Scale

File: Ex3-3.ipt
Estimated Time: 10 minutes

1. Open *ex3-3.ipt*.

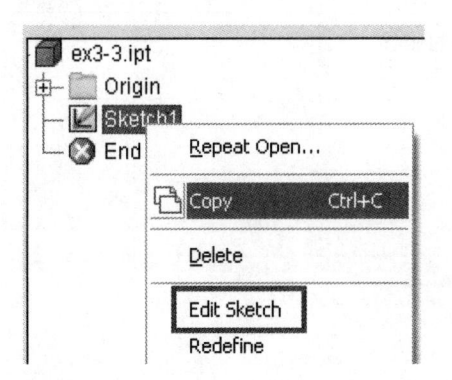

2. Highlight the **Sketch** in the Browser.
   Right-click and select **Edit Sketch**.

3. Select the **Scale** tool.

Sketch Tools

4. Press the **Select** button and window around the sketch to select it.

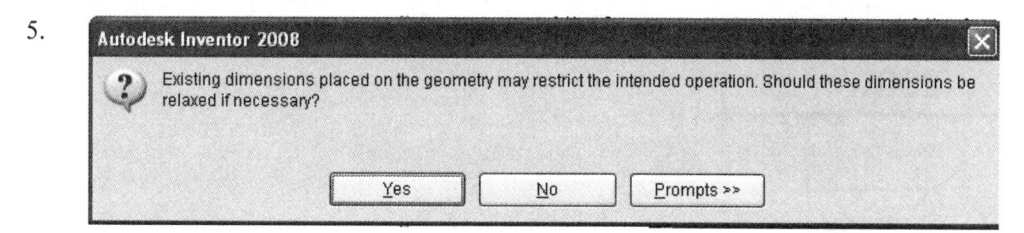

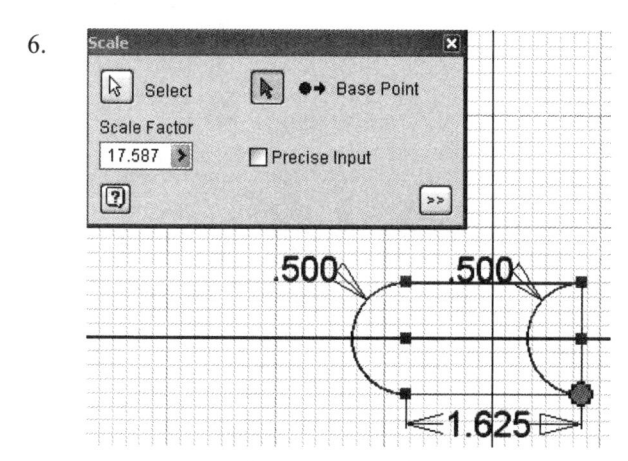

5. Press **Yes**.

6. Select the **Base Point** button and select the lower right corner of the sketch as the base point.

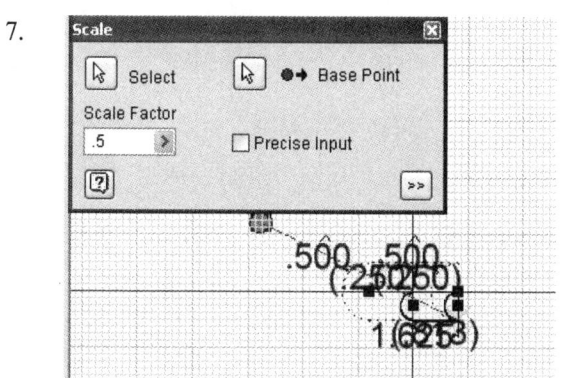

7. Type **0.5** in the Scale Factor box.

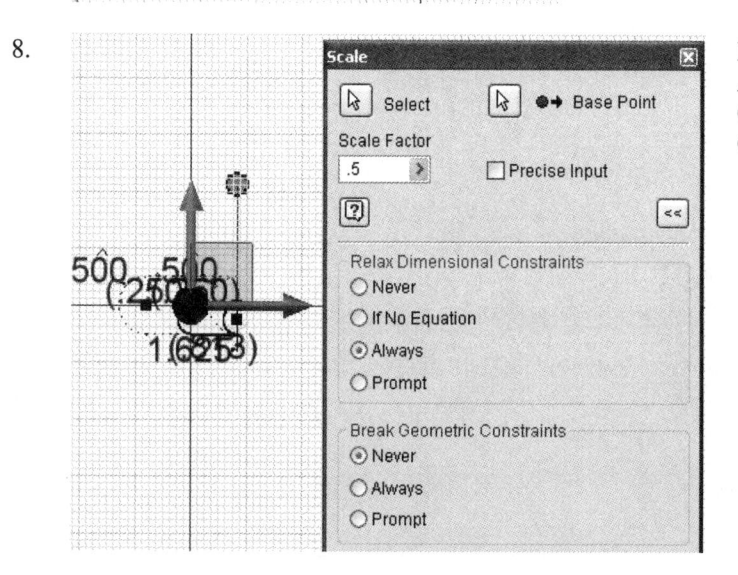

8. Press the **>> More** button and enable **Always** under Relax Dimensional Constraints and **Never** under Break Geometric Constraints.

3-21

9. Press **ENTER**.

10. The sketch is scaled.

11. Close the file without saving.

TIP: The Scale command is particularly useful when you copy geometry from AutoCAD drawings.

Exercise 3-6:
Rotate

File: Ex3-3.ipt
Estimated Time: 5 minutes

1. Open *ex3-3.ipt*.

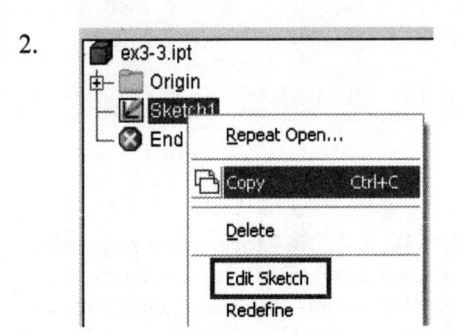

2. Highlight the **Sketch** in the Browser. Right-click and select **Edit Sketch**.

3. Select the **Rotate** tool.

Sketch Tools

4. Press the **>> More** button.

   Enable **Never** under Relax Dimensional Constraints and **Always** under Break Geometric Constraints.

   This means the size of the objects will be maintained, but the orientation may change.

5. Press the **Select** button and window around the entire sketch.

   If you projected the center point into the sketch, you need to make sure you do not select this center point or you will get an error message.

6. Press the **Center Point** button and select the center point of the left arc.

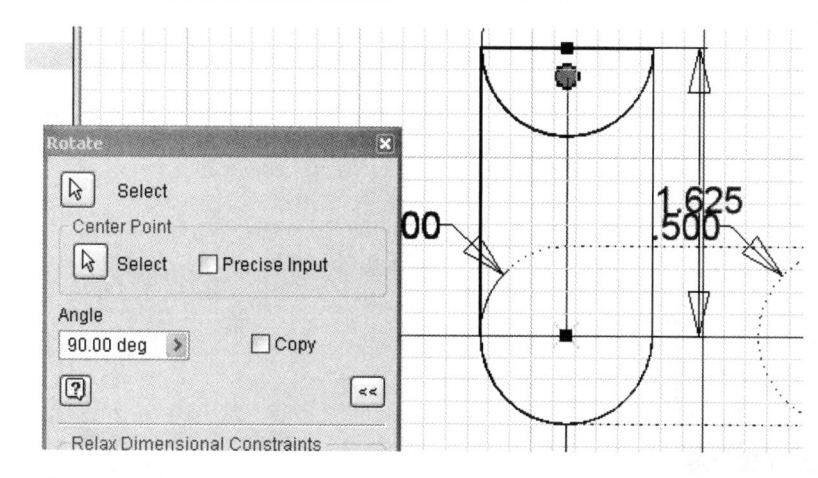

7. Change the **Angle** value to **90**.

   You can do this by moving your cursor until you see the 90 value in the dialog.

   Left-click to end.

3-23

The Power of Design: An Introduction to Autodesk Inventor

8.

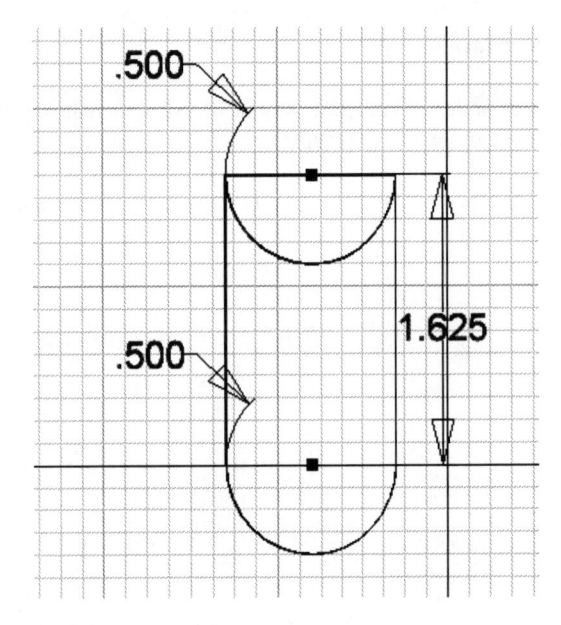

Close the dialog.

Right-click and select **Finish Sketch**.

Close the file without saving.

Exercise 3-7:
Stretch

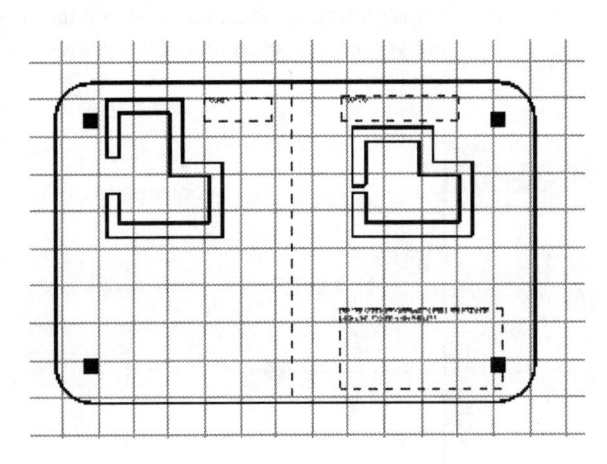

File: stretch.ipt
Estimated Time: 5 minutes

1. Open **stretch.ipt**.
   This file can be downloaded from the publisher's website at www.schroff.com/resources.

   *Note: The **Snap to Grid** option is enabled to make this exercise easier to perform.*

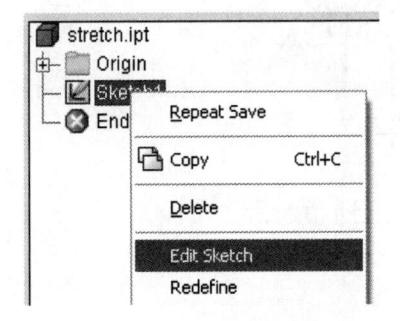

2. Highlight the **Sketch** in the Browser.
   Right-click and select **Edit Sketch**.

3-24

Sketch Tools

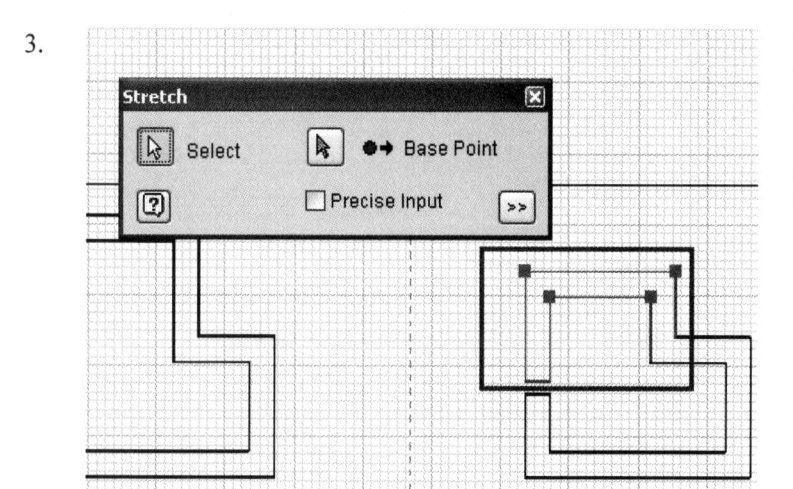

3. Press the **Select** button and window around a portion of the sketch to select it.

If you select the entire sketch, the command does not work.

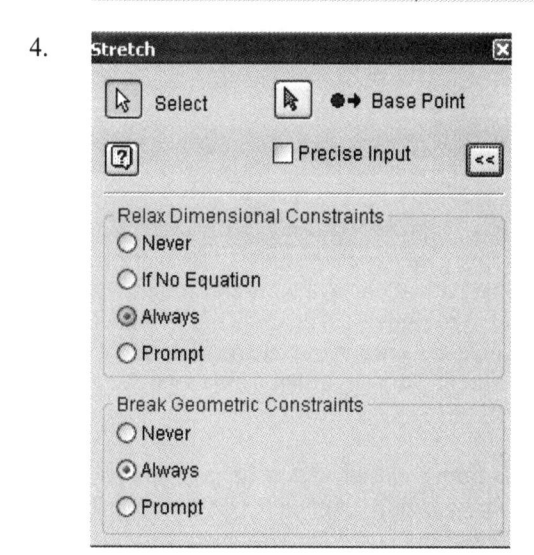

4. Press the **>> More** button.

Enable **Always** under **Relax Dimensional Constraints** and **Always** under **Break Geometric Constraints**.

This means both the size of the objects and the orientation can be changed.

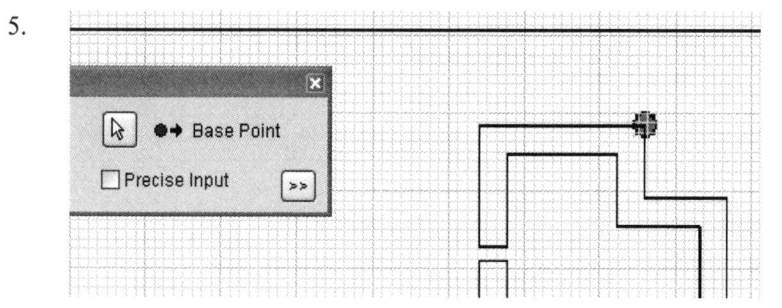

5. Select the **Base Point** button.

Pick the middle upper corner of the sketch as the base point.

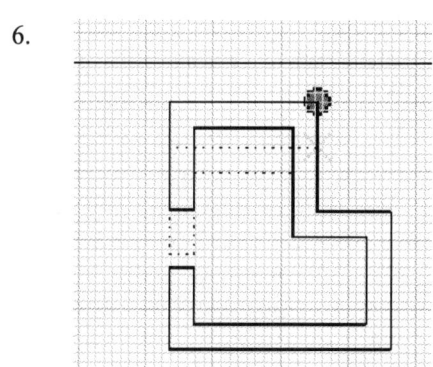

6. Drag the mouse up two grid squares so it lines up with Figure A and pick to place.

3-25

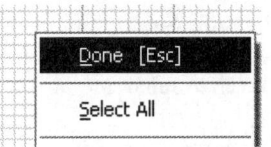

7. Right-click and select **Done**.

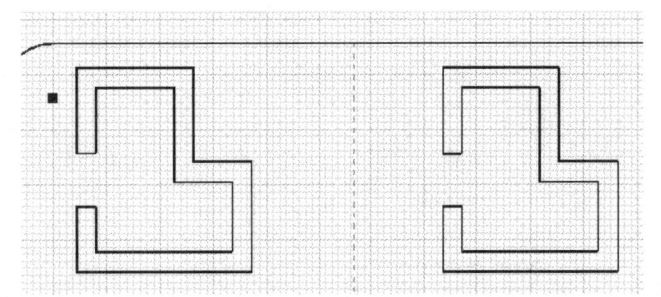

8. Your figures should match.

9. Exit sketch mode.

10. Close without saving.

Constraints

The next tool is used for adding geometric constraints. Pressing on the arrow reveals a fly-out toolbar with all the available constraints. The top row of constraints from left to right are: Perpendicular, Parallel, Tangent, Smooth, Coincident, and Concentric. The bottom row of constraints from left to right are: Collinear, Equal, Horizontal, Vertical, Fixed and Symmetric.

The **Coincident** constraint may be used to ensure that two lines form a closed angle with no overlap. The **Fixed** constraint fixes an object to a location relative to the sketch coordinate system. The other constraints are used in a similar manner to other parametric modeling software.

TIP: Press and hold CONTROL to prevent constraints from being added while sketching geometry.

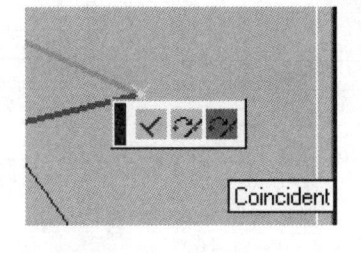

Show/Delete Constraints

To show constraints, press the **Show/Delete Constraints** tool button. Next, select the object. A small Constraint Bar will appear displaying the constraints for that object. Moving the mouse along the constraint bar will highlight each constraint.

To delete a constraint, enable the Constraint Bar. Move the mouse to the constraint to delete on the Constraint Bar. Note the highlighted objects to ensure that the correct constraint will be deleted. Right-click the mouse and the 'Delete' key will appear. Left-click the mouse to accept. If we don't wish to delete, just move the mouse off of the Constraint Bar and left-click anywhere in the window.

Sketch Tools

> **TIPS:**
> - Use the Zoom Window button on the Standard toolbar to zoom in on the area where you are working.
> - Set the grid to the spacing needed to quickly line up the sketch elements.
> - Check the Snap to Grid setting to more easily place sketch elements.
> - To select a group of sketch elements, activate the Select tool, then click in the graphics window and drag a box around the elements.
> - Use the dimension tools to set the size of sketched geometry or to add dimensions between the geometry in a sketch and elements in the underlying drawing view.
> - When you use dimensions to set the size of elements in a title block or border, the dimensions are hidden when you finish editing.

Exercise 3-8:
Adding Horizontal and Vertical Constraints

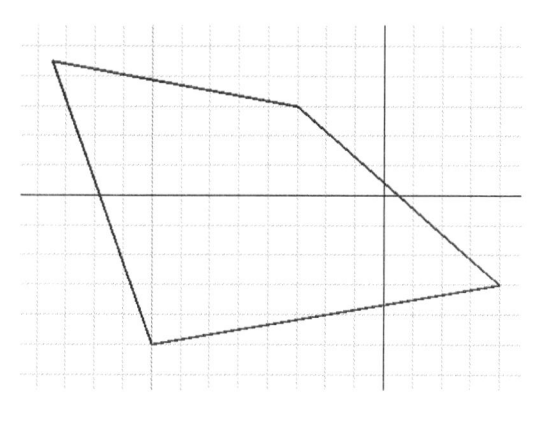

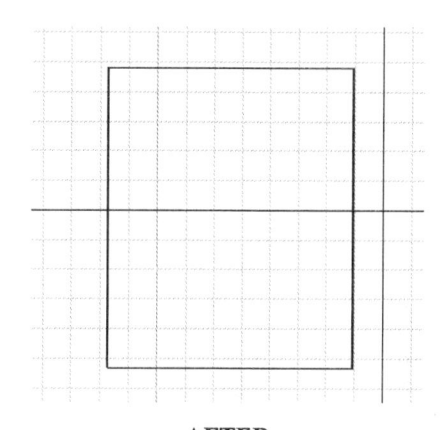

BEFORE   **AFTER**

File: constraint1.ipt
Estimated Time: 5 minutes

This exercise reinforces the following skills:

- Add Constraints

1. Open *constraint1.ipt*.
   This file can be downloaded from the publisher's website at www.schroff.com/resources.

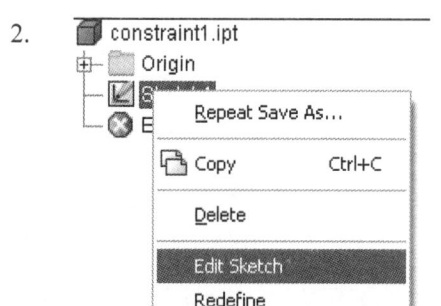

2. Highlight the **Sketch** in the Browser.
   Right-click and select **Edit Sketch**.

3-27

The Power of Design: An Introduction to Autodesk Inventor

3. Select the **Vertical** constraint tool.

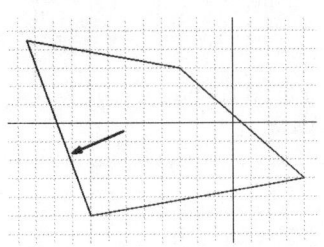

4. Select the line located on the left to apply the constraint.

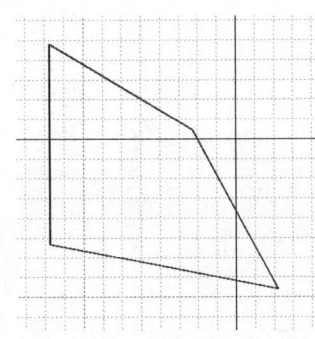

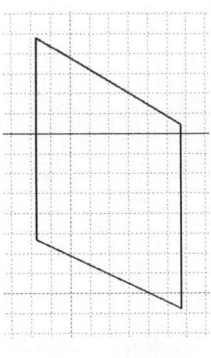

5. The object will shift as the line is constrained.

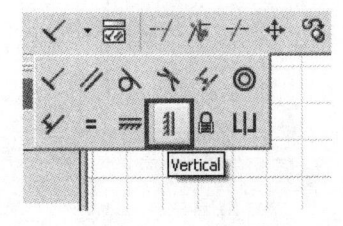

6. Select the **Vertical** constraint tool.

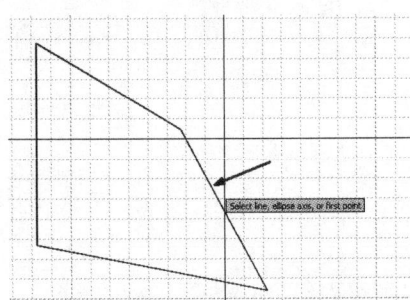

7. Select the line located on the right to apply the constraint.

8. The object will shift as the line is constrained.

Sketch Tools

9. 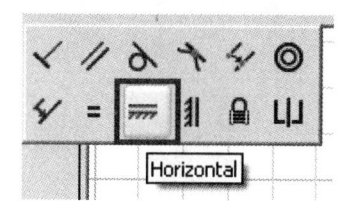 Select the **Horizontal** constraint tool.

10. 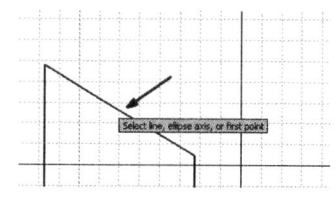 Select the angled line.

11. 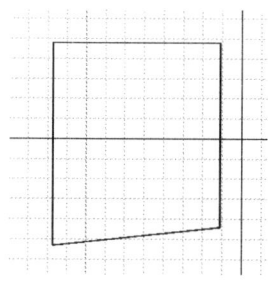 The object will shift as the line is constrained.

12. 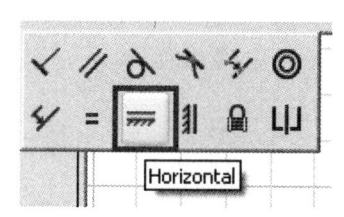 Select the **Horizontal** constraint tool.

13. 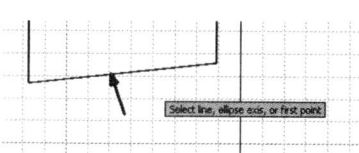 Select the line located on the bottom to apply the constraint.

14. Close without saving.

3-29

The Power of Design: An Introduction to Autodesk Inventor

Exercise 3-9:
Deleting Constraints

File: constraint2.ipt
Estimated Time: 5 minutes

This exercise reinforces the following skills:

- Show Constraints
- Hide Constraints
- Delete Constraints
- Modify Sketch

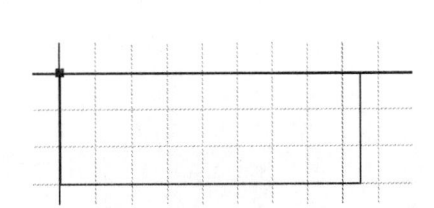

BEFORE 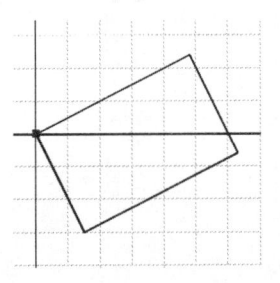   AFTER

1.  Open *constraint2.ipt*.
   This file can be downloaded from the publisher's website at www.schroff.com/resources.

2. 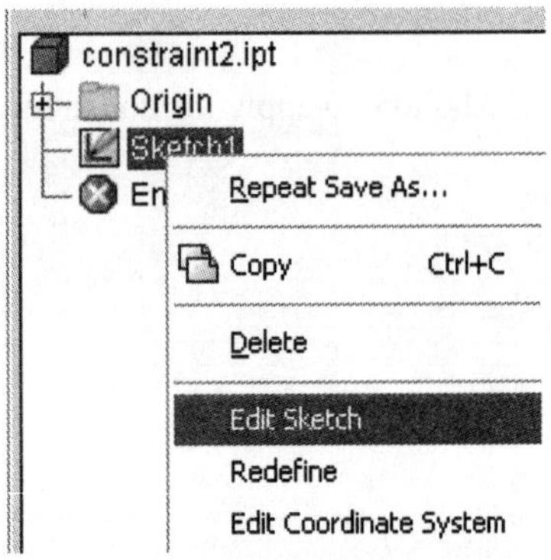 Highlight the **Sketch** in the Browser.
   Right-click and select **Edit Sketch**.

3.  Select the **Show Constraints** tool.

4. 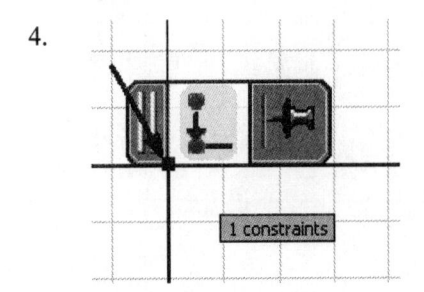 Select the upper left corner of the rectangle.

   You will see a Coincident constraint indicated.

3-30

Sketch Tools

5. Select the left vertical line.

   Right-click and select **Show All Constraints**.

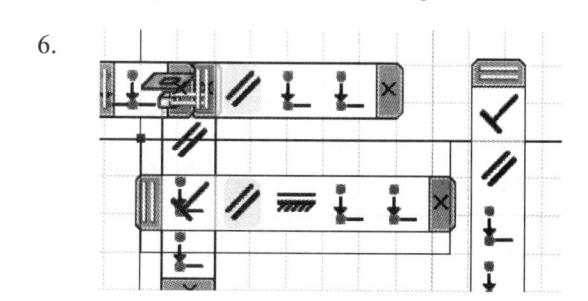

6. Mouse over each constraint.

   Note how the entities highlight to indicate where the constraints are applied.

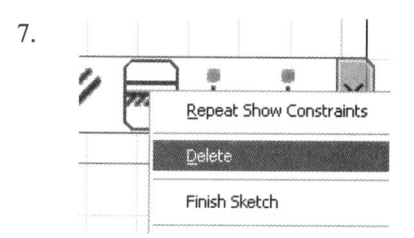

7. Locate the horizontal constraint.

   Right-click and select **Delete**.

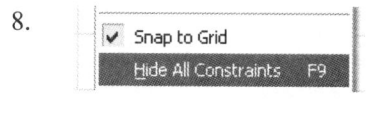

8. Right-click in the display window.

   Select **Hide All Constraints**.

9. Select the lower left corner of the rectangle.

   Use your mouse to rotate the rectangle. Notice that it remains pinned at the coincident constraint you located earlier.

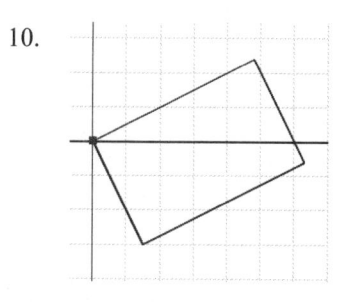

10. Rotate the rectangle and release the mouse to place.

11. Close without saving.

3-31

The Power of Design: An Introduction to Autodesk Inventor

Inventor features three projection tools: Project Geometry, Project Cut Edges, and Project Flat Pattern.

	Project Geometry Our next tool button creates reference geometry by projecting model geometry (edges and vertices), work features, or sketch geometry from another sketch onto the active sketch plane. Reference geometry can be used to constrain other sketch geometry or used directly in a profile or path sketch.
	Project Cut Edges This tool projects edges cut by the sketch plane onto the current sketch plane.
	Project Flat Pattern This tool is grayed out unless a flat pattern exists. If a flat pattern is available, the user may select a face to project it onto a selected plane.
f_x	**Parameters** The Parameters tool is used to create table-driven parts and features.
	Insert AutoCAD file Inserts an AutoCAD drawing into a sketch.
A	**Insert Text** Adds text to a sketch. It can then be extruded using the Emboss tool.
	Insert Image Adds an image to a sketch. It can be converted into a decal.

Sketch Tools

Exercise 3-10:
Inserting an AutoCAD file

File: Ta100dcd.dwg
                [can be downloaded from the publisher's website *(www.schroff.com/resources)* for free]
Estimated Time: 30 minutes

To demonstrate how it works, we use a drawing from Nidec's fan catalog, but any AutoCAD drawing will do.

This exercise reinforces the following skills:

- Insert AutoCAD file
- Measure Distance
- Trim
- Delete
- Close Loop
- Extrude

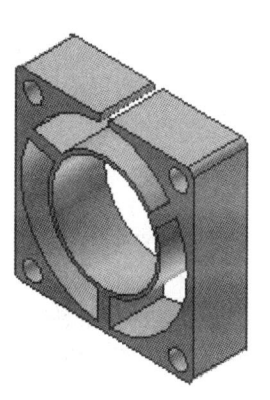

1. Start a new file using Standard units.

2. Select the **Insert AutoCAD file** tool.
   The Browser dialog will come up.

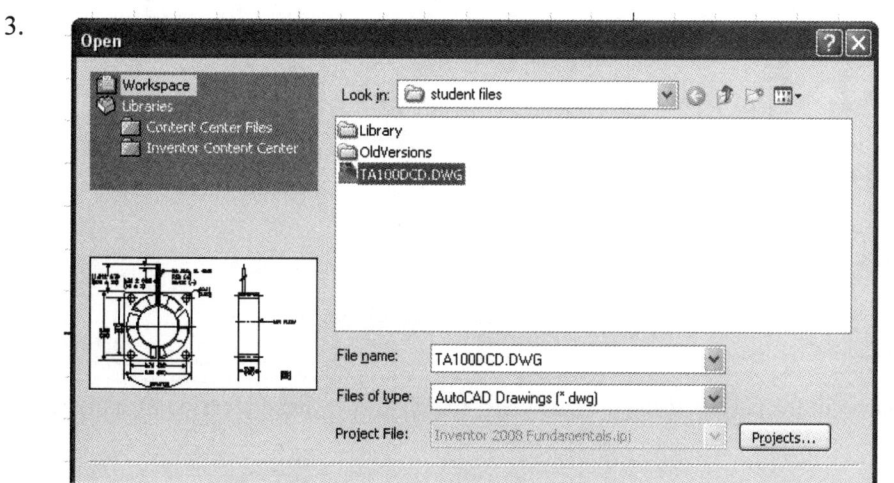

3. Locate the ***TA100DCD.dwg*** file you downloaded from the publisher's website.

   Press **Open**.

3-33

The Power of Design: An Introduction to Autodesk Inventor

4.

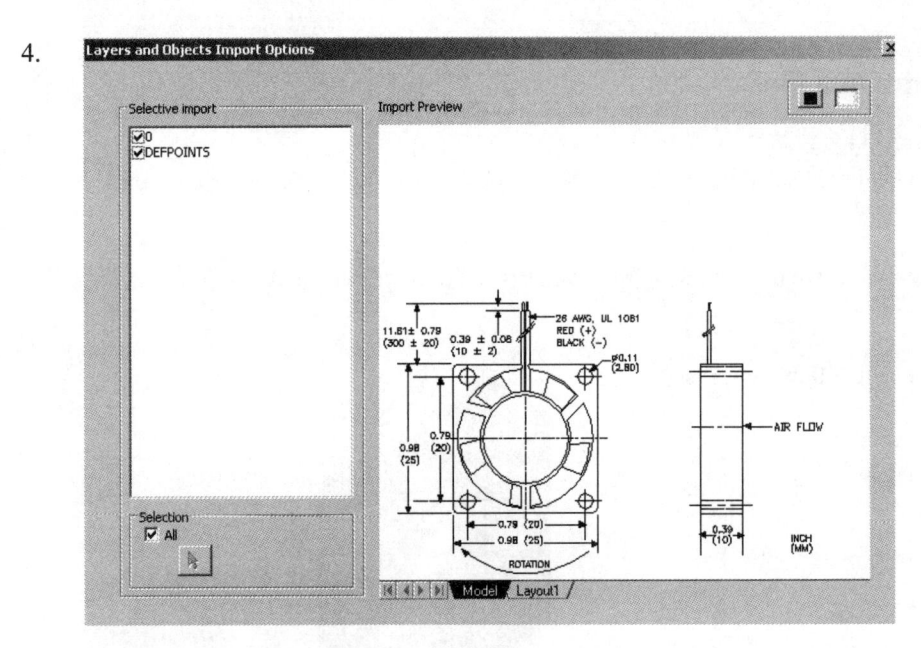

A dialog will appear to preview the drawing.
Note you can select which layers you want to import.
Press **Finish**.

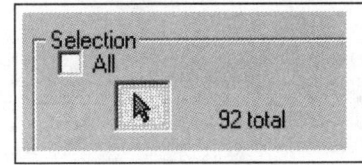

TIP: If you disable the **All** option under **Selection**, you can use **Window**, **Crossing**, or **Pick** to select the entities in the preview window you wish to import.

5. Note that the geometry is imported on a Sketch named 0, the same name as the layer used by the entities.

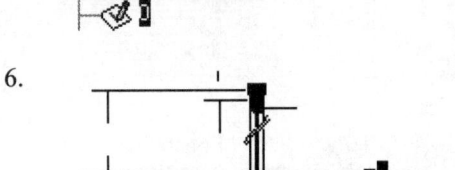

6. The AutoCAD drawing appears in your sketch.

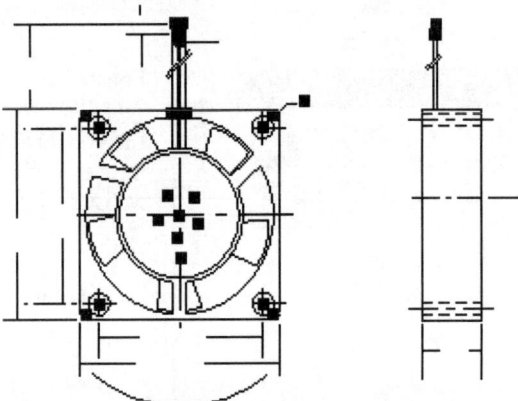

7. Measure the thickness of the fan box using the side view, so you know what dimension to apply for the extrusion.

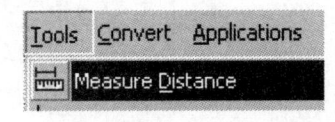

You can measure by placing a linear dimension or using the **Measure Distance** tool under the Tools menu.

3-34

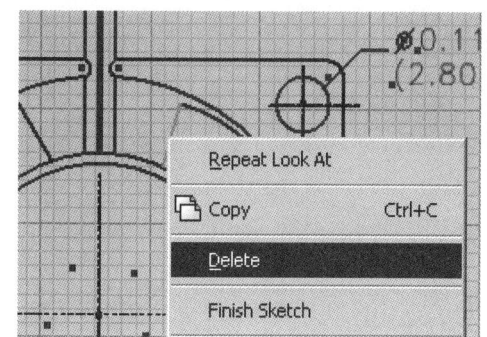

8. By holding down the Control key and picking with the left mouse, we can select all the dimension lines, then right-click and press **Delete**. We also need to delete the side view.

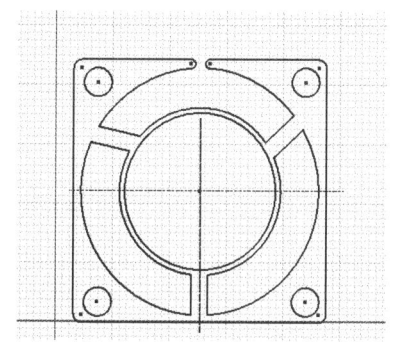

9. Continue cleaning up the sketch until you have a basic profile.

   The sketch should look like this once it has been cleaned up.

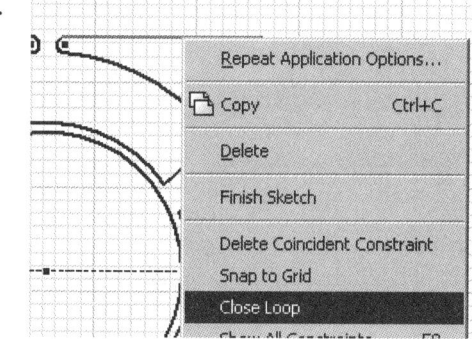

10. Use the **Sketch Doctor** to assist you in creating a closed loop profile.

    You can also use the **Close Loop** option from the shortcut menu to create your profile.

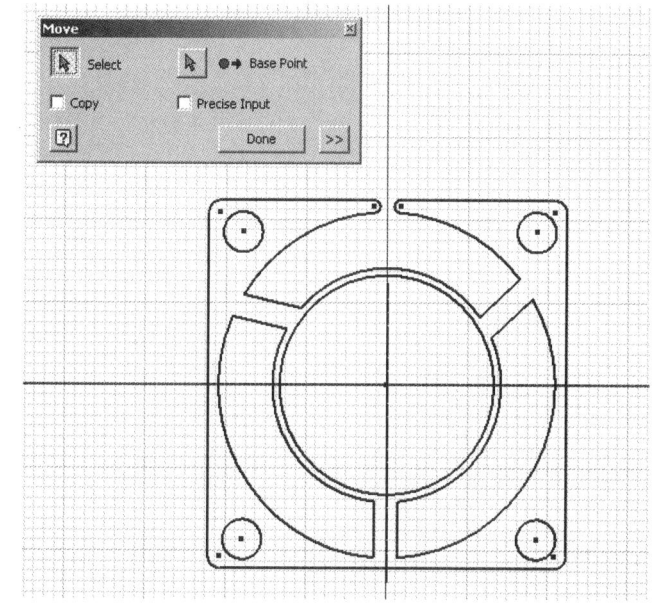

11. Project the origin into the current sketch and using the center point of the fan body sketch, move the profile so it is centered at the origin point.

12.

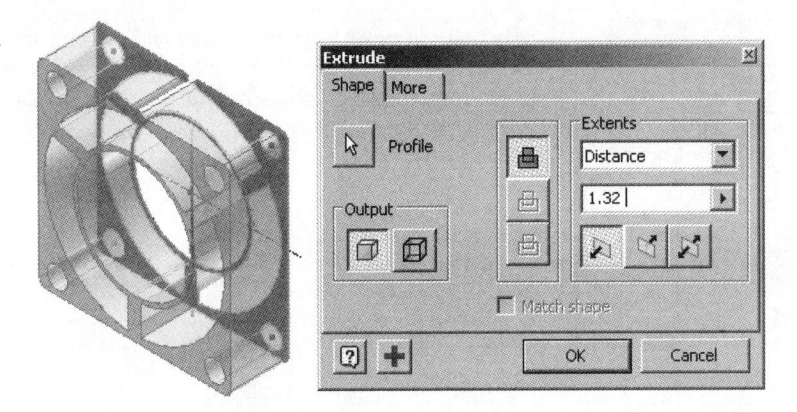

Select **Extrude**.

Set the **Distance** to **1.32 mm**.

13. We have successfully transformed an AutoCAD 2D drawing into a 3D Parametric part in minutes.

    Save the file as *ex3-10.ipt*.

Cursor Cues

As we create our sketches, we see cursor cues telling us how Inventor is interpreting what we are drawing. By watching for the visual feedback Inventor provides, we can create sketches faster and with fewer edits required.

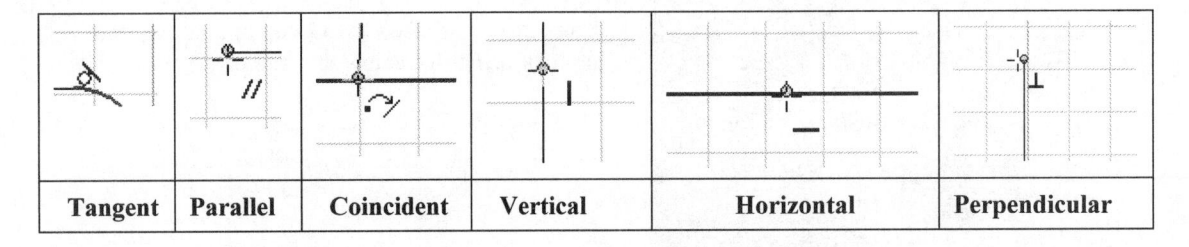

Tangent	Parallel	Coincident	Vertical	Horizontal	Perpendicular

Exercise 3-11:
Creating Etched Text

File: Ex3-10.ipt
Estimated Time: 15 minutes

This exercise reinforces the following skills:

- Create Point
- Text
- Rotate Sketch
- Extrude

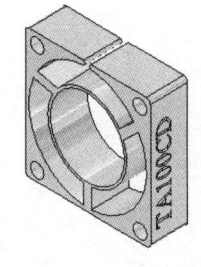

1. Open or continue working in file *ex3-10.ipt*.

3-36

Sketch Tools

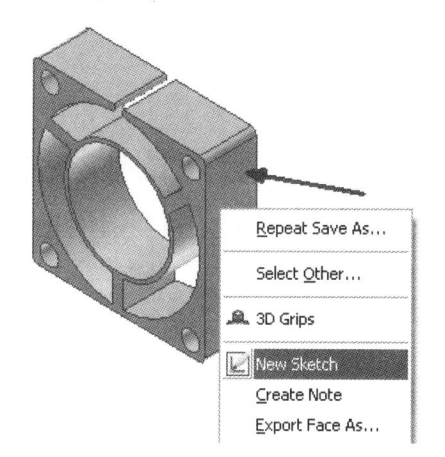

2. Select the right vertical side of the fan box as shown.

   Right-click and select **New Sketch**.

3. In order to place the text, it is a good idea to place a point to use to align your text.

   Select the **Point** tool and place a point as shown.

   Dimensions are in mm.

   You can go to Tools→Document Settings to change the units at any time.

4. Select the **Text** tool.

   Pick the point to bring up the **Format Text** dialog.
   This will act as the insertion point for the text.

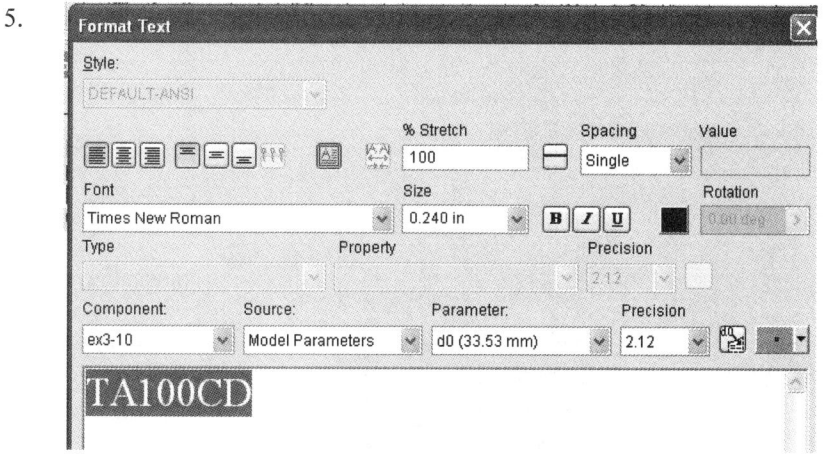

5. Select **Times New Roman** for the font to be used.
   Set the **Font Size** to **0.240 in**.
   Type *TA100CD* in the text field area.

   Press **OK**.

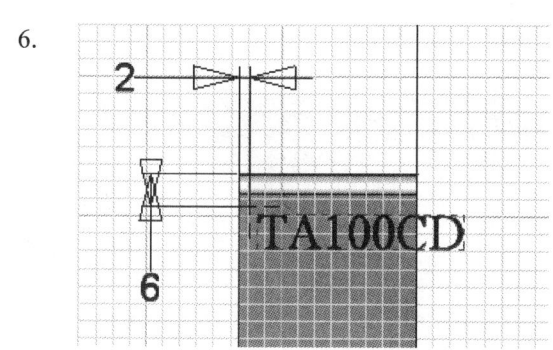

6. Your text appears with a rectangle around it.

3-37

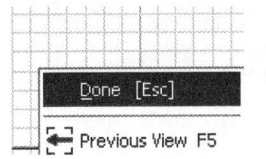

The Power of Design: An Introduction to Autodesk Inventor

Right-click and select **Done**.

You can use the rectangle to locate your text.

Simply pick the upper left corner of the text rectangle and drag it to the point.

You can also add a Coincident constraint between the point and the upper left corner of the rectangle to locate the text.

7. Select the **Coincident** tool.

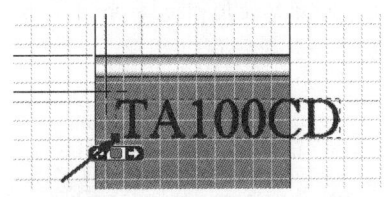

   Select the lower left corner of the text box.

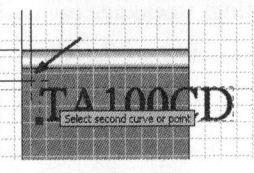

   Select the point that was placed.

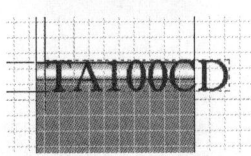

   The text shifts position.

8. Select the **Rotate** tool to rotate the sketch 90 degrees.

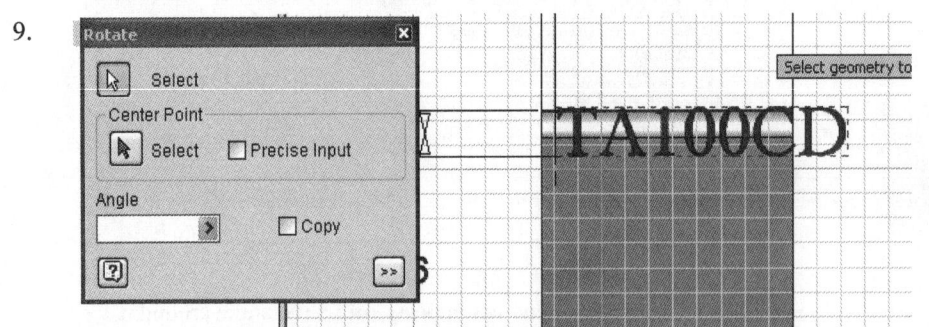

9. Select the text.

   Select the bottom left corner as the center point.

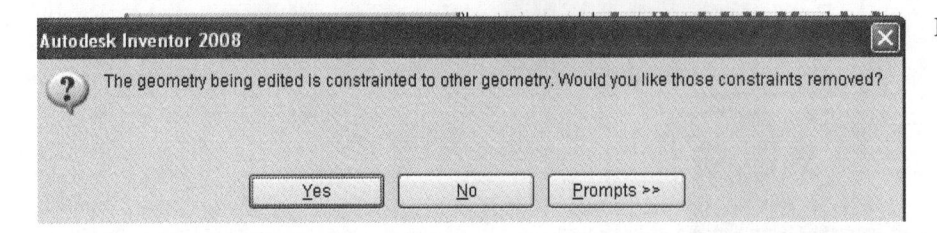

   Press **Yes**.

3-38

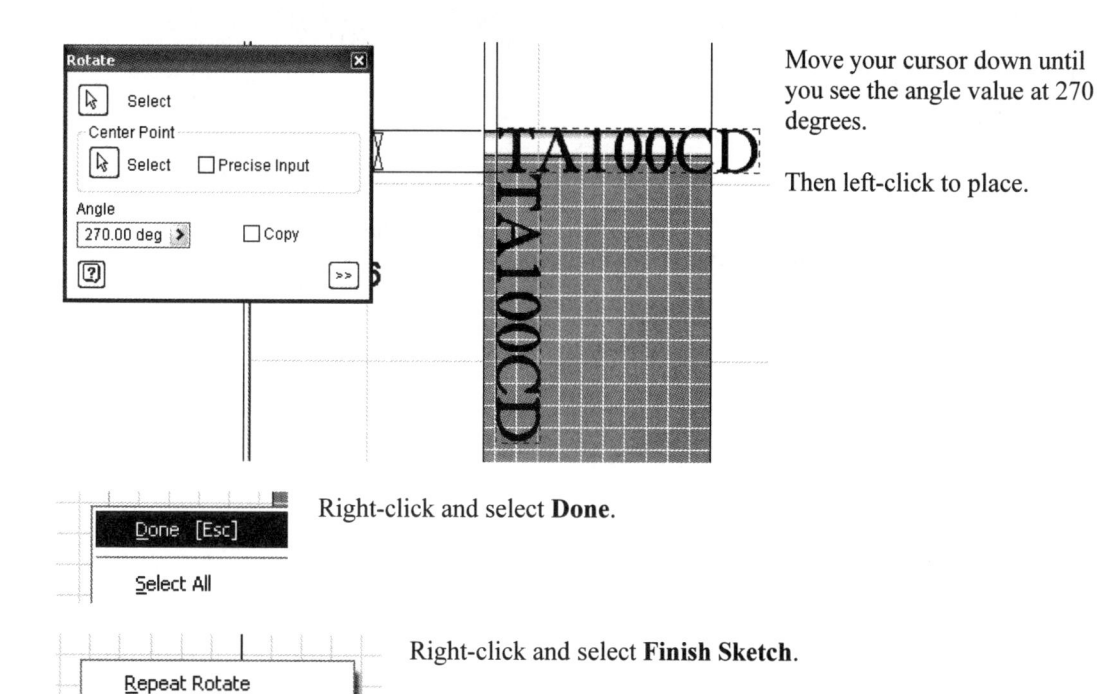

Sketch Tools

Move your cursor down until you see the angle value at 270 degrees.

Then left-click to place.

Right-click and select **Done**.

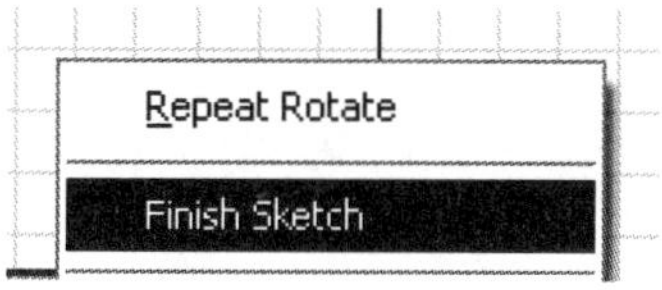

Right-click and select **Finish Sketch**.

10.  Select the **Extrude** tool.
    You can also use the **Emboss** tool. The Emboss tool is best for curved faces.

11. 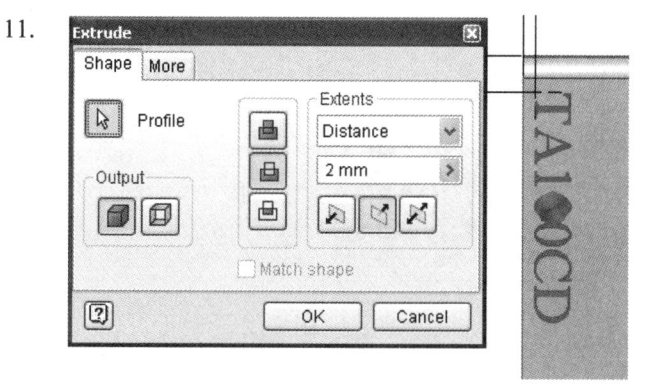 Set the **Cut** option and set the Distance to **2 mm**.

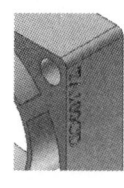

The text is etched into the fan body.

12. Save the file as *ex3-11.ipt*.

3-39

Insert Image

Images can be inserted into a sketch and then applied as a decal to a face using the **Insert Image** tool.

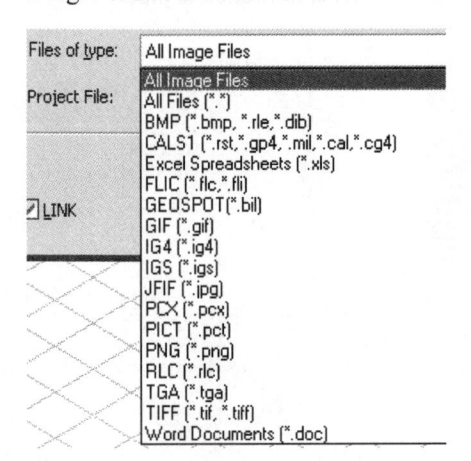

You can use any image file, as well as *.doc and *.xls files with the **Image** tool.

Exercise 3-12:
Insert Image

File: Ex3-11.ipt, Nidec-logo.bmp (downloaded from the publisher's website.)
Estimated Time: 15 minutes

This exercise reinforces the following skills:

- Insert Image
- Decal

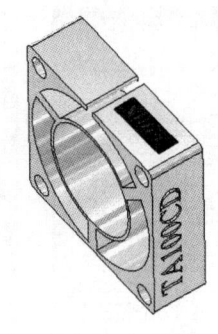

1. Open or continue working in file *ex3-11.ipt*.

2. Select the top face of the fan body.
   Right-click and select **New Sketch**.

3. Download the *nidec-logo.bmp* file from the publisher's website, or use any bmp file of your choice.

4. Select the **Insert Image** tool.

3-40

Sketch Tools

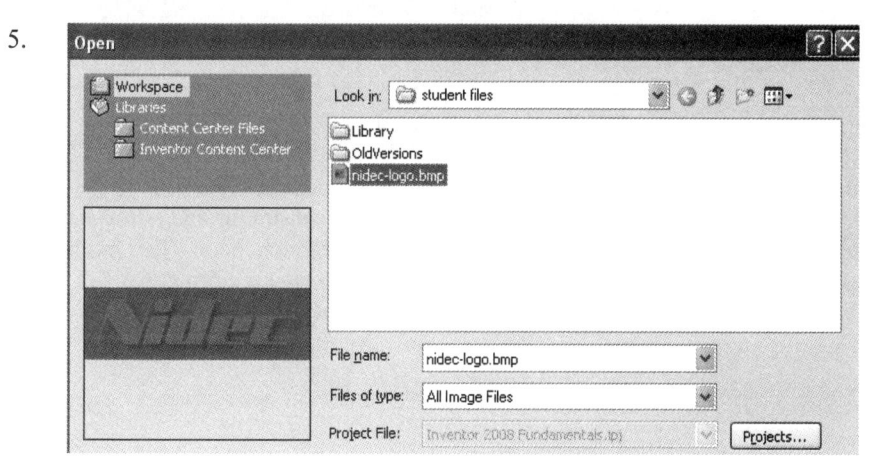

5. Locate the *nidec-logo.bmp* file you downloaded.

Note that you can see a preview of the image in the preview window on the left.

Press **Open**.

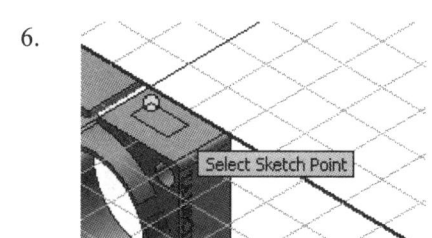

6. Pick to place in your sketch.

Right-click and select **Done**.

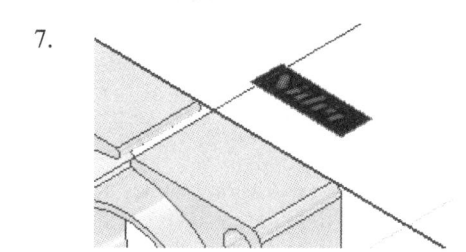

7. You can scale the image by picking on the corners and holding down the left mouse button until you have the desired size.

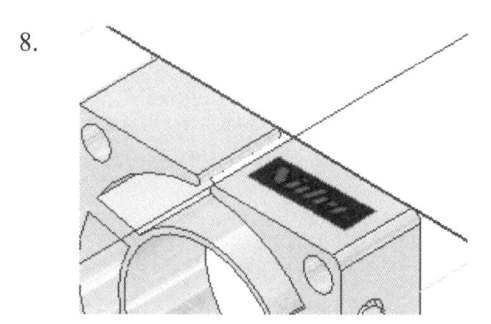

8. To position the image, window around the entire image to select and drag into place.

9. Select the **Decal** tool.

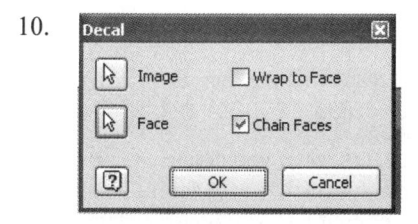

10. Select the Image. Then, select the face where you placed the Image.

Press **OK**.

Rotate the part around to see how the image appears from different angles.

Remember you can only see the Decal in Shaded Display Mode.

11. Save the file as *ex3-12.ipt*.
Close the file.

3-41

Edit Coordinate System

This tool allows you to redefine the Coordinate System of a sketch. The Coordinate System controls the orientation of features. Modifying the Coordinate System can affect applied constraints and reference geometry. The Edit Coordinate System tool does not work when you are in Sketch mode. This has been a bug for the past few releases. If you delete features, you may get an error message requiring that you redefine the coordinate system. Use this method to repair any existing sketches.

Exercise 3-13:
Edit Coordinate System

File: Ex3-8.ipt
Estimated Time: 10 minutes

This exercise reinforces the following skills:

- Edit Coordinate System
- Use of the Sketch Doctor

1. Open *ex3-8.ipt*.

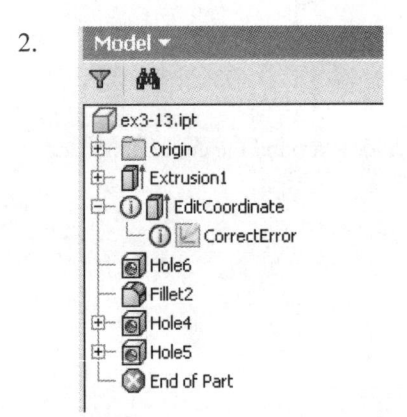

2. If you look in the Browser, you will see an extrusion that has an ⓘ symbol next to it. The symbol indicates that there is an error that needs to be corrected.

3. If you look in the Standard toolbar, you will see a Red Cross ✚ symbol. This is the symbol for the **Sketch Doctor**, indicating a sketch needs to be repaired.

    Select the **Red Cross** symbol.

4. A list of errors is displayed. Note that the primary error is an **Invalid axis or origin**.

    This error usually occurs when you delete an existing feature and a dependent sketch is affected. The error can be corrected by editing the coordinate system for that sketch.

    Close the Sketch Doctor.

Sketch Tools

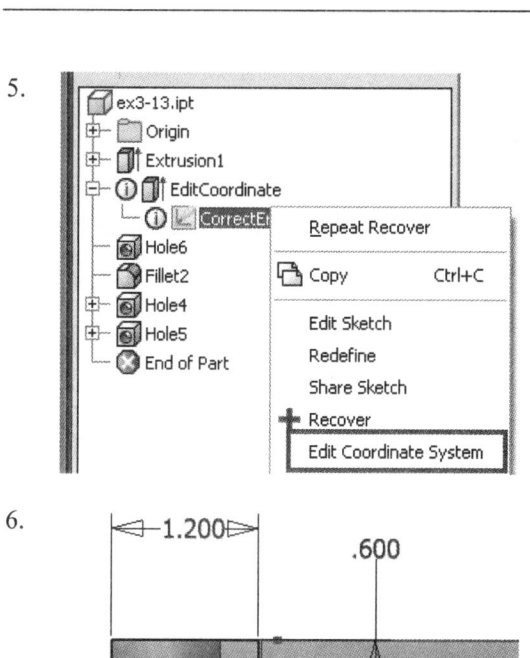

5. Highlight the sketch called **CorrectError**.

   Right-click and select **Edit Coordinate System**.

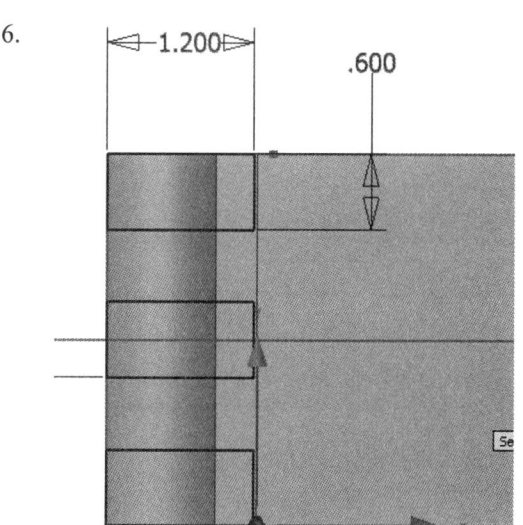

6. Select the **Red Origin Point**.

   Select the lower corner to relocate the coordinate system.

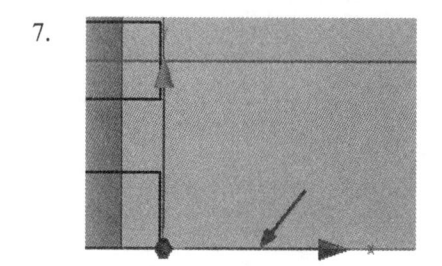

7. Select the **X-axis**.

   Select the edge of the part as the new X-axis.

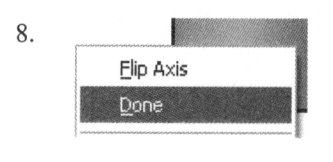

8. Right-click and select **Done**.

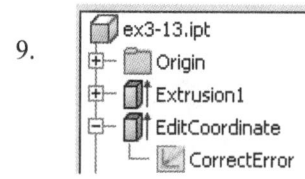

9. Note that the error symbols are now gone.

10. Note that the Red Cross symbol is now gray, indicating no sketch errors.

11. Close without saving.

3-43

Import Points

This tool allows you to add points into a sketch using an Excel spreadsheet. You may import either 2D or 3D point sets (X,Y) or (X,Y,Z). In order to import 3D point sets, you must be in 3D Sketch mode.

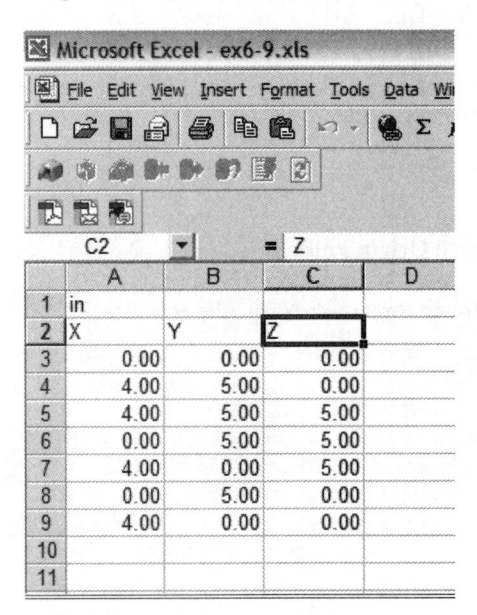

In order to import the points, the Excel spreadsheet must follow a specific format. If your spreadsheet does not meet the format, you will get an error message.

The first row and cell (A1) should state the units (in or mm).

The second row should be the header row.

The remaining rows are the point values.

Exercise 3-14:
Import Points

File: Ex3-9.xls
Estimated Time: 5 minutes

This exercise reinforces the following skills:

- Import Points

1. Start a new file using Standard.

2. Close the active sketch.

3. Select **3D Sketch** from the sketch toolbar.

4. Select the **Import Points** tool.

5. You can use the spreadsheet available for download from the publisher's website, or use your own.

   Press **Open**.

6. The points are placed.

7. Close without saving.

Review Questions

A. ▨ B. ▨ C. ▨ D. ▨

Identify the geometric constraint

1. Vertical
2. Fixed
3. Parallel
4. Coincident

5. The Spline tool is located under this drop-down:

    A. Line
    B. Arc
    C. Circle
    D. Rectangle

6. The three types of arc options are:

    A. 3 Point, Tan-Tan-Tan, Start End Radius
    B. 3 Point, Star t Direction Radius, Start End Radius
    C. 3 Point, Center Two Ends, Tangent
    D. 3 Point, Center Radius, Start End Radius

7. To draw a construction line or circle:

    A. Use the Style drop-down and select Construction
    B. Use the Construction Line/Circle tool
    C. Select the Line/Circle, Right-click and enable 'Construction'
    D. While drawing the line, hole down the CONTROL button.

8. To switch to arc mode while using the Line tool:

    A. Hold down the CONTROL key
    B. Left-click and Hold down the left mouse button
    C. Hold down the TAB Key
    D. Right-click and select ARC from the menu.

9. To switch from TRIM mode to EXTEND mode:

    A. Hold down the CONTROL key
    B. Press and hold SHIFT
    C. Right-click and select EXTEND from the menu.
    D. Hold down the TAB Key

10. To modify a dimension:

    A. Double click on top of the dimension
    B. Select the Edit Dimension tool
    C. Select the Dimension in the Browser, right-click and select Edit
    D. Select Edit Text from the Modify menu.

11. In order to see a Decal, your model must be in:

    A. Wireframe Display mode
    B. Hidden Edges Display mode
    C. Shaded Display mode
    D. It makes no difference

12. You cannot move and copy sketch geometry in the same operation.

    A. True
    B. False

13. When mirroring a sketch, the mirror line can be selected as part of the set to be mirrored.

    A. True
    B. False

14. When you add text to a sketch, it can be Extruded or Embossed to create a feature.

    A. True
    B. False

15. Inserted images cannot be scaled once they are placed in a sketch.

    A. True
    B. False

ANSWERS: 1) D; 2) B; 3) A; 4) C; 5) A; 6) C; 7) A; 8) B; 9) B; 10) A, 11) C; 12) B; 13) B; 14) A; 15) B

NOTES:

Lesson 4
The Features Toolbar

The Features toolbar has six sections.

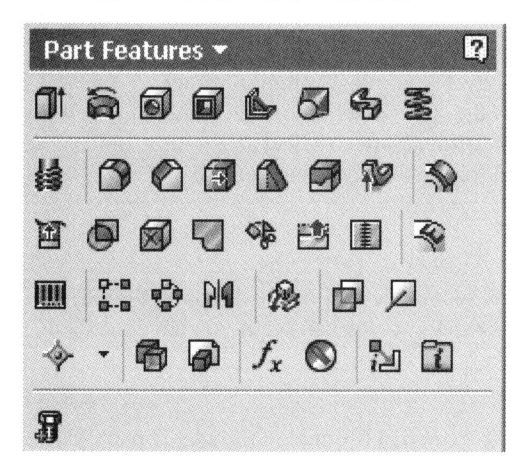

Sketched Feature Tools:
    Extrude, Revolve, Hole, Shell, Rib, Loft, Sweep, Coil
Thread Tool
Placed Feature Tools:
    Fillet, Chamfer, Move Face, Face Draft, Part Split, Bend, Thicken/Offset, Replace Face, Sculpt, Delete Face, Boundary Patch, Trim Surface, Extend Surface, Stitch Surface, Emboss, Decal
Pattern Feature Tools:
    Rectangular Array, Circular Array, Mirror Feature
Place Feature from Content Center
Work Features:
    Work Plane, Work Axis, Work Point
Copy Object, Derived Component
Parameters, iMate
IFeature, iFeature Catalog
Component Authoring

The Sketched feature tools use sketched geometry to build features.

The Thread tool maps the image of a thread onto a cylindrical face.

The Placed feature tools add finishing touches to existing features.

Pattern feature tools add multiple instances of existing features.

Content Center allows the user to place hardware components into assemblies.

Work features create work planes, work axes, or work points that you can use to define other features in a model.

iFeature tools create or add predefined features to a model.

You'll notice that when you start your part, most of the Features toolbar is grayed out. That's because Inventor automatically disables any tools you don't need until you need them. Those tools will become available once you have created the appropriate features.

Sketched Feature Tools

The eight options of this section are Extrude, Revolve, Hole, Shell, Rib, Loft, Sweep and Coil.

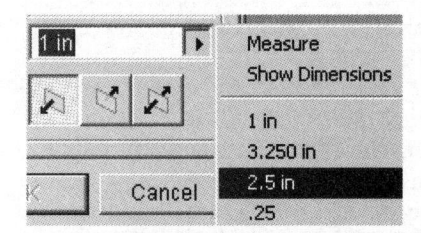

Inventor automatically saves your most common values. You can also measure objects on the fly to determine what value you want to use. You can use **Show Dimensions** to select an existing dimension on the model to be used for the feature being defined.

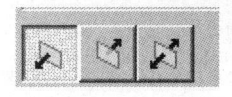

The buttons on the bottom of the **Extents** section determine the direction for the extrusion. Inventor will automatically preview, so you can see if you have selected the correct direction.

The **Taper Angle** sets a taper angle of up to 180 degrees for the extrusion (normal to the sketch plane). The taper extends equally in both directions. If a taper angle is specified, a symbol in the graphics window shows the fixed edge and direction of taper.

TIP: To taper a feature in only one direction, create an extruded feature with no draft, then use the Face Draft tool to add draft to a specific face.

Exercise 4-1:
Performing a Revolve

File: New (Standard.ipt using Inches)
Estimated Time: 15 minutes

This exercise reinforces the following skills:

- Project Geometry
- Line
- Arc
- Dimension
- Revolve

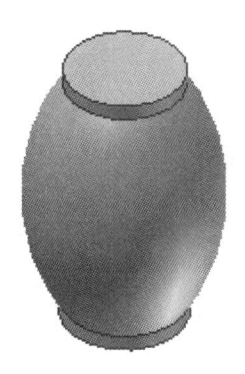

1. Go to **File→New**.

2. Select the **Standard.ipt** icon on the Defaults tab.

3. Select the **Project Geometry** tool.

4. Select the **Center Point** to add to the current sketch.

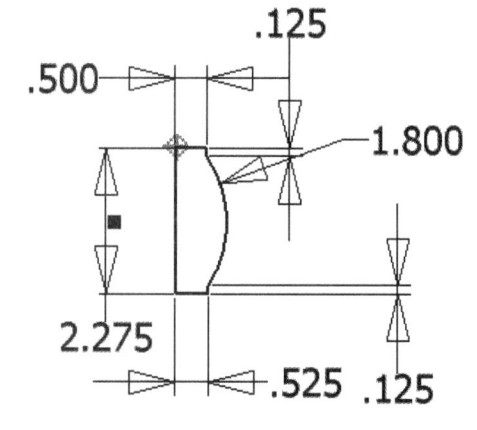

5. Create this sketch.

   The upper left corner of the sketch should be coincident with the center point.

   Right-click and select **Finish Sketch**.

6. Select the **Revolve** tool.

The Power of Design: An Introduction to Autodesk Inventor

7.

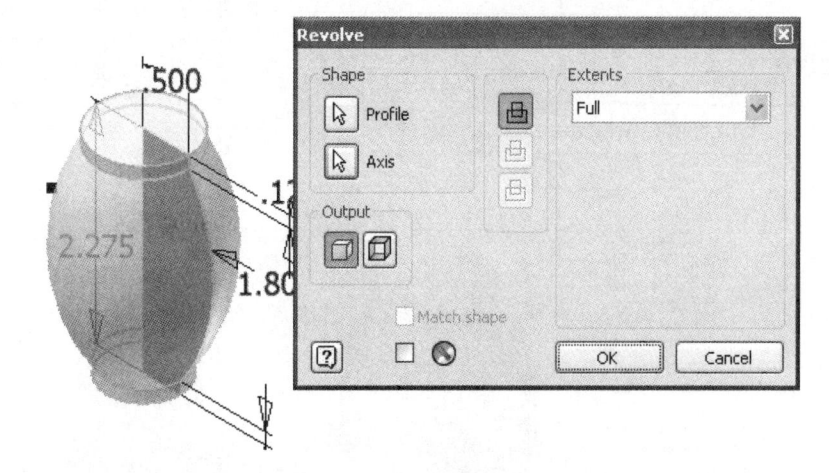

Select the vertical line as the axis.
Press **OK**.

8.

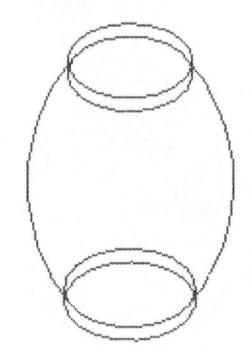

The completed revolve is shown here.

Save as *ex4-1.ipt*.

4-4

The Features Toolbar

Exercise 4-2:
Adding a Hole using Linear

File: Ex1-1.ipt
Estimated Time: 15 minutes

This exercise reinforces the following skills:

- Hole using the Linear method

1. Open *ex1-1.ipt*.

2. Select the **Hole** tool.

3. Set the Placement to **Linear**.
   With Face enabled, select the face indicated.

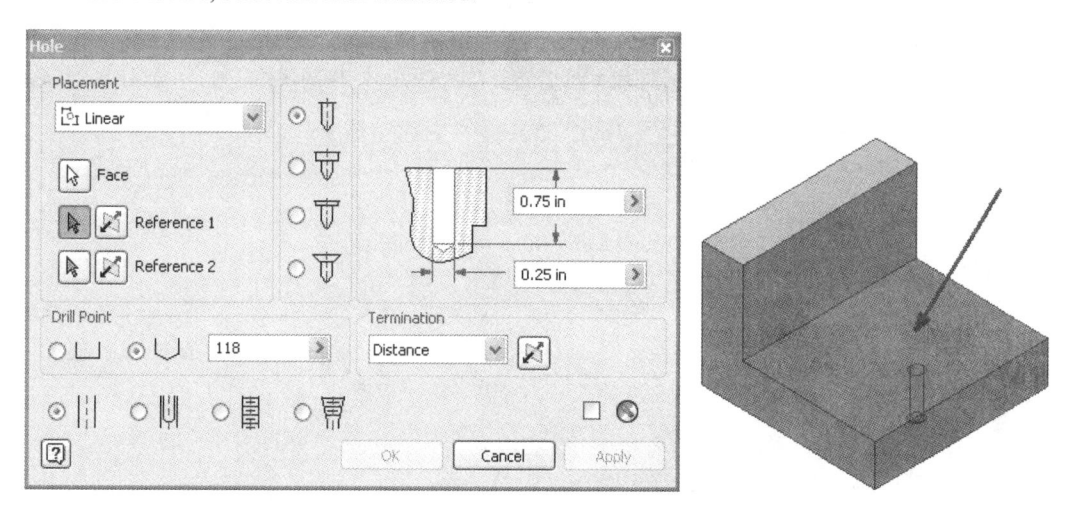

4. Enable **Reference 1**.

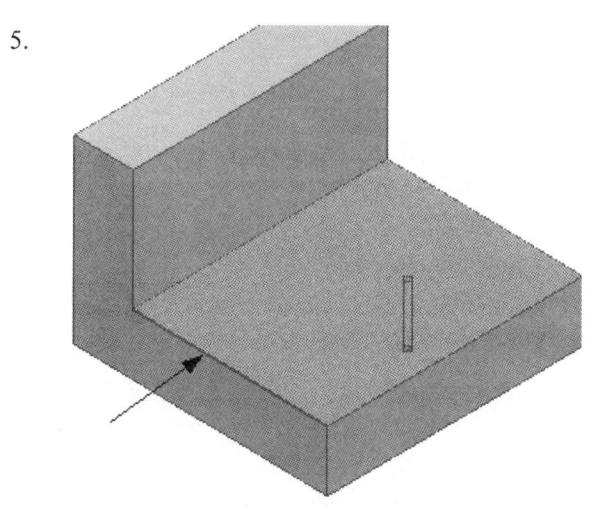

5. Select the edge indicated.

4-5

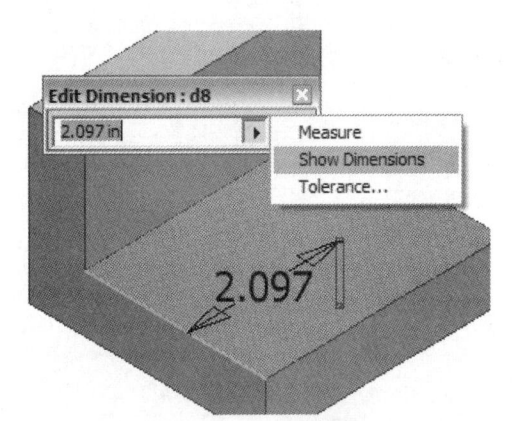

The Power of Design: An Introduction to Autodesk Inventor

6. When the dimension appears, select the **Show Dimensions** option.

Left-click the part to display the dimensions.

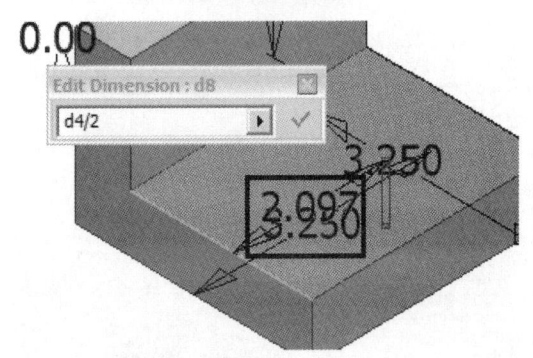

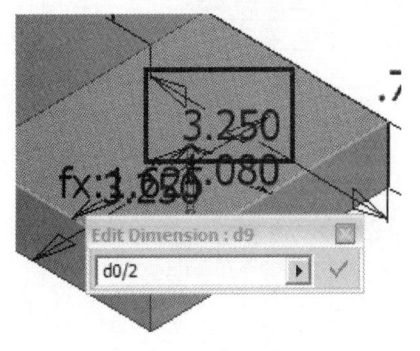

7. Select the **3.25** dimension corresponding to the selected edge.

The dimension name should appear in the dimension text box.

Add **/2** to the text box to place the hole in the middle of the face.

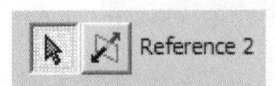

8. Reference 2 should now be enabled.

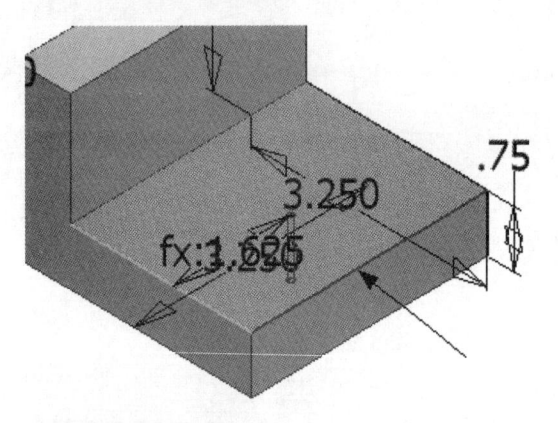

9. Select the edge indicated as the second Reference.

10. Select the **3.25** dimension corresponding to the selected edge.

The dimension name should appear in the dimension text box.

Add **/2** to the text box to place the hole in the middle of the face.

4-6

The Features Toolbar

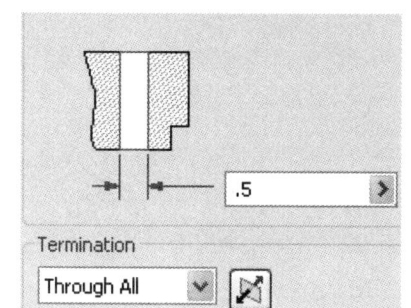

11. Set the **Termination** to **Through All**.

Set the **diameter** to **.50**.

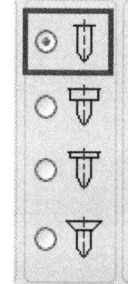

12. Set the hole as a simple hole.

13. Press **OK**.

14. Save As *ex4-2.ipt*.

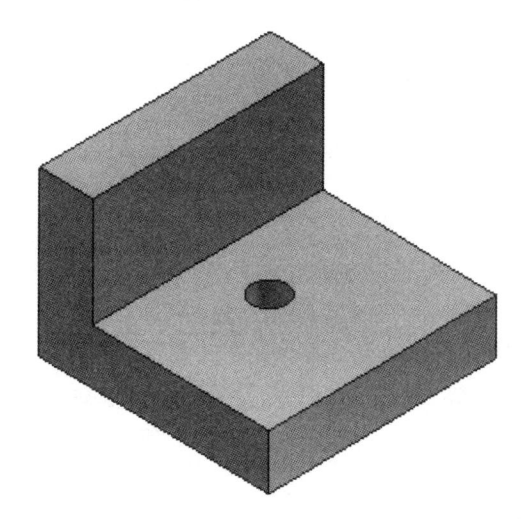

4-7

The Power of Design: An Introduction to Autodesk Inventor

Exercise 4-3:
Adding a Hole using Concentric

File: Ex4-1.ipt
Estimated Time: 15 minutes

This exercise reinforces the following skills:

- Hole using the Concentric method

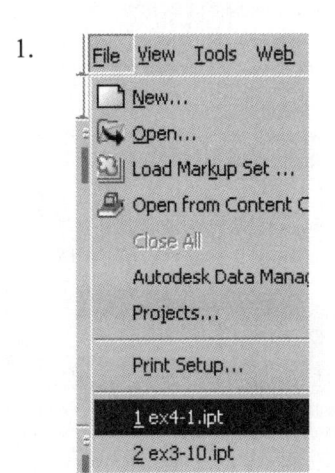

1. You can open *ex4-1* by locating it on the list of recently used files under the File menu.

   Open *ex4-1.ipt*.

2. Select the **Hole** tool.

3. Set the **Placement** to **Concentric**.

4. The first selection is the face where the hole will be placed.

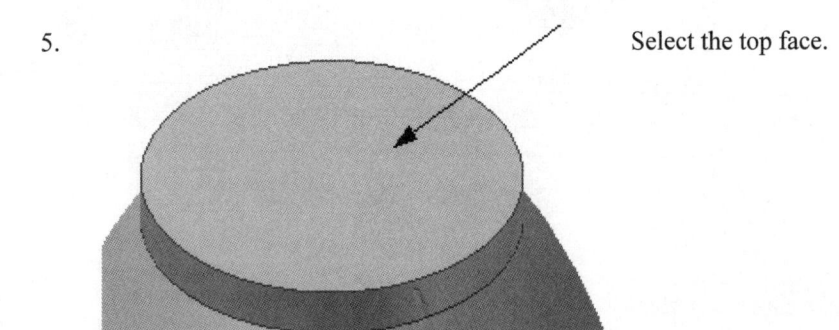

5. Select the top face.

6. The second selection is the cylindrical feature used for concentricity.

The Features Toolbar

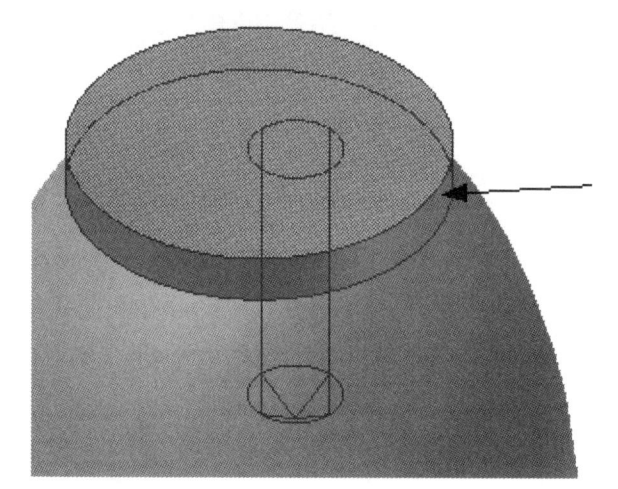

7. Select the cylinder adjacent to the selected face.

   The hole preview will automatically shift to the correct position.

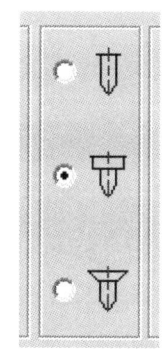

8. Set the **hole type** to be **Counterbore**.

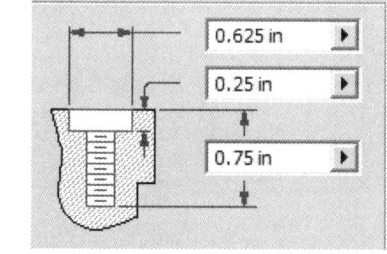

9. Set the **counterbore diameter** to **0.625**.
   Set the **counterbore depth** to **0.25**.
   Set the **shaft depth** to **0.75**.

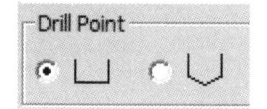

10. Set the **Drill Point** to **flat**.

11. Set the **hole type** to **Tapped**.

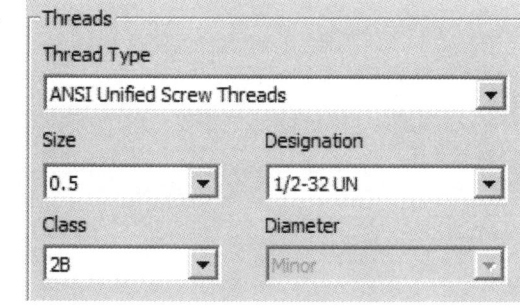

12. Set the **Thread Type** to **ANSI Unified Screw Threads**.
    Set the **Size** to **0.5**.
    Set the **Designation** to **½-32 UN**.
    Set the **Class** to **2B**.

4-9

13. Enable **Full Depth**.

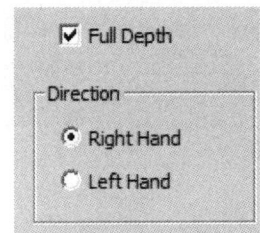

   Enable **Right Hand**.

14. Press **OK**.

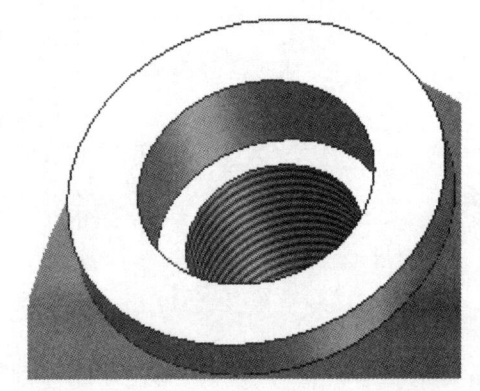

15. Rotate the part to inspect the placed hole.

   Save as *ex4-3.ipt*.

Exercise 4-4:
Adding a Hole using Sketch

File: Ex4-2.ipt
Estimated Time: 15 minutes

This exercise reinforces the following skills:

- Hole using the Sketch method

1. Open *ex4-2.ipt*.

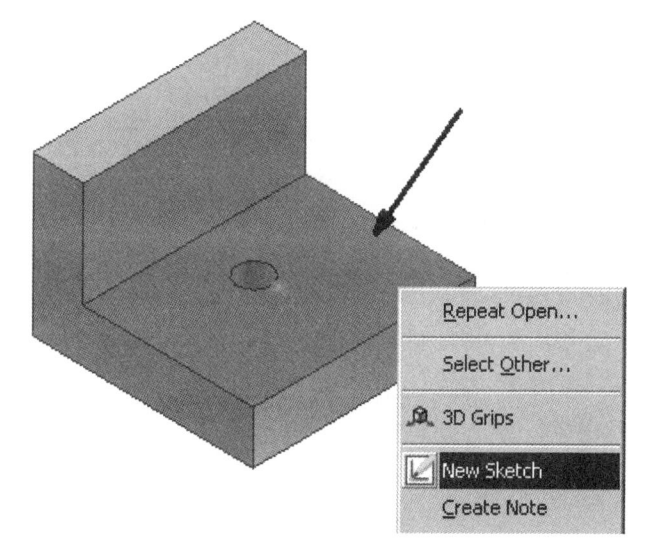

2. Select the face indicated.

    Right-click and select **New Sketch**.

3. Draw a rectangle and dimension.

    Add three points at the midpoints on three sides of the rectangle.

4. Select the **Hole** tool.

5. Set the **Placement** to **From Sketch**.

    Centers should be enabled.

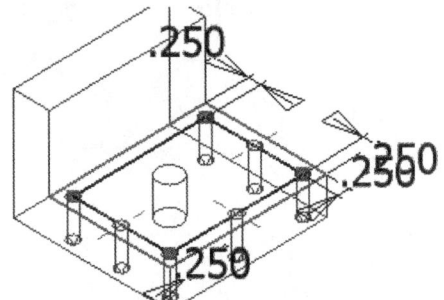

6. The points are automatically selected.

   Pick the four vertices of the rectangle.

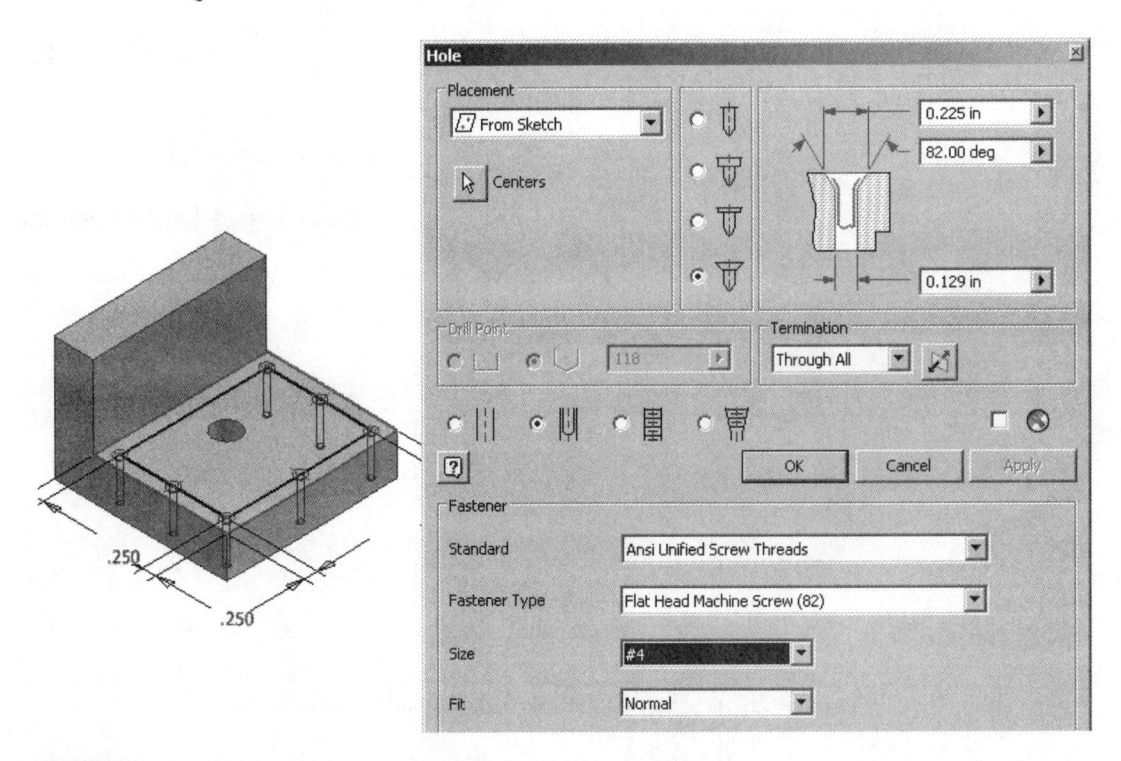

7. Set the **hole type** as **Countersink**.

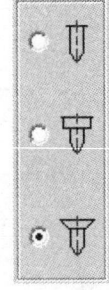

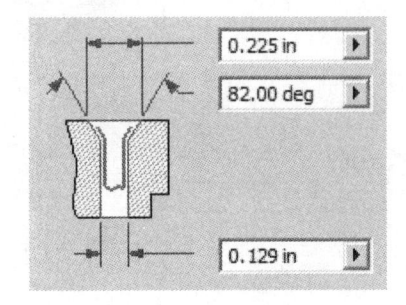

8. Set the **countersink diameter** to **0.225**.
   Set the **countersink angle** to **82.00 degree**.
   Set the **clearance hole diameter** to **0.129**.

The Features Toolbar

9. Set the **Termination** to **Through All**.

10. Set the hole type to **Clearance**.

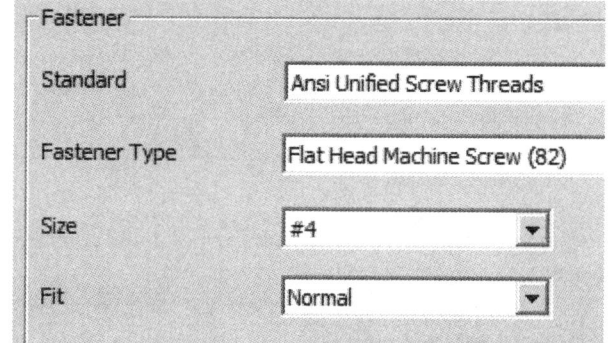

11. Set the **Standard** to **ANSI Unified Screw Threads**.

    Set the **Fastener Type** to **Flat Head Machine Screw (82)**.

    Set the **Size** to **#4**.

    Set the **Fit** to **Normal**.

    Press **OK**.

12. Save As *ex4-4.ipt*.

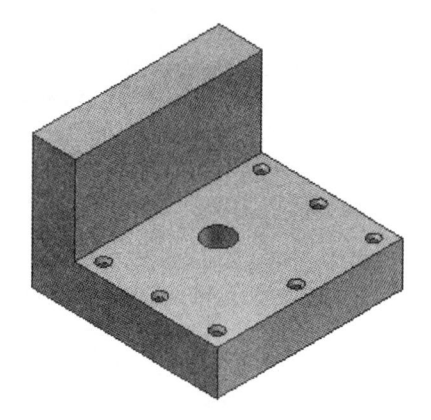

TIP: To place more than one hole using different methods without having to exit the **Hole** dialog, use **Apply** and then define the next hole type.

4-13

Customizing Hole Data

Autodesk Inventor uses an Excel spreadsheet to manage thread and tapped hole data. By default, the spreadsheet is located in the Program Files\Autodesk\Inventor<*version*>\Design Data folder. The file name is *thread.xls*.

The spreadsheet contains some common industry standard thread types and standard tapped hole sizes. You can modify the spreadsheet to:

- Include more standard thread sizes.
- Include more standard thread types.
- Create custom thread sizes.
- Create custom thread types.

You can type new thread data in the spreadsheet, but you are less likely to make errors if you copy existing rows and edit the values. Before you do, always make a copy of the current spreadsheet.

When you modify the *thread.xls* file, you need to close and re-launch Autodesk Inventor in order to access the new thread definitions.

To add a new thread type, create a new sheet in the spreadsheet, then copy the data from an existing sheet to preserve the spreadsheet columns. Edit the columns and rows as necessary to define the new thread type.

In general, copy an inch-based sheet if you want to create a new inch type, or a metric-based sheet for a new metric type.

NOTE: *If you do not need as much data as an entire sheet, be sure to copy at least the first three rows to preserve the correct format of the spreadsheet. Cell A1 sets the thread family to be inch or mm, and straight or tapered.*

Exercise 4-5:
Customizing the Thread Data

File: Thread.xls
Estimated Time: 15 minutes

This exercise reinforces the following skills:

- Threads
- Hole

1. Thread.xls — Locate the **Thread.xls** file.
   The file should be located in the *Program Files\Autodesk\Inventor<version>\ Design Data* folder.

Notice that the spreadsheet is divided into sheets to make it easy to locate specific thread types.

inch		External						
			Major Dia		Pitch Dia		Minor Dia	
Size	Thread Designation	Class	Max	Min	Max	Min	Max	Min
0.06	0-80 UNF	2A	0.0595	0.0563	0.0514	0.0496	0.0446	
		3A	0.06	0.0568	0.0519	0.0506	0.0451	
0.073	1-64 UNC	2A	0.0724	0.0686	0.0623	0.0603	0.0538	
		3A	0.073	0.0692	0.0629	0.0614	0.0544	
	1-72 UNF	2A	0.0724	0.0689	0.0634	0.0615	0.0559	
		3A	0.073	0.0695	0.064	0.0626	0.0565	
0.086	2-56 UNC	2A	0.0854	0.0813	0.0738	0.0717	0.0642	
		3A	0.086	0.0819	0.0744	0.0728	0.0648	
	2-64 UNF	2A	0.0854	0.0816	0.0753	0.0733	0.0668	
		3A	0.086	0.0822	0.0759	0.0744	0.0674	
0.099	3-48 UNC	2A	0.0983	0.0938	0.0848	0.0825	0.0734	

You can then add data by inserting rows and saving the file.

2. Activate each tab, so you can see the data on each sheet.

3. Close the file without saving.

Exercise 4-6:
Creating a Shell

File: New
Estimated Time: 15 minutes

This exercise reinforces the following skills:

- Project Geometry
- Extrude
- Point, Hole Center
- Hole
- Shell

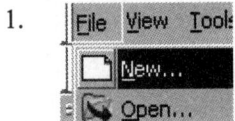

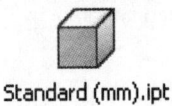

1. Go to **File→New**.

2. Select the **Metric** tab.

   Select the *Standard (mm).ipt* template.

   Press **OK**.

3. Select the **Project Geometry** tool.

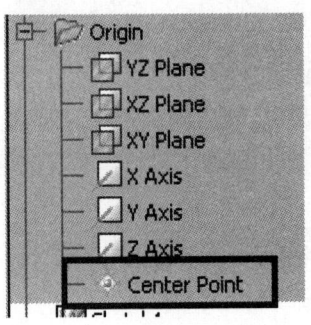

4. Select **Center Point** from the Browser.

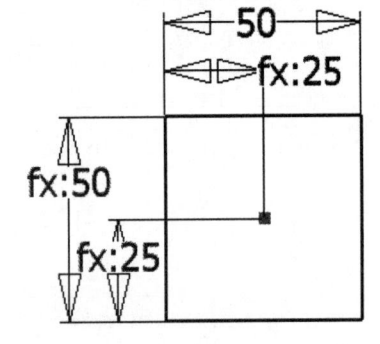

5. Draw a square using the **Rectangle** tool.

   To center the square on the center point, dimension from an endpoint to the center point.

   You can create a relationship between the length of the side and its location relative to the center point by defining an equation.

4-16

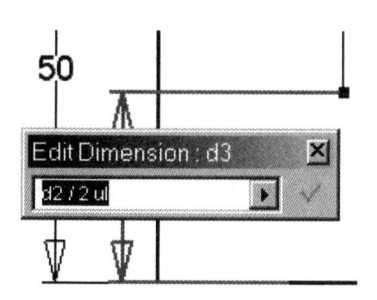

Simply erase the contents of the dimension field.

Select the dimension you wish to use for the equation and type **/2**.

6. Select the **Extrude** tool.

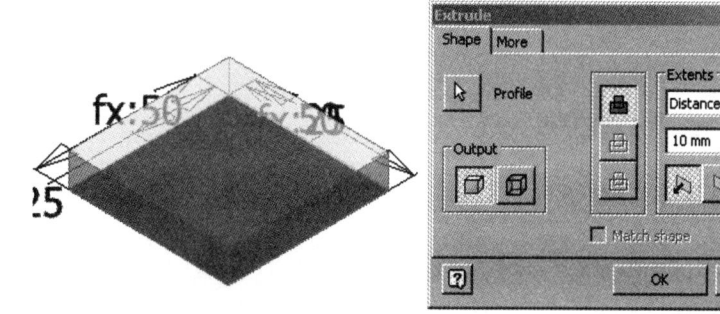

Enter a Distance of **10 mm**.

Press **OK**.

7. Select the top face.

Right-click and select **New Sketch**.

8. Draw a rectangle and center it on the face as shown.

9. Select the **Hole, Point Center** tool.

4-17

The Power of Design: An Introduction to Autodesk Inventor

10. Right-click and select the **Midpoint** option to locate the midpoint of each side of the rectangle.

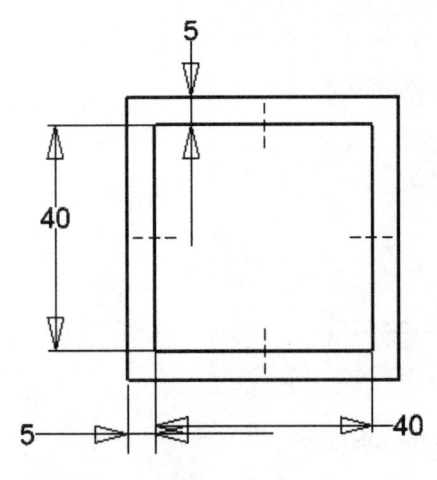

11. Place four points so they are located at the midpoint of each side.

    Right-click and select **Finish Sketch**.

12. Select the **Hole** tool.

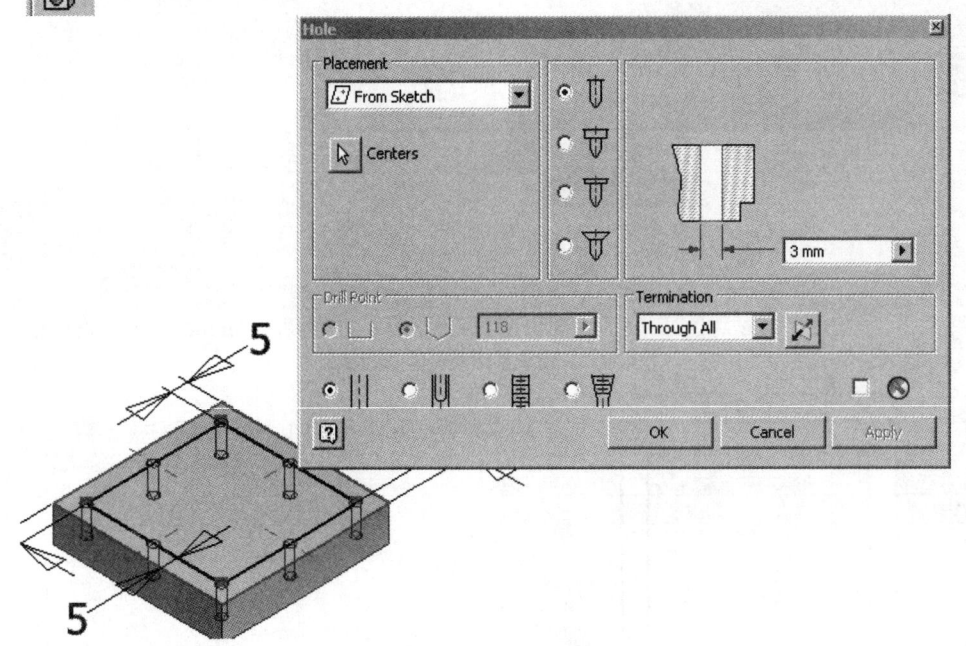

    Select the corners of the rectangle to be used as hole centers. The dialog will automatically select all the placed points.

    Set the **Termination** to **Through All**.
    Set the **diameter** to **3 mm**.
    Press **OK**.

13. Select the **Shell** tool.

14.

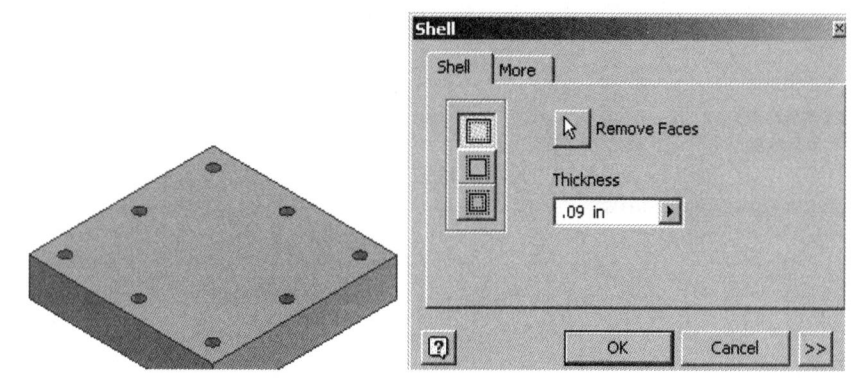

Select the top face for the Remove Faces selection.
Set the **Thickness** to **.09 in**.

Press **OK**.

15.

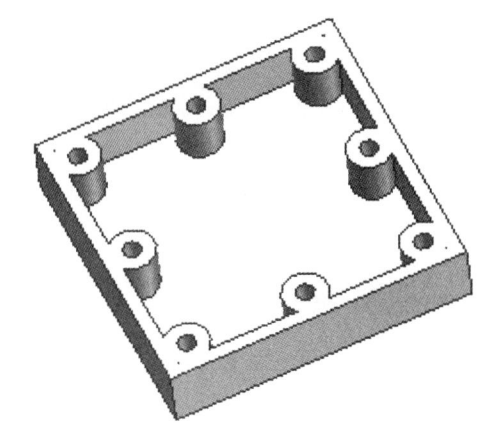

Save as *ex4-6.ipt*.

TIP: To reclaim a face, press and hold Ctrl and select the face.

The Power of Design: An Introduction to Autodesk Inventor

Exercise 4-7:
Creating a Rib

File: Ex4-6.ipt
Estimated Time: 15 minutes

This exercise reinforces the following skills:

- Rib
- Share Sketch

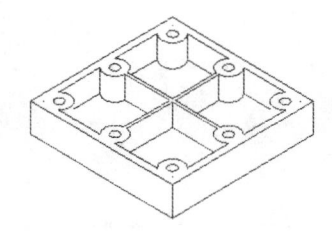

1. Open or continue working in *ex4-6.ipt*.

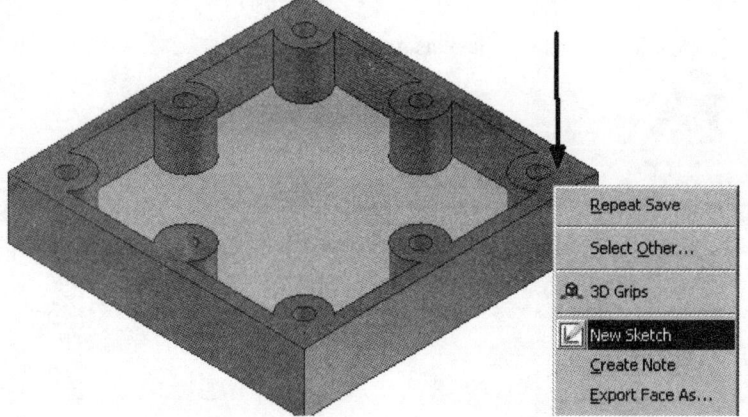

2. Select the top face of the extrusion.

   Right-click and select **New Sketch**.

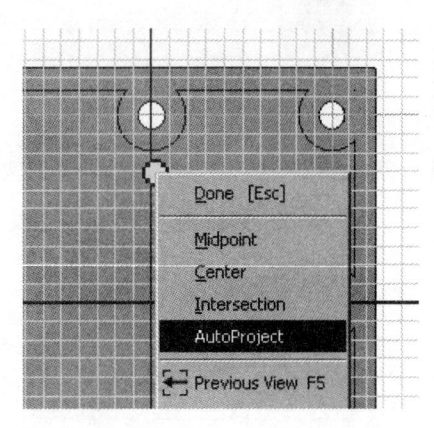

   Draw a horizontal line between the two middle posts using the **Auto-Project** option.

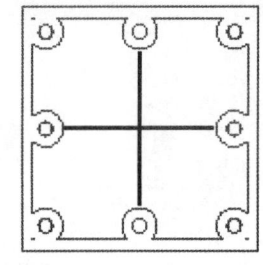

3. Draw a vertical line between the two middle posts using the Auto-Project option.

4. Select the **Rib** tool.

4-20

The Features Toolbar

5.

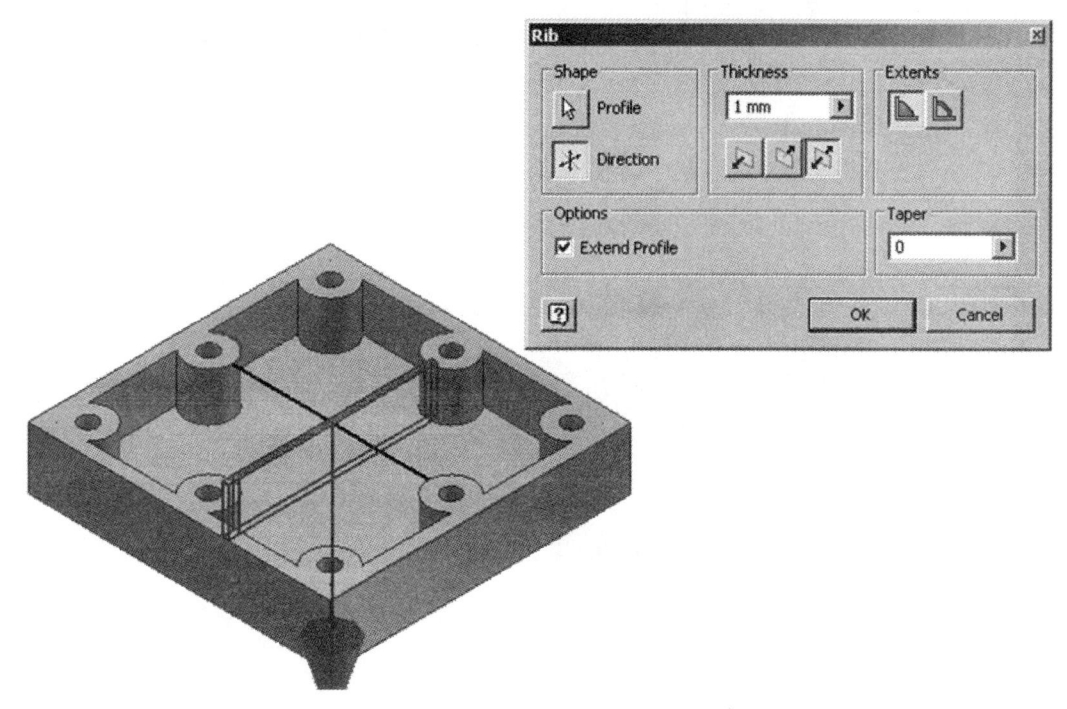

Set the **Thickness** to **1 mm**.

Set the **Direction** into the bottom face.

Enable **Extend Profile**.

Press **OK**.

The first rib is created.

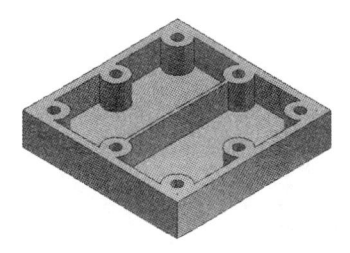

6. 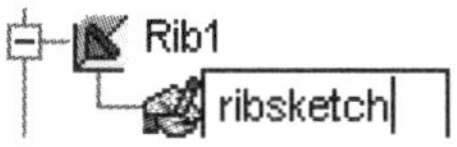 Expand the Rib definition in the Browser.
Rename the sketch to **ribsketch** to make it easier to identify.

To rename, simply click on the sketch name to activate the edit box.

7. 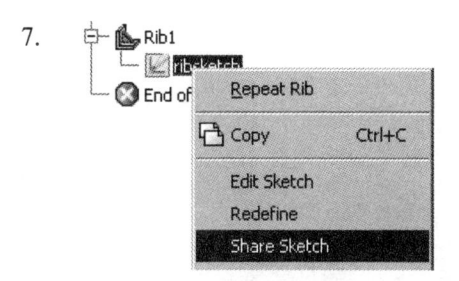 Highlight the **ribsketch**.
Right-click and select **Share Sketch**.

8.  Select the **Rib** tool.

4-21

9.

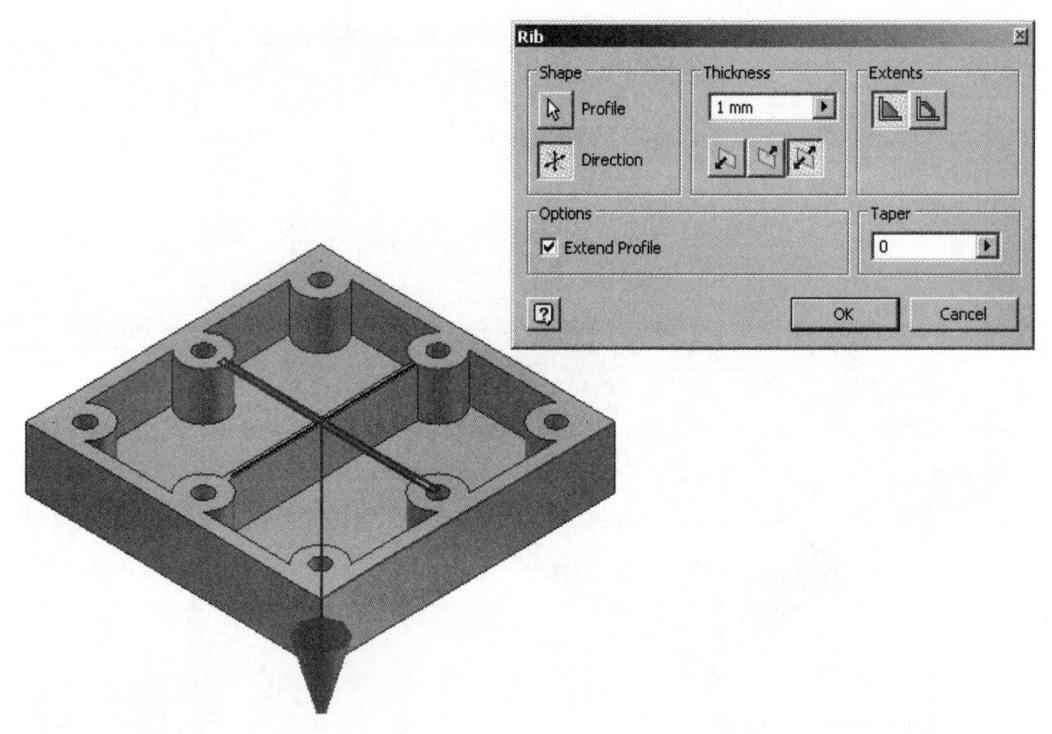

Select the other line for the second rib.

Set the **Thickness** to **1 mm**.

Set the **Direction** into the bottom face.

Enable **Extend Profile**.

Press **OK**.

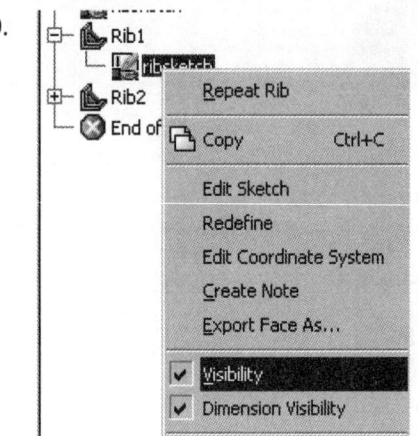

10. Highlight the **ribsketch** in the Browser. Right-click and disable **Visibility**.

11. Save As *ex4-7.ipt*.

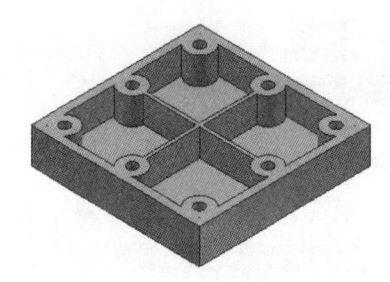

> ☑ Autoproject part origin on sketch create
>
> **TIP:** You can set to autoproject the part origin in all new sketches in the Application Options dialog under the Sketch tab.

Exercise 4-8:
Loft

File: New.ipt
Estimated Time: 15 minutes

This exercise reinforces the following skills:

- Loft
- Project Geometry
- Work Plane
- Circle
- Ellipse

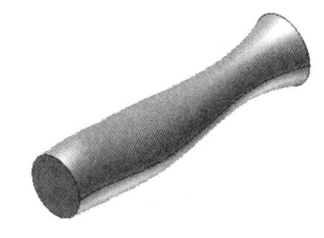

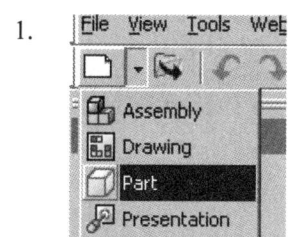

1. Start a new **Part** file.

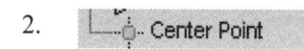

2. Project the **Center Point** on the XY workplane.

3. Draw a circle with its center point coincident with the projected centerpoint.

   Exit the sketch.

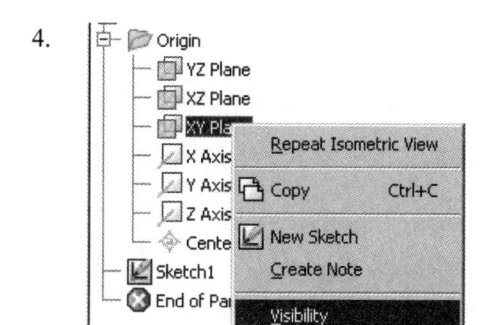

4. Turn on the **Visibility** of the **XY** Plane.

4-23

The Power of Design: An Introduction to Autodesk Inventor

5.  Select the **Work Plane** tool.

6. 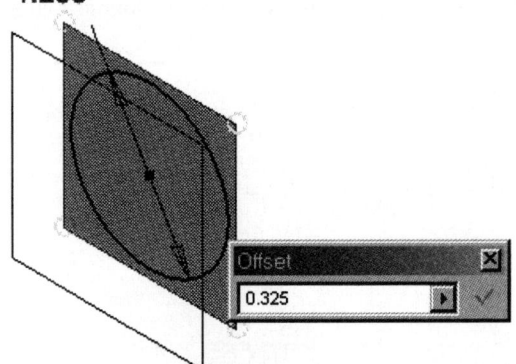 **1.200**   Create an **Offset** workplane a distance of **0.325** from the XY workplane.

7. 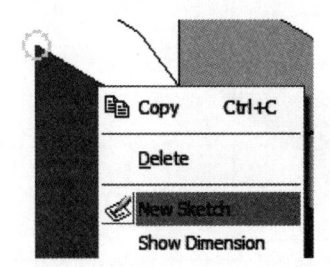 Select the new work plane.

   Right-click and select **New Sketch**.

8. Turn **off** the **Visibility** of the first sketch to make it easier to see what you are drawing.

9. 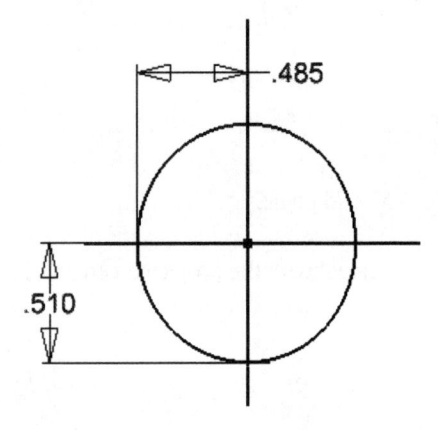 Create an ellipse.
   The **Ellipse** tool is available under the circle drop-down.

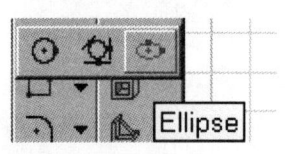

   Right-click and select **Finish Sketch**.

   Turn **off** the **Visibility** of the sketch.

10. 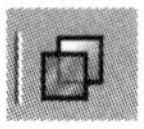 Select the **Work Plane** tool.

11. 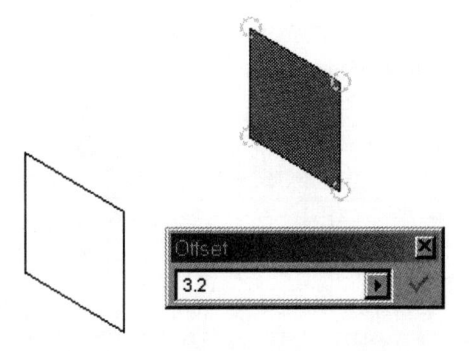 Create an offset workplane a distance of **3.2** in from the first offset workplane.

4-24

12. Rename the workplane **Middle Workplane**.

13. Select the **Middle Workplane**.
    Right-click and select **New Sketch**.

14. Create an ellipse on the Middle Workplane.

    The centerpoint of the ellipse should be located .100 below the part center point.

    Use a **Vertical** constraint to align the center points.

    Right-click and select **Finish Sketch**.

    Rename the sketch **MiddleSketch**.

    Turn **off** the **Visibility** of the sketch.

15. Turn **on** the **Visibility** of the first workplane.

16. Select the **Work Plane** tool.

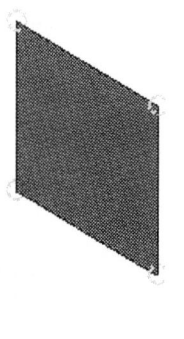

17. Create an offset workplane a distance of **4.8 in** from the first offset workplane.

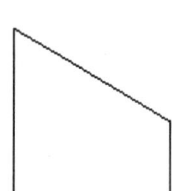

18. Rename the workplane **EndWorkPlane**.

19. Highlight the **EndWorkPlane** in the Browser.
    Right-click and select **New Sketch**.

20. Draw a circle with a center point coincident with the part's centerpoint.

    Dimension with a **diameter** of **1.050 in**.

21. Select the Work Plane tool.

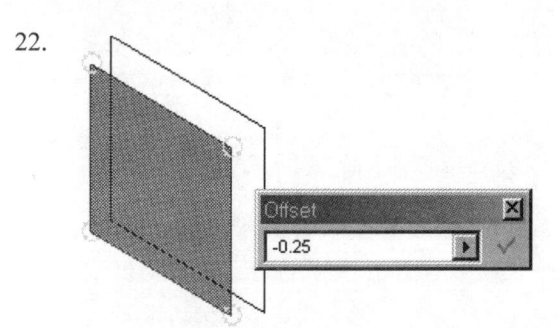

22. Create an offset workplane a distance of **-0.25 in**. from the **EndWorkPlane**.

23. Rename the new workplane **B4endworkplane**.

24. Highlight the **B4endworkplane** in the Browser.
    Right-click and select **New Sketch**.

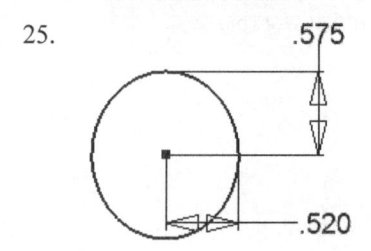

25. Create an ellipse with a center point coincident with the part's centerpoint.

26. Turn **on** the **Visibility** of all the sketches.

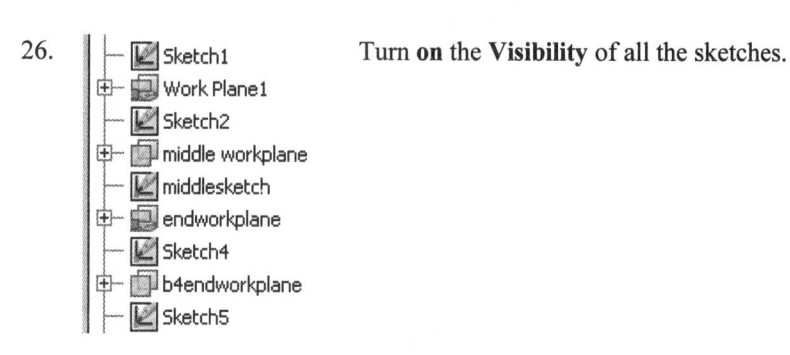

27. Select the **Loft** tool.

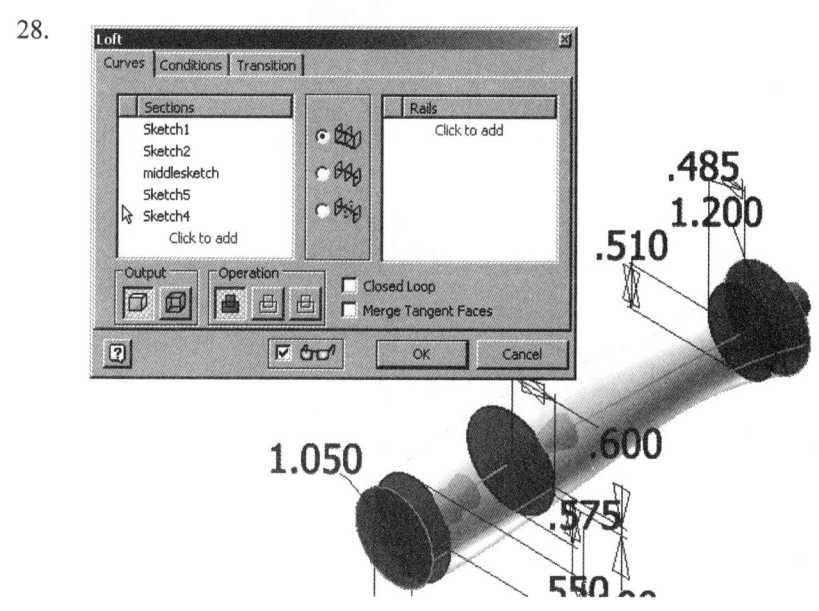

28. Select the sketches. Sketch5 will be selected before Sketch4 because it precedes it positionally.

    The second sketch may require you to select regions because the axes bisect the sketch.

    Press **OK**.

29. Save the file as *ex4-8.ipt*.

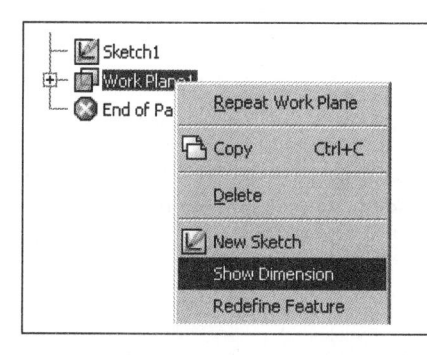

TIP: If you need to change the distance for an offset work plane, simply select the work plane in the browser. Right-click and select **Show Dimension**. This will bring up the edit dimension dialog box and you can modify the offset value.

Exercise 4-9:
Loft with Rails

File: New.ipt
Estimated Time: 30 minutes

This exercise reinforces the following skills:

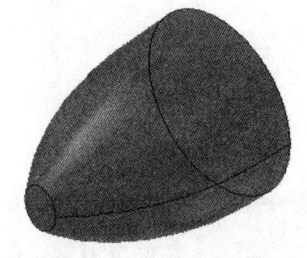

- Loft
- Project Geometry
- Work Plane
- Circle

1. 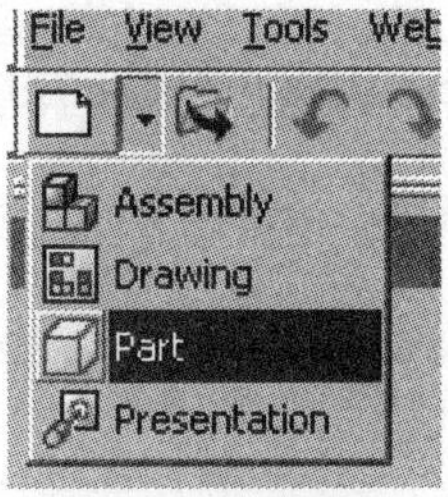 Start a new **Part** file.

2. 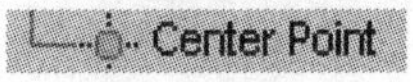 **Project** the **Center Point** on the XY workplane.

3. 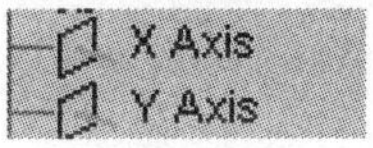 **Project** the **X** and **Y axes**.

4. 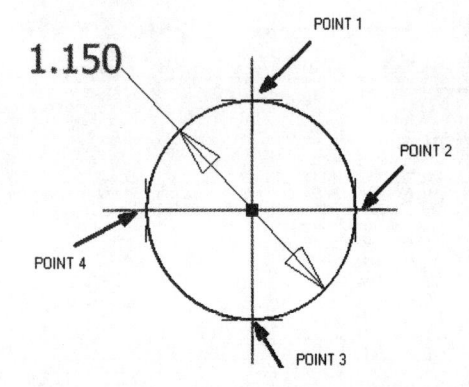 Draw a circle with its center point coincident with the projected centerpoint.

    Place a point at the each of the four quadrants of the circle.

    To place the points, project the X and Y axes. Then use the **Intersect** option to locate the intersection point between the axis and the circle. A yellow X will appear to assist you in placing the point.

    Exit the sketch.

5. 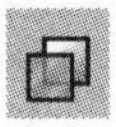 Select the **Work Plane** tool.

TIP: A yellow X symbol will appear next to your cursor when you are placing the point at the intersection between the axis and the circle.

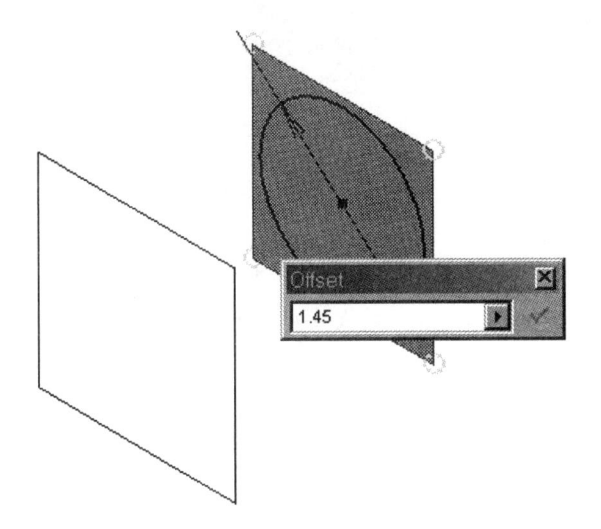

6. Create an offset workplane **1.45** from the XY workplane.

7. Select the offset workplane.

   Right-click and select **New Sketch**.

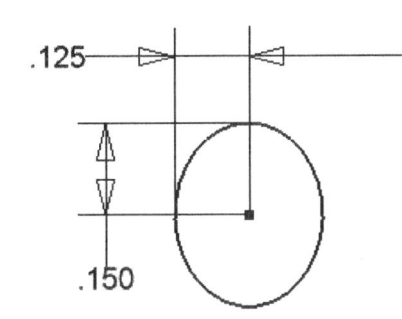

8. Create an ellipse with a center point coincident with the part's centerpoint.

   .125
   .150

9. Place a point at each quadrant of the ellipse. Project the X and Y axes to assist you in locating the intersect points.

   Exit Sketch mode.

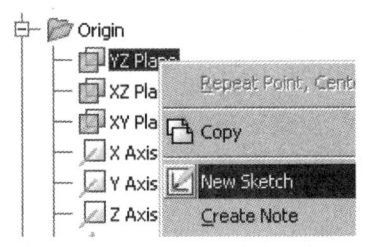

10. Highlight the **YZ** workplane.

    Right-click and select **New Sketch**.

4-29

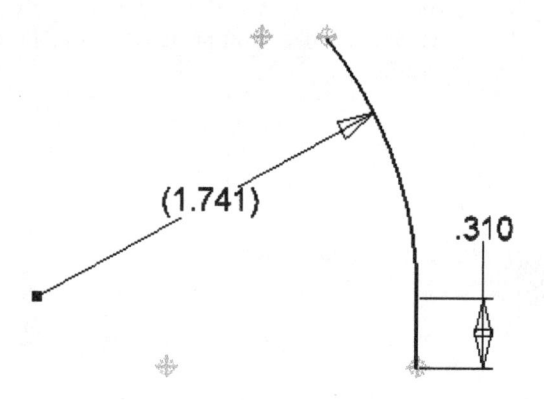

11. Project a quadrant point from each sketch on the right of the axis.

    Create the sketch as shown.

    Add a tangent constraint between the arc and the line.

    Right-click and **Finish Sketch**.

    Rename this sketch **Rail1**.

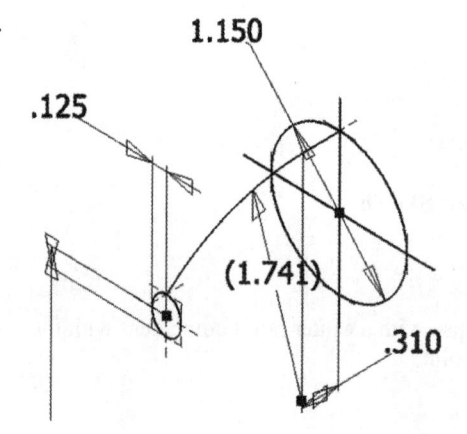

12. If you turn on the visibility of all the sketches, you can see how the rail connects the first two sketches.

13. Highlight the **YZ** workplane.
    Right-click and select **New Sketch**.

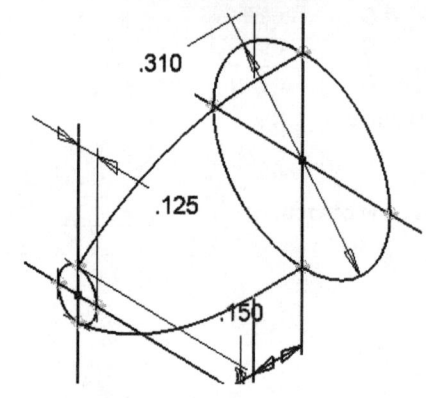

14. Create a sketch that mirrors the previous sketch using the quadrant points on the left side of the axis.

    Right-click and **Finish Sketch**.

    Make sure the rail sketch is coincident at each endpoint to the two profiles.

    Rename this sketch ***Rail2***.

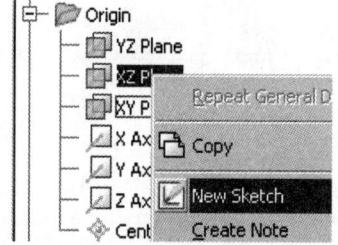

15. Highlight the **XZ** Workplane.
    Right-click and select **New Sketch**.

16. Project the two quadrant points to the right of the Y axis.

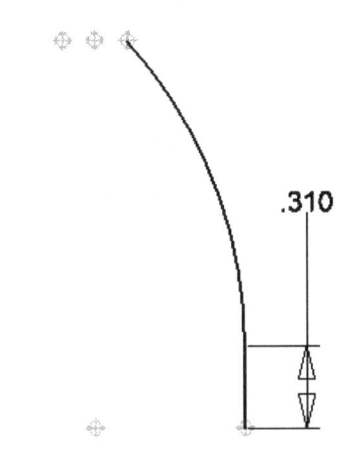

   Create the sketch and name it *Rail3*.

   Right-click and select **Finish Sketch**.

17. Highlight the YZ workplane.
    Right-click and select **New Sketch**.

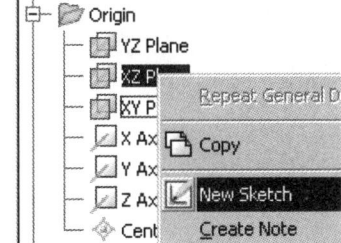

18. Project the two quadrant points to the left of the Y axis.

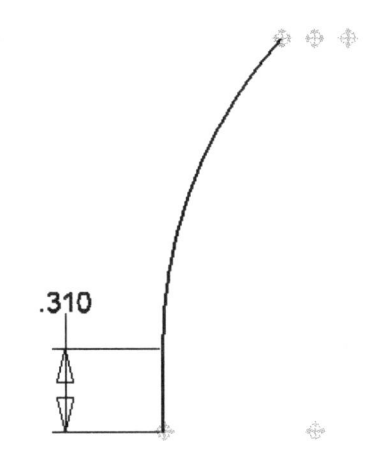

   Create the sketch and name it *Rail4*.

   Right-click and select **Finish Sketch**.

19. Turn **on** the **Visibility** of all the sketches.

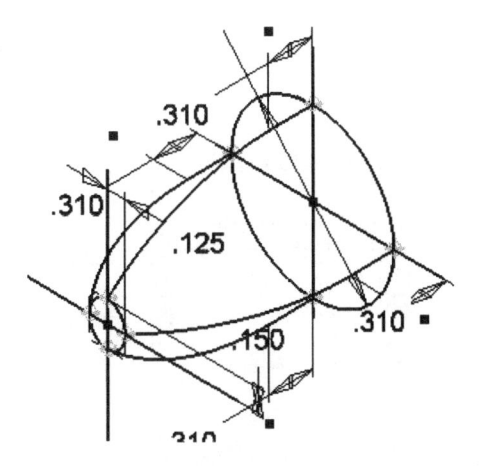

4-31

The Power of Design: An Introduction to Autodesk Inventor

20. Select the **Loft** tool.

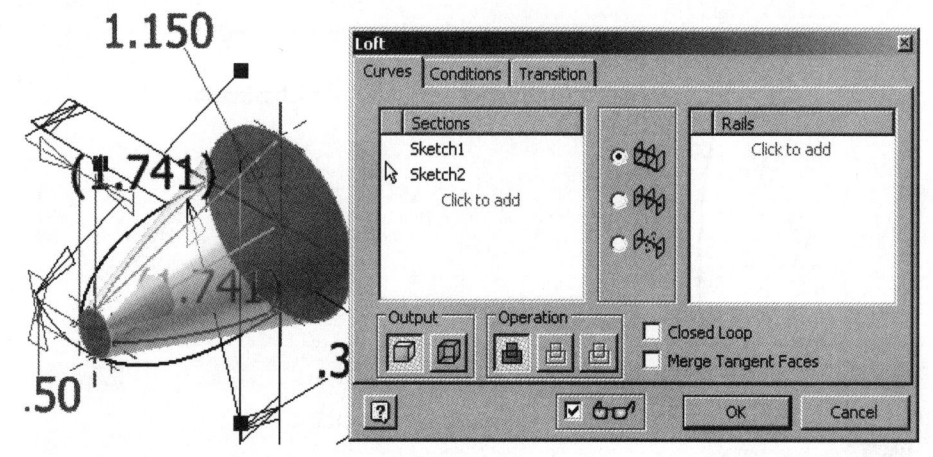

21. Select the two sketches.
DO NOT SELECT THE RAILS.
Be sure to select each region of each sketch.

Press **OK**.

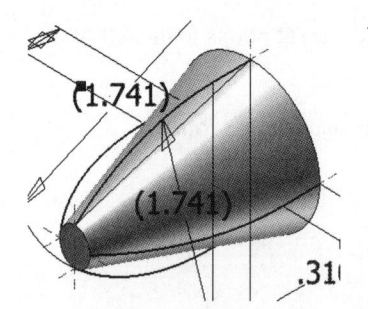

You see what the loft would look like without using rails.

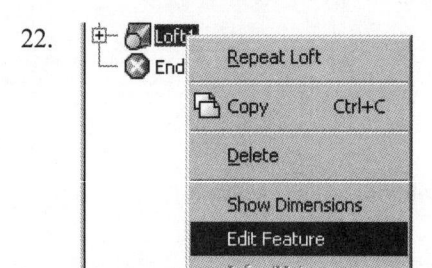

22. Highlight the Loft in the Browser.

Right-click and select **Edit Feature**.

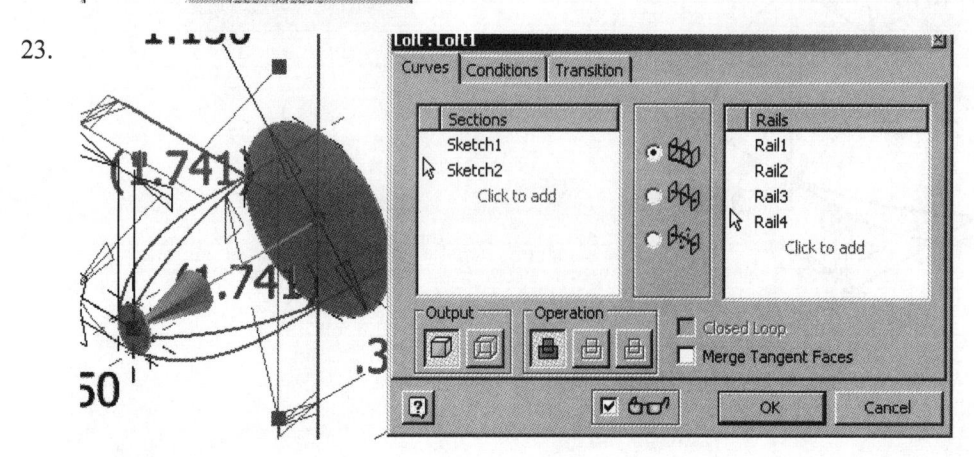

23. Select the Rails and add in the Rails dialog.

Press **OK**.

4-32

24. Save as *ex4-9.ipt*.

    Close the file.

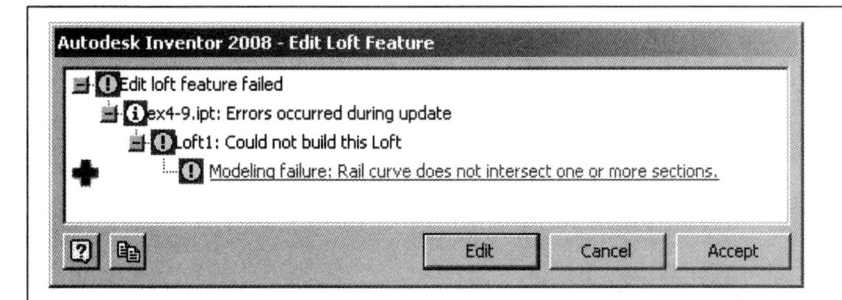

Autodesk Inventor 2008 - Edit Loft Feature

- Edit loft feature failed
  - ex4-9.ipt: Errors occurred during update
    - Loft1: Could not build this Loft
      - Modeling failure: Rail curve does not intersect one or more sections.

[Edit] [Cancel] [Accept]

If you see this error, it means that you are missing a **Coincident constraint** between the endpoint of one of the rails and one of the quadrant points on the sketches.

To troubleshoot which rail is the problem, delete all the rails from the loft and add one rail at a time.

Also, check if you placed the quadrant points so that they are actually coincident to the circle or ellipse.

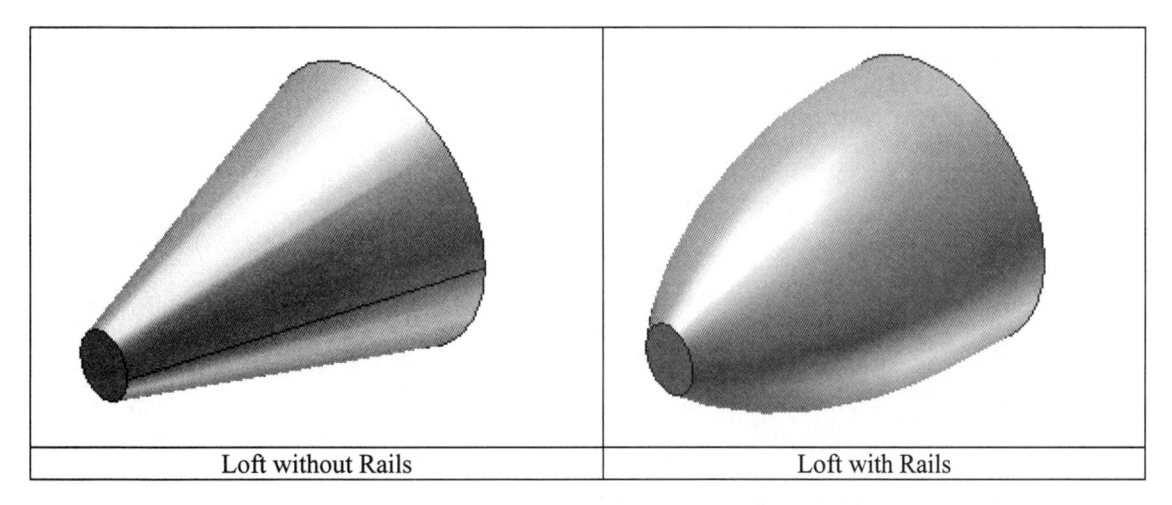

Loft without Rails	Loft with Rails

Compare the results of the two lofts. Note how the use of rails can control the loft and allow you to better define the desired shape.

The Power of Design: An Introduction to Autodesk Inventor

Exercise 4-10:
Sweep

File: Ex4-8.ipt
Estimated Time: 20 minutes

This exercise reinforces the following skills:

- Shell
- Workplane
- Extrude
- Sweep
- Project Geometry

1. Open *ex4-8.ipt*.
   Save as *ex4-10.ipt*.

2. Select the **Shell** tool.

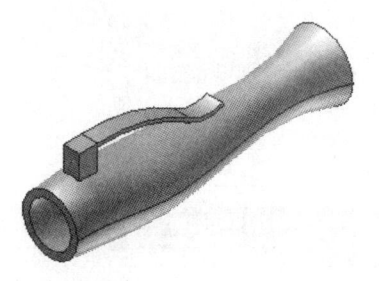

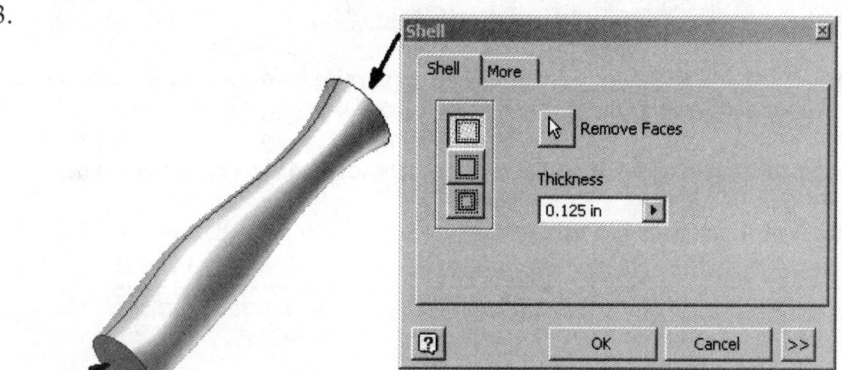

3. Select the two end faces to be removed.

   Set the **Thickness** to **0.125**.

   Press **OK**.

4. Create an offset workplane from the XZ plane.

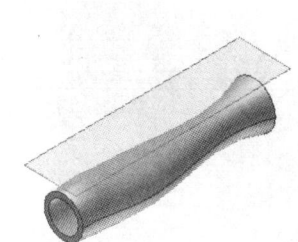

   Set the **offset distance** to **1 in**.

   Rename the workplane *ClipPlane*.

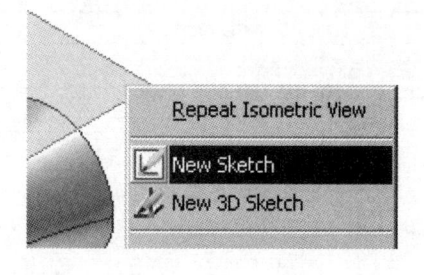

5. Highlight the **ClipPlane** in the Browser.
   Right-click and select **New Sketch**.

4-34

The Features Toolbar

6. 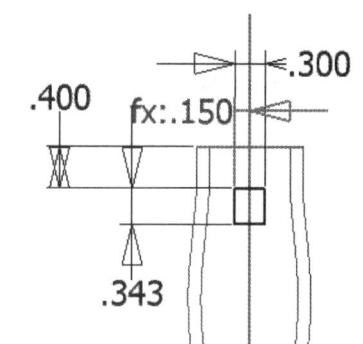 Create the sketch shown.

7. 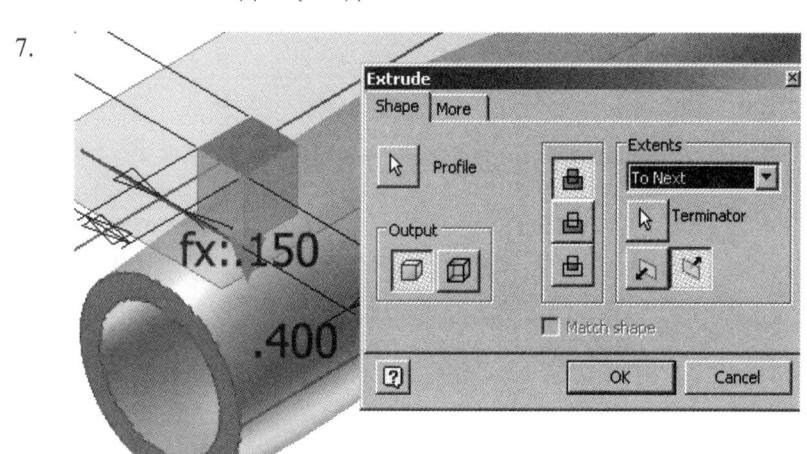 Select the **Extrude** tool.

   Set the **Extents** to **To Next**.

   Select the top surface of the loft as the terminator.

   Press **OK**.

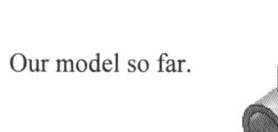

Our model so far.

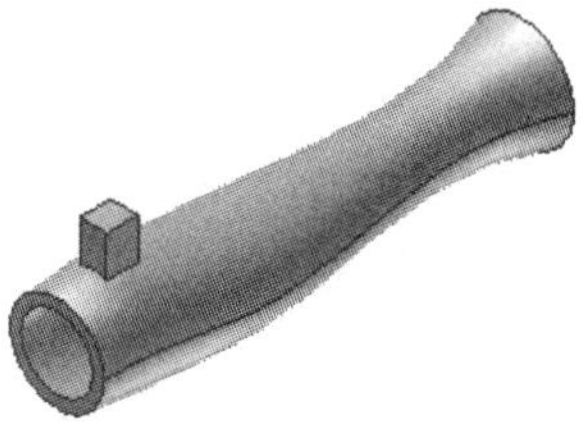

8.  Highlight the **YZ** workplane. Right-click and select **New Sketch**.

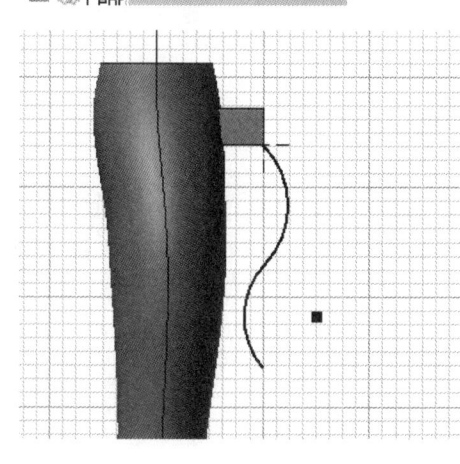

Draw two arcs.

4-35

9. Dimension as shown.

Rename the sketch **Path**.

Look for the **Fully Constrained** message in the lower right corner of the screen to help you figure out how many dimensions are still needed in the sketch.

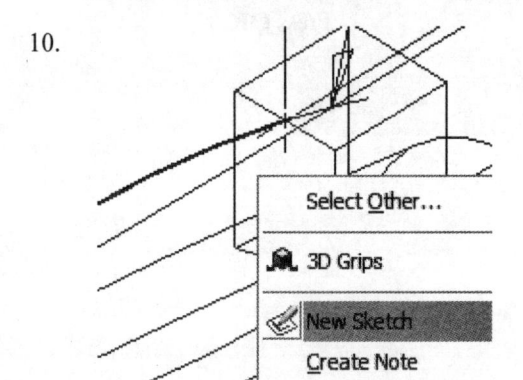

10. Select the inside face of the extrusion.
Right-click and select **New Sketch**.

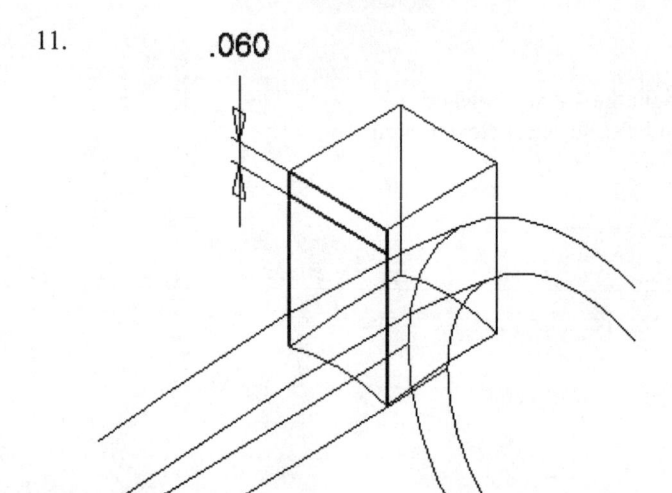

11. Draw a rectangle with the width of **.060**.

Use a **Collinear** constraint to define the top and side edges.

Rename this sketch **SweepProfile**.

12. Select the **Sweep** tool.

13. The profile will automatically be selected.

    Select the path sketch and press **OK**.

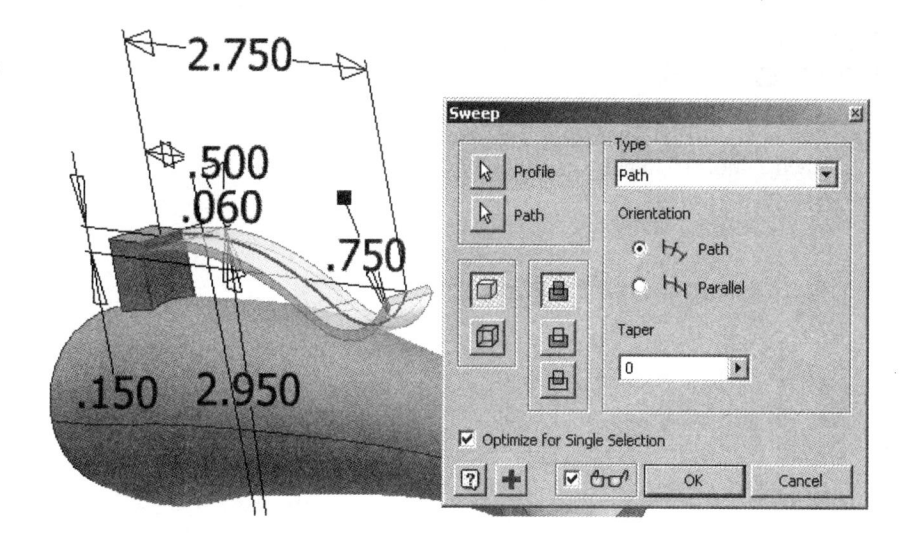

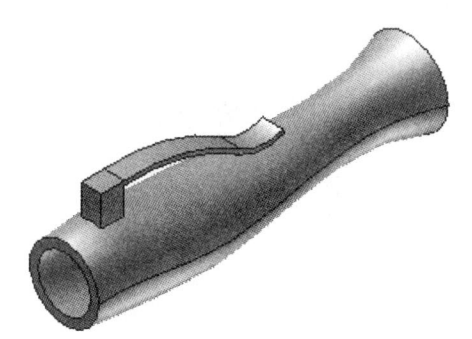

14. Your sweep is created.

    Save your file as *ex4-10.ipt*.

The Power of Design: An Introduction to Autodesk Inventor

Exercise 4-11:
Creating a Coil

File: Ex4-10.ipt
Estimated Time: 15 minutes

This exercise reinforces the following skills:

- Extrude
- Coil
- Work Axis

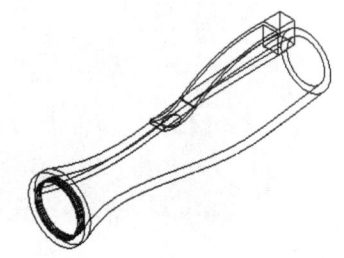

1. Open *ex4-10.ipt*.
   Save as *ex4-11.ipt*.

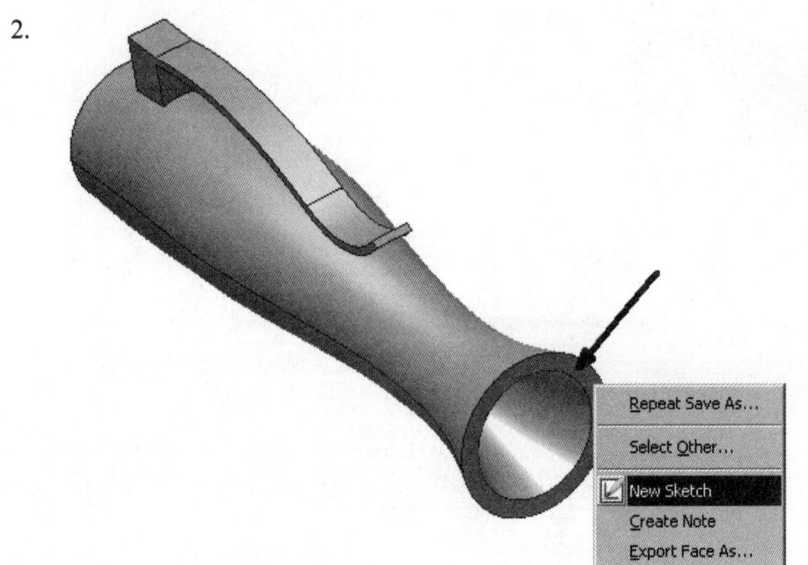

2. Select the end face indicated.

   Right-click and select **New Sketch**.

3. Select the **Project Geometry** tool and select the two circular edges.

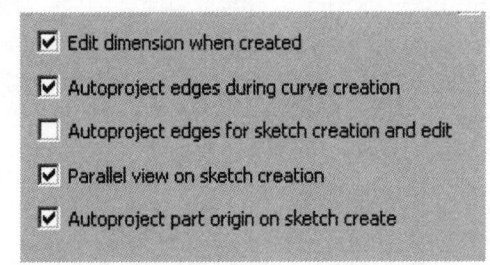

   This step is only necessary if you have turned OFF the Autoproject edges under Sketch Options.

   I usually have this option turned off to eliminate unnecessary elements in my sketches.

4. Select the **Extrude** tool.

4-38

5. Extrude the sketch **.125 in**.

   Press **OK**.

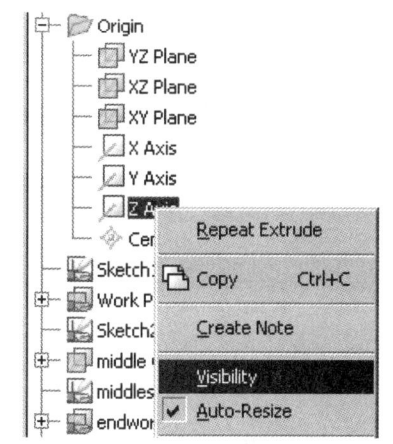

6. Turn **on** the **Visibility** of the Z axis.

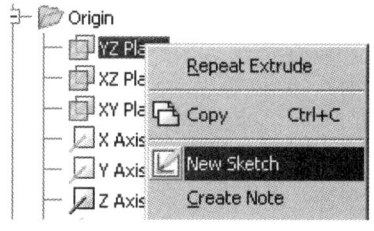

7. Highlight the **YZ** Plane.

   Right-click and select **New Sketch**.

8. Switch to a **Wireframe** view.

9. Create a circle with a diameter of **.030 in**.

   Place it **0.466** from the z-axis.

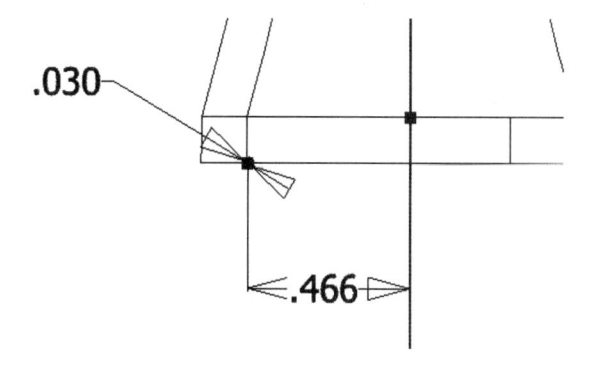

10. Select the **Coil** tool.

4-39

The Power of Design: An Introduction to Autodesk Inventor

11.	Select the **Z-axis** as the axis.

Change the direction of the coil so it goes into the part using the **Rotation** button.

Select the **Coil Size** tab.

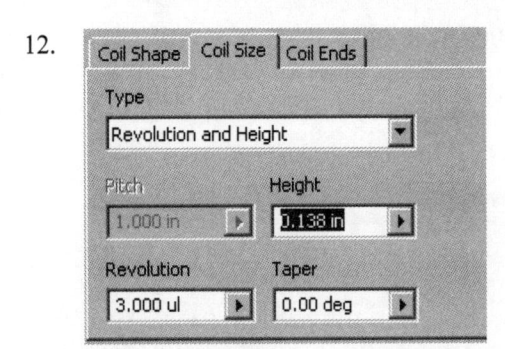

12.	Set the **Type** to **Revolution and Height**.

Set the **Height** to **.138**.

Set the **Revolution** to **3**.

Select the **Coil Ends** tab.

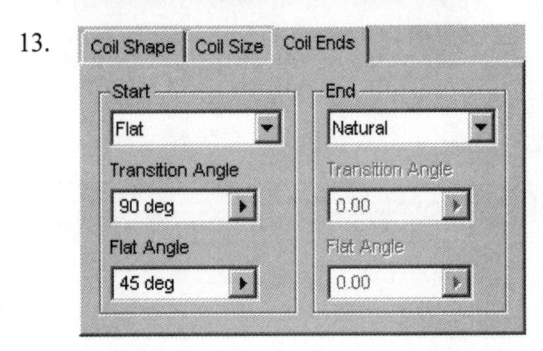

13.	Set the **Start** to **Flat**.

Set the **Transition Angle** to **90 deg**.

Set the **Flat Angle** to **45 deg**.

Press **OK**.

14.	Save the file.

4-40

Exercise 4-12:
Adding a Thread

File: Ex4-3.ipt
Estimated Time: 15 minutes

This exercise reinforces the following skills:

- Shell
- Thread

1. Open *ex4-3.ipt*.
   Save as *ex4-12.ipt*.

2. Select the **Thread** tool.

   There may be a slight pause while Inventor loads the Excel thread data.

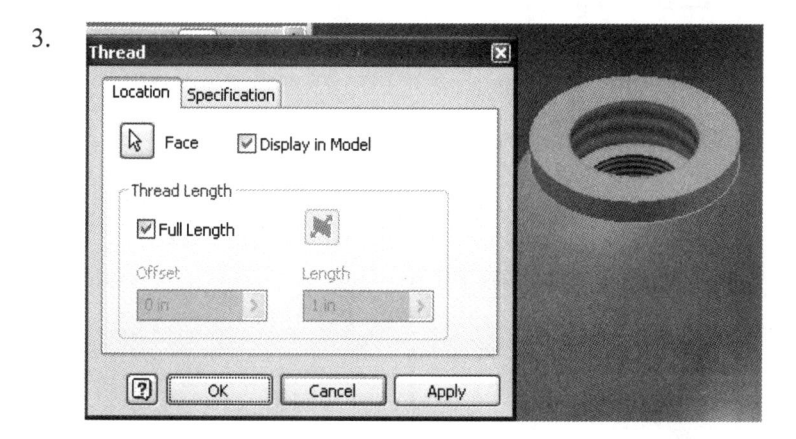

3. Select the lip of the model.
   Enable **Full Length**.
   Enable **Display in Model**.
   Select the **Specification** tab.

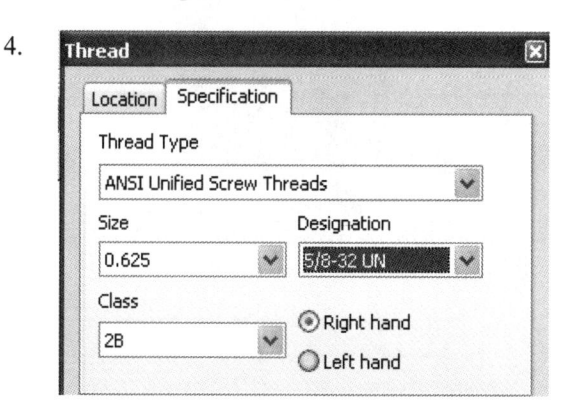

4. Set the **Pitch** to **5/8-32 UN**.

   Press **OK**.

4-41

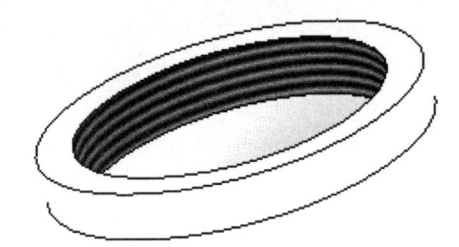

5. You will only be able to see the threads in Shaded mode.

Save the file and close.

Exercise 4-13:
Adding a Fillet

File: Ex4-10.ipt
Estimated Time: 15 minutes

This exercise reinforces the following skills:

* Fillet

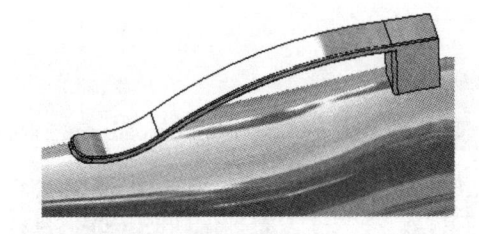

1. Open *ex4-10.ipt*.
   Save as *ex4-13.ipt*.

2. Select the **Fillet** tool.

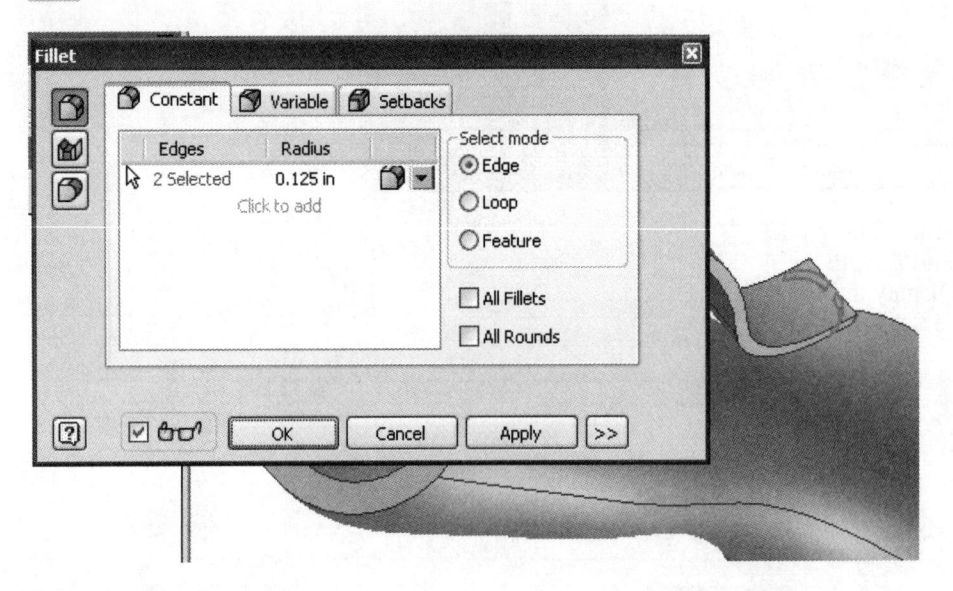

3. Select the two edges of the sweep.
   Set the **radius** to **0.125**.
   Press **OK**.

4. Select the **Fillet** tool.

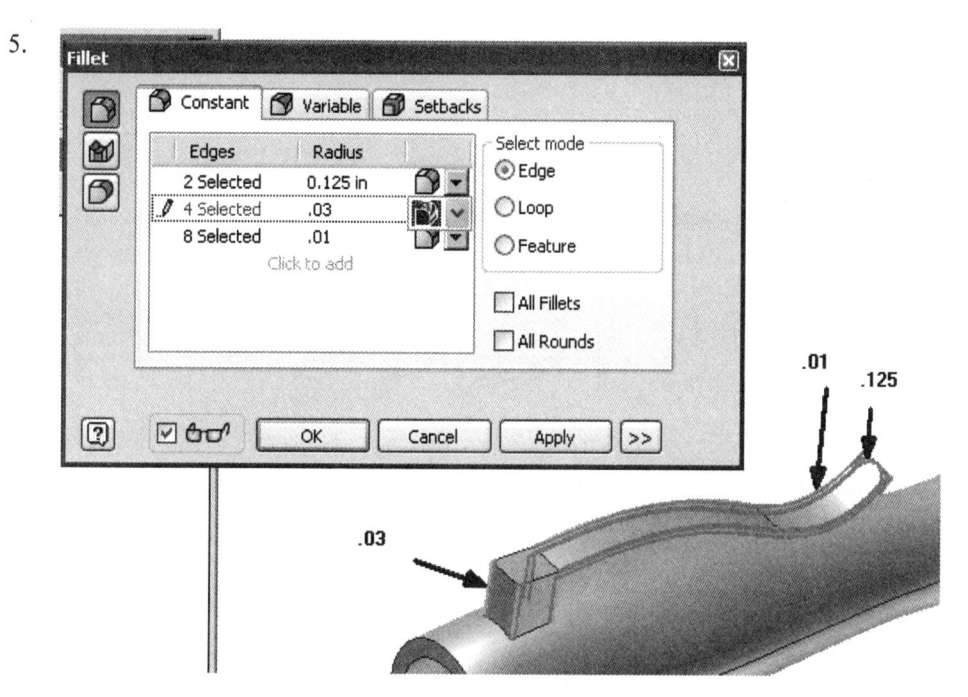

5.

Set the **Radius** to **0.01** for the top and bottom edges of the sweep.
Set the **Radius** to **0.03** for the inside edges of the extrusion.
For the 0.03 fillets, select the **Smooth (G2) Fillet** option.
Press **OK**.

6. Save the file.

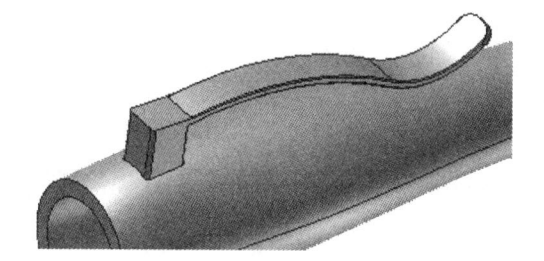

The Power of Design: An Introduction to Autodesk Inventor

Exercise 4-14:
Adding a Chamfer

File: Ex4-7.ipt
Estimated Time: 15 minutes

This exercise reinforces the following skills:

- Chamfer

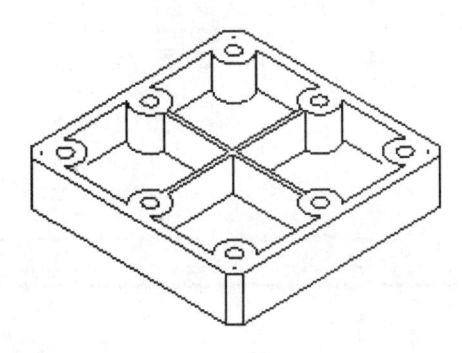

1. Open *ex4-7.ipt*.
   Save as *ex4-14.ipt*.

2. Select the **Chamfer** tool.

3.

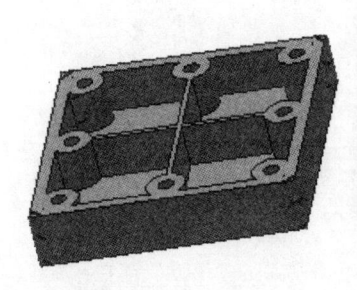

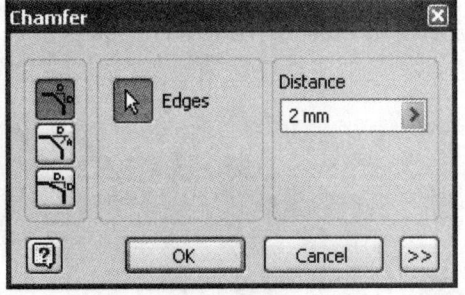

   Select the four corners.
   Set the **Distance** to **2 mm**.
   Press **OK**.

4. Save the file and close.

TIP: The first edge selected when using the Draft tool determines the curves you can select for draft. For example, if the first edge you select is linear, you cannot select edges with continuous tangency.

Exercise 4-15:
Applying a Face Draft

File: Ex4-13.ipt
Estimated Time: 5 minutes

This exercise reinforces the following skills:

- Face Draft

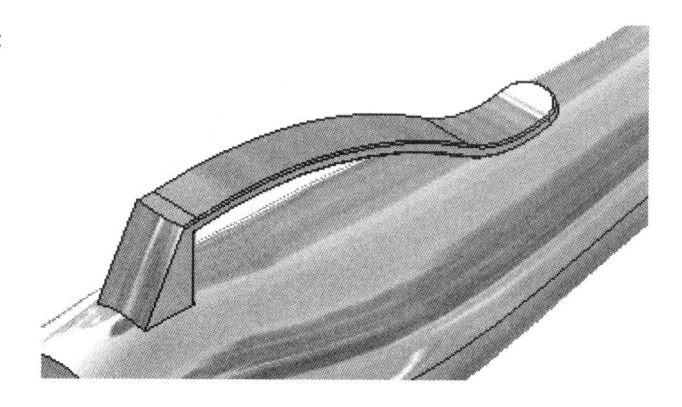

1. Open *ex4-13.ipt*.
   Save a copy as *ex4-15.ipt*.

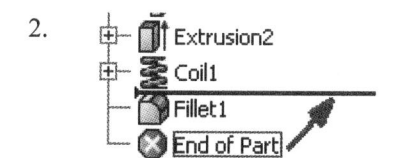

2. Drag the **End of Part** to above the fillet.

   This will suppress the fillets.

3. Select the **Face Draft** tool.

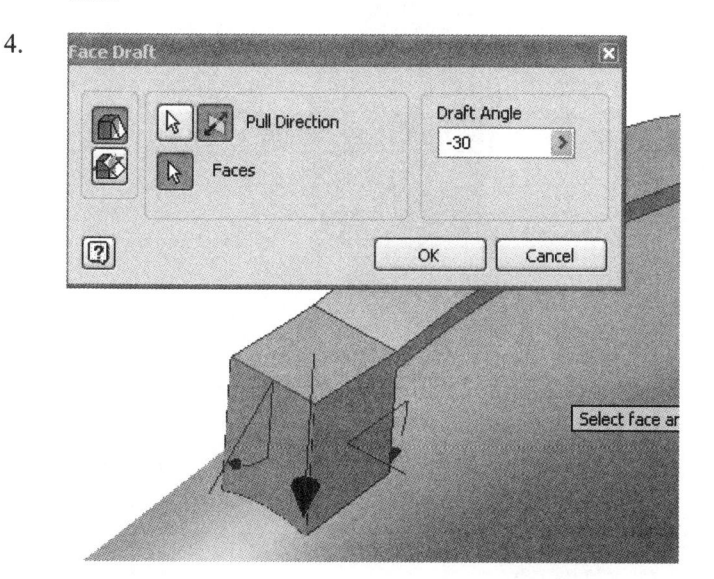

4. Select the top of the clip to indicate the pull direction.

   Flip the **Pull Direction** so that the face draft is going into the part.

   Select the back face of the clip.
   Set the **Draft Angle** to **-30 deg**.

   Press **OK**.

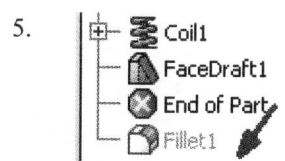

5. Drag the **End of Part** below the fillet feature.

   The fillets will be restored.

6. Save the file.

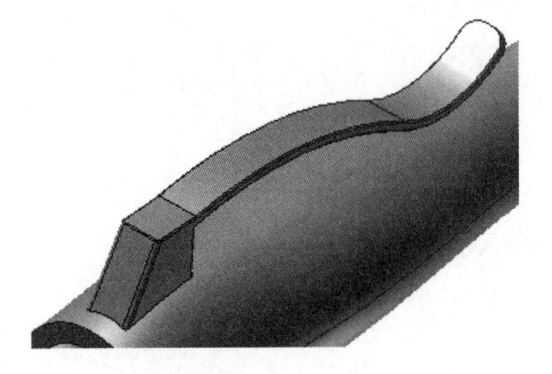

> **TIP:** When filleting: To select part of an edge that already has filleted corners, clear the check box. If a face is tangent to other faces, all tangent faces are highlighted.

Exercise 4-16:
Splitting a Part

File: Ex4-3.ipt
Estimated Time: 10 minutes

This exercise reinforces the following skills:

- Edit Feature
- Visibility of Work Planes
- Part Split

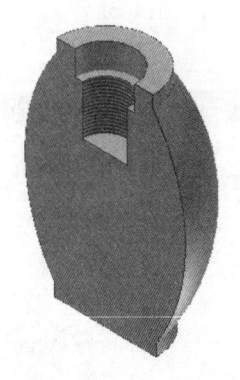

1. Open *ex4-3.ipt*.
   Save as *ex4-16.ipt*.

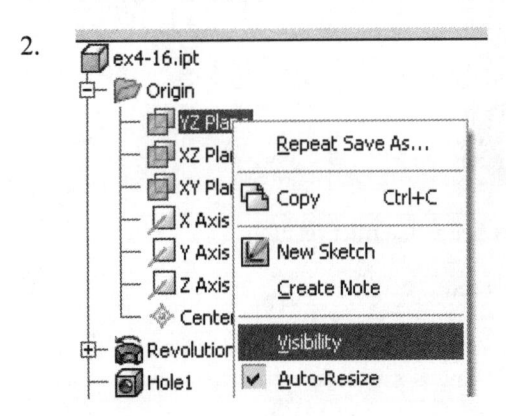

2. Highlight the **YZ plane**.
   Right-click and enable **Visibility**.

4-46

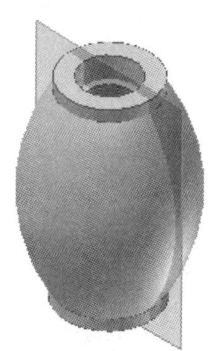

3. We see the work plane bisects the part.

4. Select the **Split** tool.

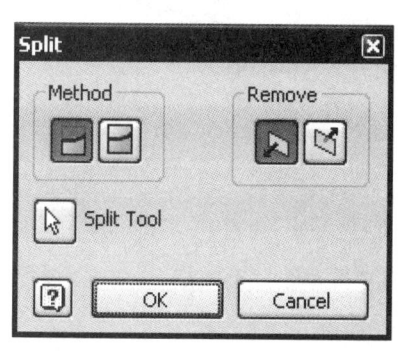

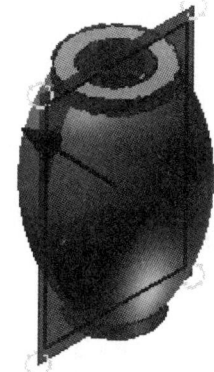

5. Enable the **Part Split** button to split the part.

    Select the **YZ** plane as the split tool.

    An arrow appears to indicate the section of the part that will be removed.

    Since the part is symmetrical, it doesn't really matter which part we remove.

6. Press **OK**.

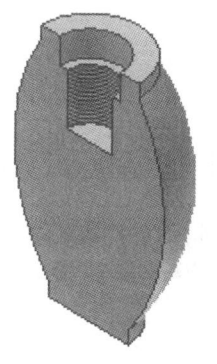

7. Save the file and close.

Exercise 4-17:
Emboss

File: Ex4-15.ipt
Estimated Time: 10 minutes

This exercise reinforces the following skills:

- New Sketch
- Offset Workplane
- 2D Sketch Text
- Insert Text
- Edit Coordinate System
- Emboss

1. Open *ex4-15.ipt*.
   Save as *ex4-17.ipt*.

2. Create an offset workplane 0.06 above the Clip Plane.

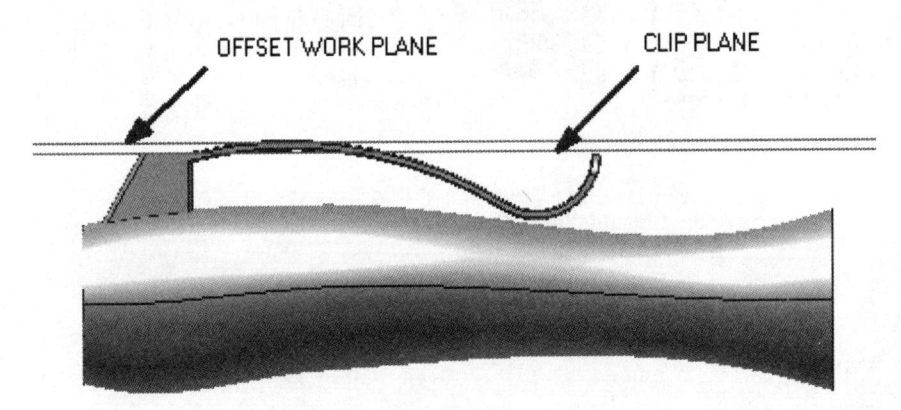

3. Highlight the offset workplane in the Browser. Right-click and select **New Sketch**.

4. Select the **Text** tool from the **2D Sketch Panel**.

4-48

The Features Toolbar

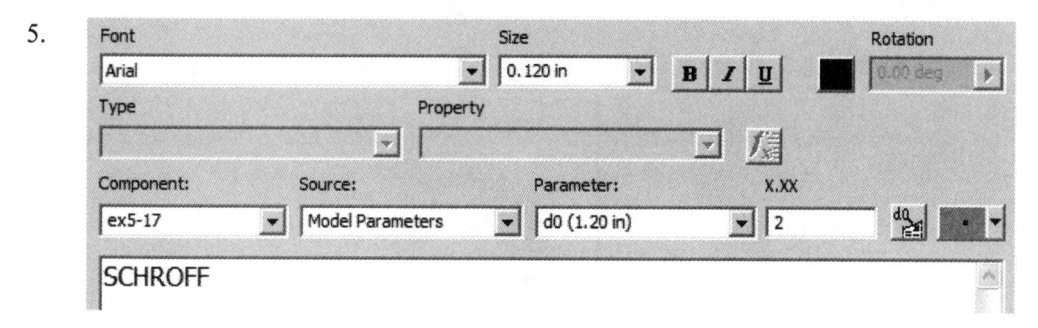

5. Set the **Font** to **Arial**.

    Type **SCHROFF** in the text field.

    Press **OK**.

6. Select the **Rotate** tool.

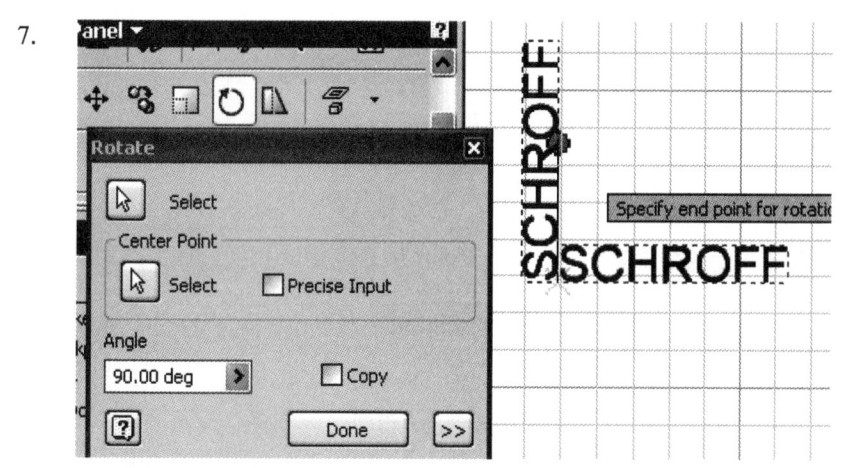

7. Select the text by windowing around it.

    Select the **Center Point** button.
    Select the lower left corner of the text box.
    Set the **Angle** to **90**.

    Press **Done**.

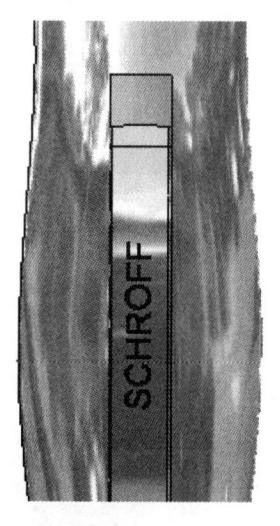

8. Drag the text so it is centered on the clip.

    Right-click and select **Finish Sketch**.

9. Change the **sketch name** to **Text**.

10. Select the **Emboss** tool.

4-49

11.

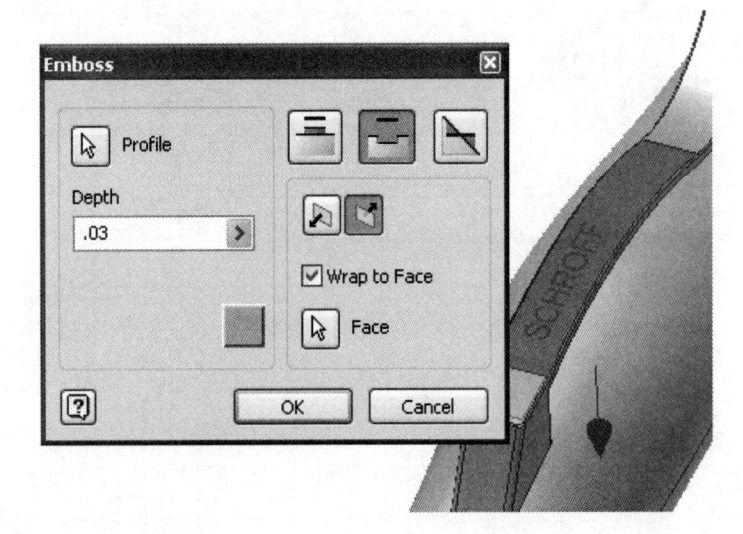

Select the **Text** sketch.
Set the **Depth** to **0.03 in**.
Set the option to **Engrave from Face**.

Enable **Wrap to Face**.
Select the curved clip face.

Select the **Color** button.

12. Set the **Color to Red (Flat)**.

Press **OK**.

13. Save and close the file.

Exercise 4-18:
Decal

File: Ex4-17.ipt, ex4-18.doc
Estimated Time: 10 minutes

This exercise reinforces the following skills:

- New Sketch
- Insert Image
- Decal

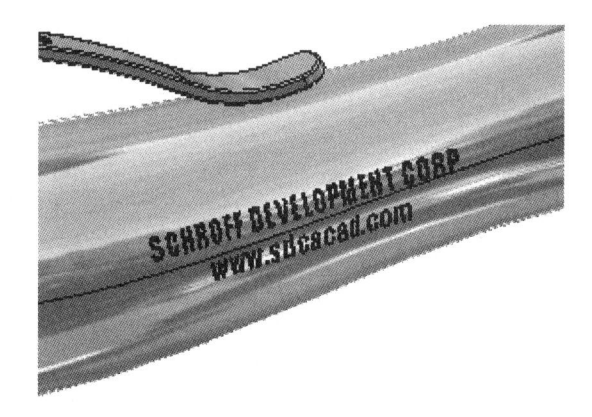

1. Open *ex4-17.ipt*.
   Save as *ex4-18.ipt*.

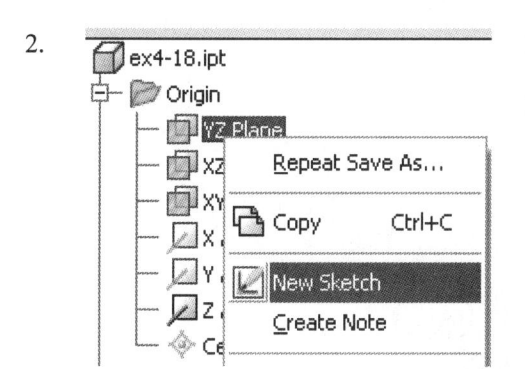

2. Highlight the **YZ** workplane.

   Right-click and select **New Sketch**.

3. Select the **Insert Image** tool from the **2D Sketch Panel** bar.

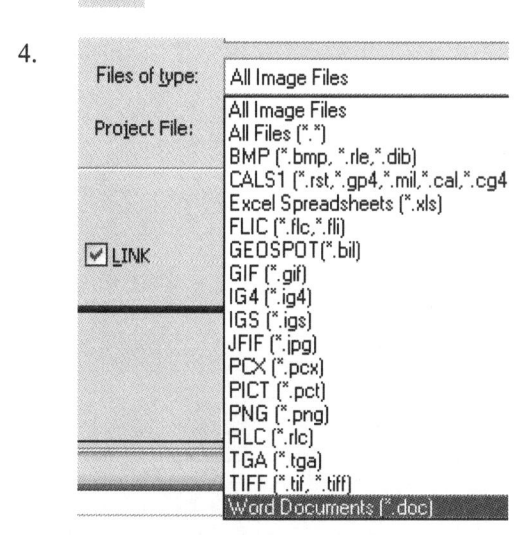

4. Select the ***.doc** option under **Files of type**.

4-51

The Power of Design: An Introduction to Autodesk Inventor

5. Locate the *ex4-18.doc* file downloaded from the publisher's website or use a Word document of your own choosing.

   Press **Open**.

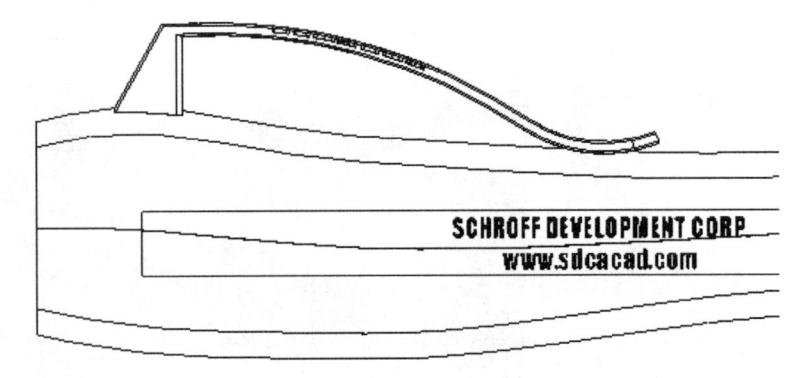

6. Rotate the document and align with the body.

   Right-click and select **Finish Sketch**.

7. Select the **Decal** tool.

8. Select the **Image**.
   Then select the lofted **Face**.

   Press **OK**.

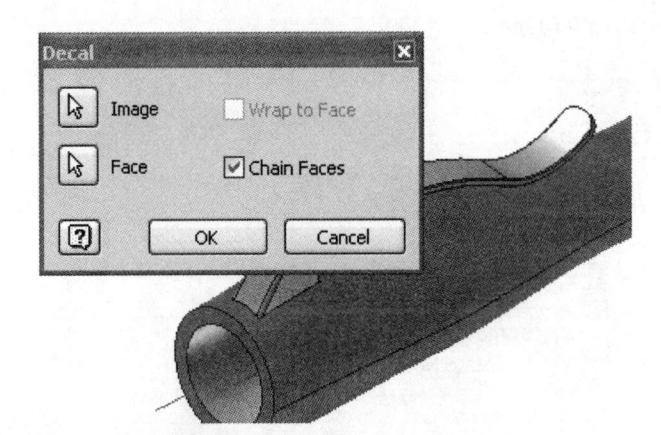

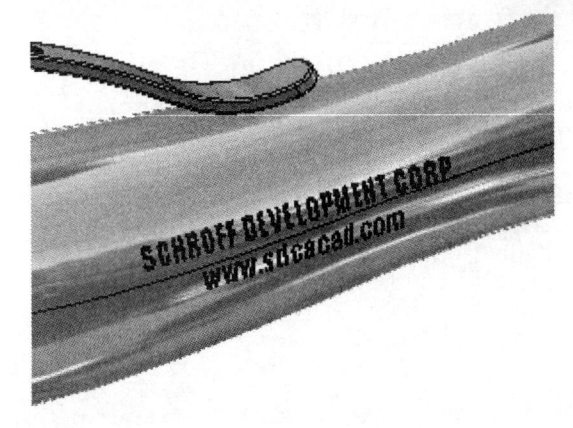

9. Save the file and close.

TIP: Decals are only visible in Shaded mode. If you want the decal to be visible in a drawing view, the drawing view also must be set to Shaded.

4-52

Exercise 4-19:
Circular Pattern

File: Ex4-3.ipt
Estimated Time: 15 minutes

This exercise reinforces the following skills:

- Workplane
- Extrude
- Circular Pattern

1. Open *ex4-3.ipt*.
   Save as *ex4-19.ipt*.

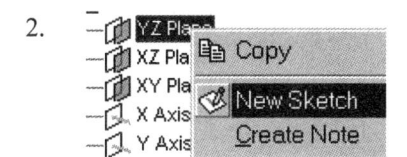

2. Select the **YZ Plane**.

   Right-click and select **New Sketch**.

3. Draw a rectangle as shown.

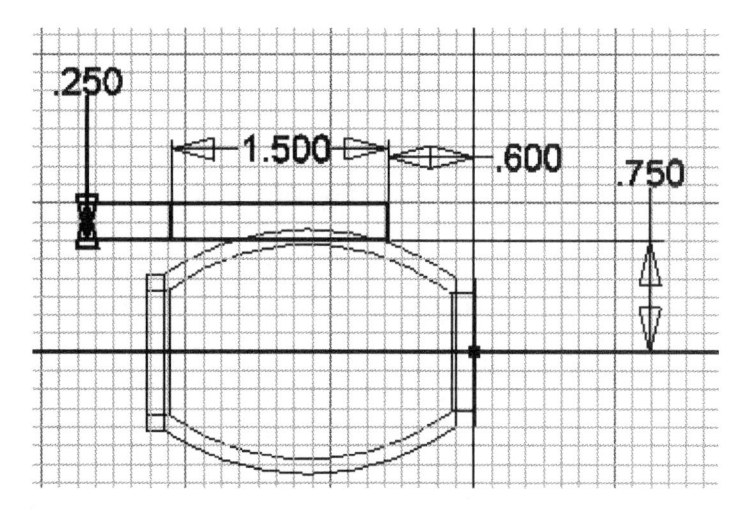

4. Select the **Extrude** tool.

5.

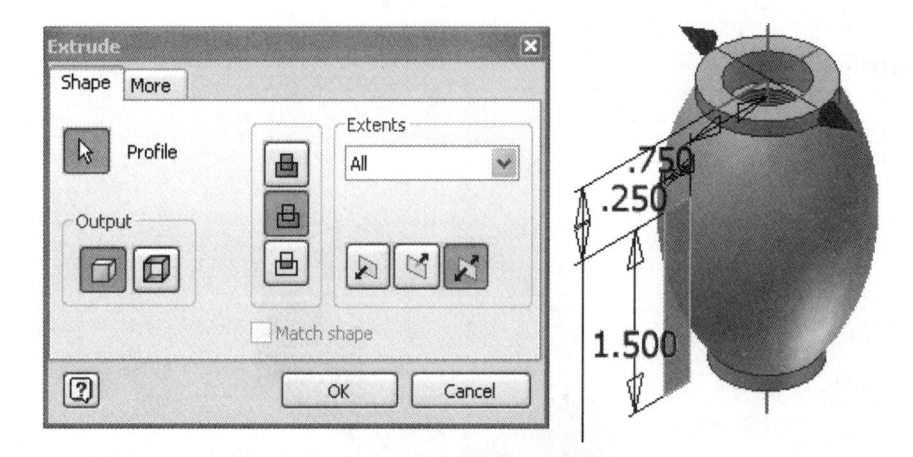

Extrude as a **Cut**.
Set **Extents** to **All**.
Set the **Direction** to **Mid-Plane**.

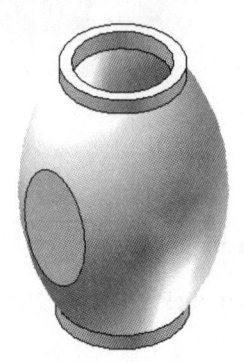

6. Select the **Circular Pattern** tool.

7.

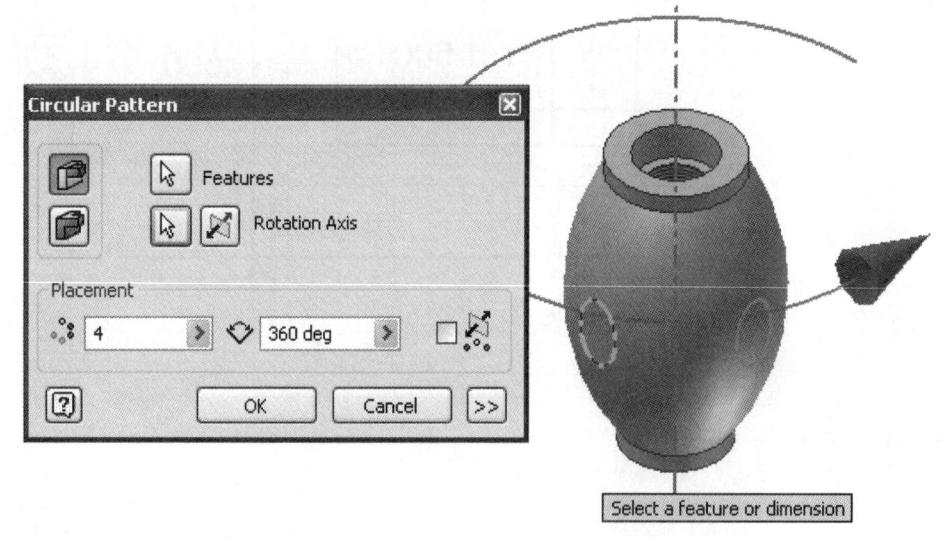

Enable the **Pattern Feature** option.
Select the **Extruded Cut** you just created as the feature to be patterned.
Select the **Z-axis** from the Browser as the Rotation Axis.
(It does not have to have the visibility turned on to be selected.)

Set the **Placement** to **4**.
Set the **Angle** to **360 deg**.

Press **OK**.

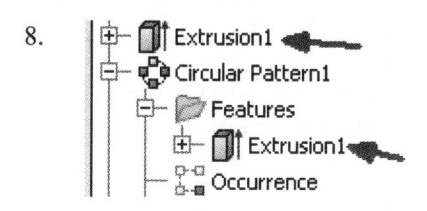

8. In the Browser, we see the Extrusion we patterned listed twice: once under the Circular Pattern and once above.

Turn off the visibility of any of the work axes that are displayed.

9. Save and close the file.

TIP: Features mirrored with the Identical method calculate faster than the Adjust to Model method. Using Adjust to Model, the mirrored feature terminates if it encounters a planar face, and may result in a feature whose size and shape differs from the original.

Exercise 4-20:
Bend Part

File: ex3-13.ipt
Estimated Time: 5 minutes

This exercise reinforces the following skills:

♦ Bend Part

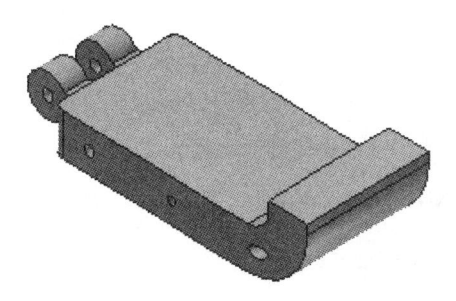

1. Open the *ex3-13.ipt* file.
   Save as *ex4-20.ipt*.

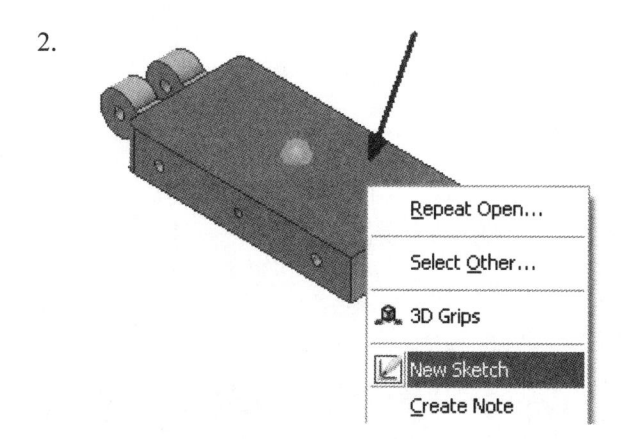

2. Select the top of the part.

   Right-click and select **New Sketch**.

4-55

3. Draw a single line on the face.

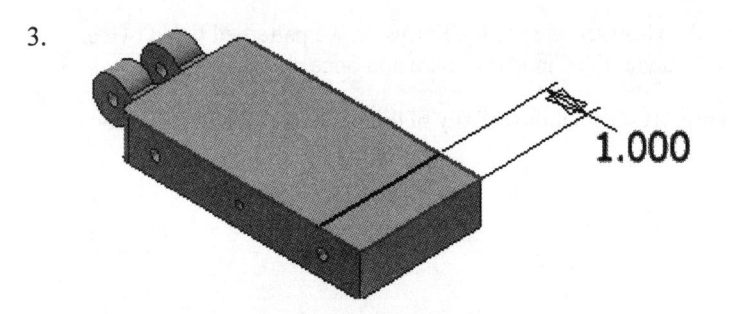

   Project the top and bottom edges and set the line end points coincident to the edges.

   Add the dimension.

   Right-click and **Finish Sketch**.

4.  Select the **Bend Part** tool.

5. Select the line.
   Set the **Radius** to **0.5**.
   Press **OK**.

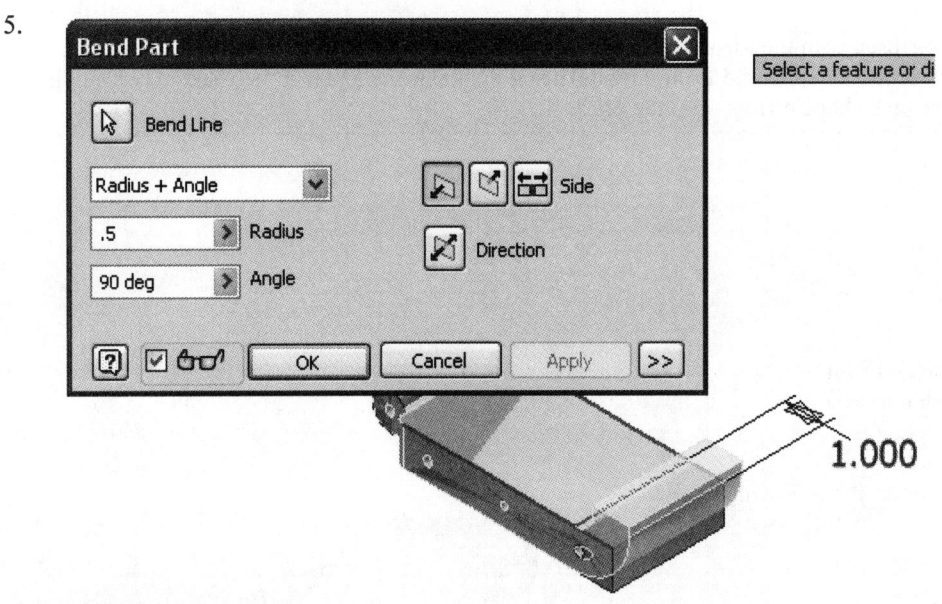

6. Save and close the file.

Exercise 4-21:
Mirror Feature

File: Ex4-9.ipt
Estimated Time: 5 minutes

This exercise reinforces the following skills:

- Mirror Feature

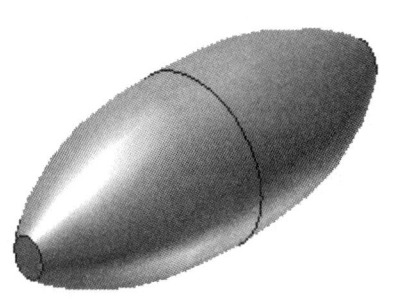

1. Open the *ex4-9.ipt* file.
   Save as *ex4-21.ipt*.

2. Select the **Mirror Feature** tool.

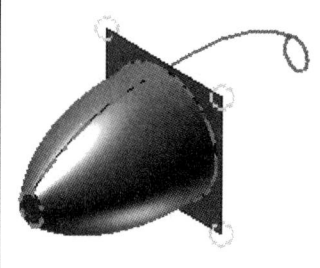

3.
   Select the **Loft** in the Browser.
   Pick the **Mirror Plane Select** button.
   Select the **XY Plane** in the Browser. (The plane does not need to be visible.)
   Press **OK**.

4. Save and close.

The Power of Design: An Introduction to Autodesk Inventor

Exercise 4-22:
Sculpt

File: New
Estimated Time: 5 minutes

This exercise reinforces the following skills:

- Sculpt

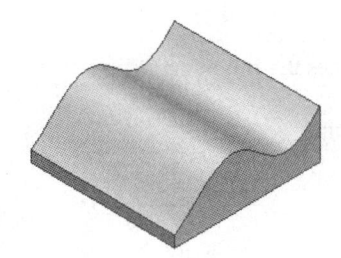

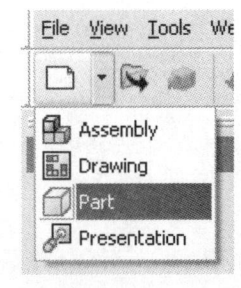

1. Start a new part.

2. Create a 2 in x 2 in x 1 in box using **Extrude**.

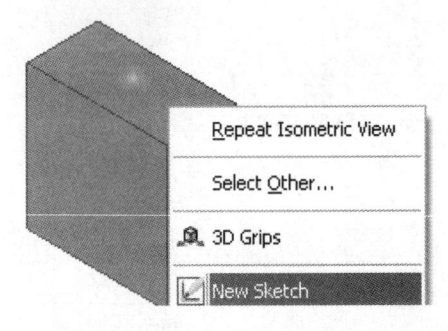

3. Select the top of the box.

   Right-click and select **New Sketch**.

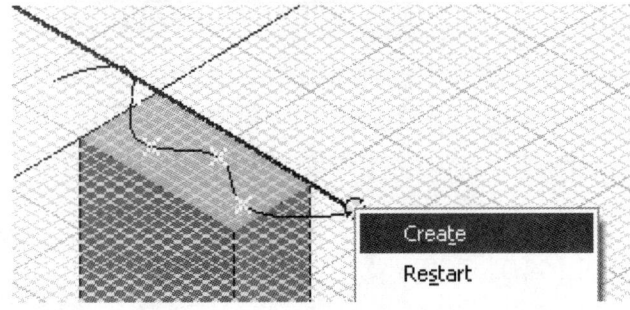

4. Draw a free-form spline.

4-58

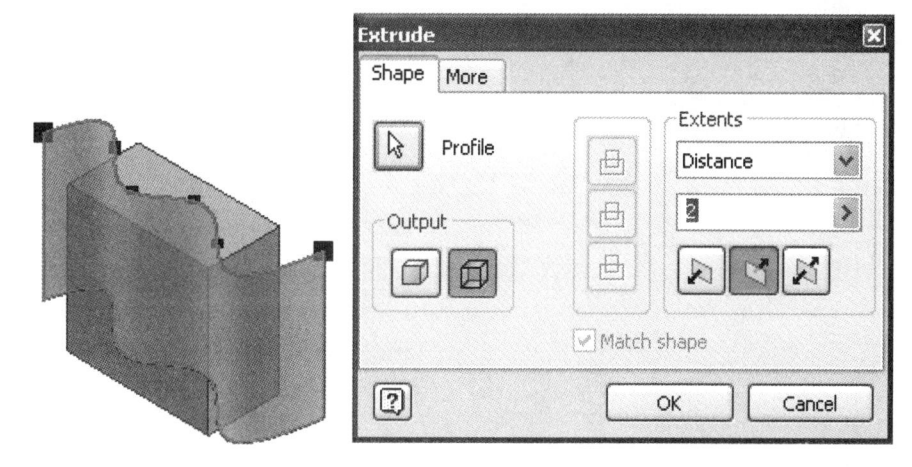

5. Extrude as a surface through the block.

6. Select the **Sculpt** tool.

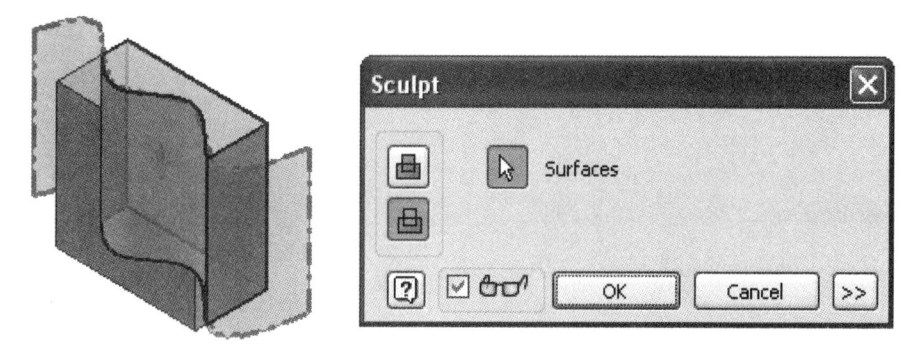

7. Enable **Cut**.
    Select the surface as the cutting tool.
    Press **OK**.

8. Close the file without saving.

The Power of Design: An Introduction to Autodesk Inventor

Review Questions

Identify the icon.

1. Hole
2. Extrude
3. Coil
4. Loft

Identify the icon.

5. Part Splitting
6. Face Draft
7. Fillet
8. Chamfer

9. Sculpt can be used to:

    A. Cut away an extruded feature
    B. Create statues or clay figures
    C. Create a part with an irregular face
    D. All of the above

10. The three options for creating a chamfer are:

    A. Equal distance, distance-angle, two distances
    B. Variable, constant, radial
    C. Equal Distance, Variable Distance, Distance-Angle
    D. Two angles, Two Distances, Equal Distance

11. When you create a circular pattern, you select one or more features to pattern. What happens to the original features selected?

    A. They are deleted.
    B. They are copied into the pattern.
    C. They are suppressed.
    D. They become iFeatures.

12. You can select more than one profile to extrude at a time.

    A. True
    B. False

13. When placing holes you can use a Point, Hole Center or any vertex points on a sketch.

    A. True
    B. False

14. Base features require an unconsumed sketch.

    A. True
    B. False

15. To deselect a selected profile:

    A. Press down the Control key and pick.
    B. Press down the Shift key and pick.
    C. Press down the Tab key and pick.
    D. Press down the Alt key and pick.

16. Select the item listed that is NOT considered a DEPENDENT feature:

    A. Fillet
    B. Hole
    C. Pattern
    D. Base Extrude

17. You extrude a cylindrical surface on top of a base extrude. You then add a hole to the cylindrical surface. Next, you delete the cylindrical surface. What happens to the hole?

    A. It is automatically deleted.
    B. A dialog box comes up to allow you to retain the hole and place it somewhere else.
    C. An error message comes up advising that you need to delete the hole first.
    D. A fatal error occurs causing the software to crash.

ANSWERS: 1) B; 2) A; 3) D; 4) C; 5) D; 6) C; 7) A; 8) B; 9) C; 10) A; 11) B; 12) A; 13) A; 14) A; 15) A; 16) D; 17) B

NOTES:

Lesson 5
Drawing Management

Learning Objectives:

Upon completion of this lesson, the user will be familiar with:

- Creating Base Views
- Creating Orthographic Views
- Creating Auxiliary Views
- Creating Section Views
- Creating Detail Views
- Creating Broken Views
- Creating Sheets
- Creating Title Blocks
- Modifying Title Blocks
- Managing Views
- Managing Sheets

We do not see either the Drawing Management toolbar or the Drawing Annotation toolbar unless we are in the drawing layout environment.

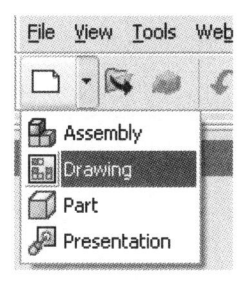

To get there, we select **Drawing** under the **New File** pull-down.

Exercise 5-1:
Creating a Base View

File: Ex5-1.ipt
Estimated Time: 15 minutes

1. Open a new drawing file.

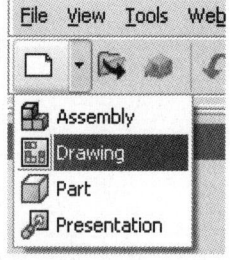

There is more than one way to create drawing views in Inventor.

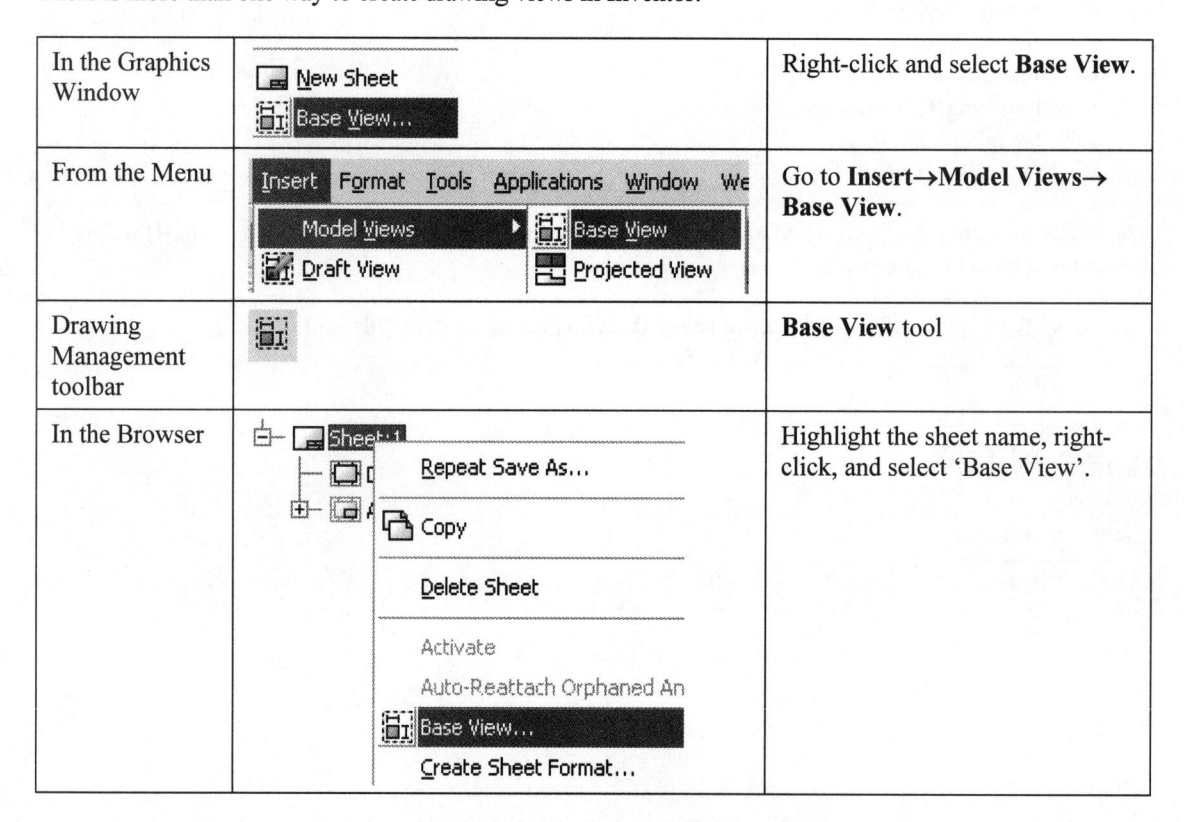

In the Graphics Window		Right-click and select **Base View**.
From the Menu		Go to **Insert→Model Views→Base View**.
Drawing Management toolbar		**Base View** tool
In the Browser		Highlight the sheet name, right-click, and select 'Base View'.

2. Select the **Base View** tool from the **Drawing Management** toolbar.

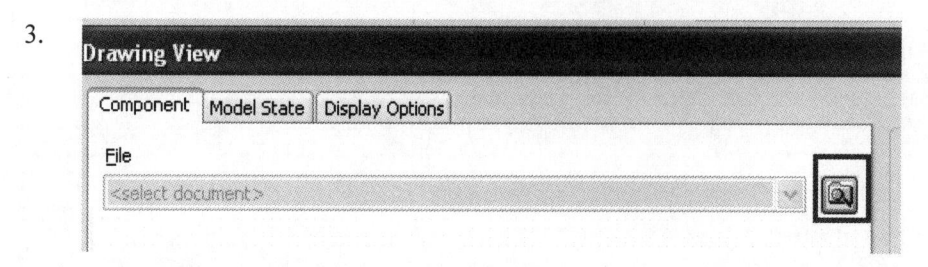

3. Select the **Browse** button to select a file.

The dialog File drop-down list will show any open part, assembly or presentation files.

4. Locate *ex5-1.ipt*.

This file may be downloaded from the publisher's website at www.schroff.com/resources.

Note that you will see a preview of the part in the dialog.

Select **Open**.

5. Verify that the **Front** view is highlighted in the dialog box.

You will also see a preview of the view at the end of your cursor.

6. Select **Hidden Line** as the **Style**.

7. Set the **Scale to 1:1**.

NOTE: If you want the view to be labeled with a scale, click on the light bulb to turn on the visibility.

8. Select the **Display Options** tab.

9. Enable **All Model Dimensions**.

   Enable **Tangent Edges**.

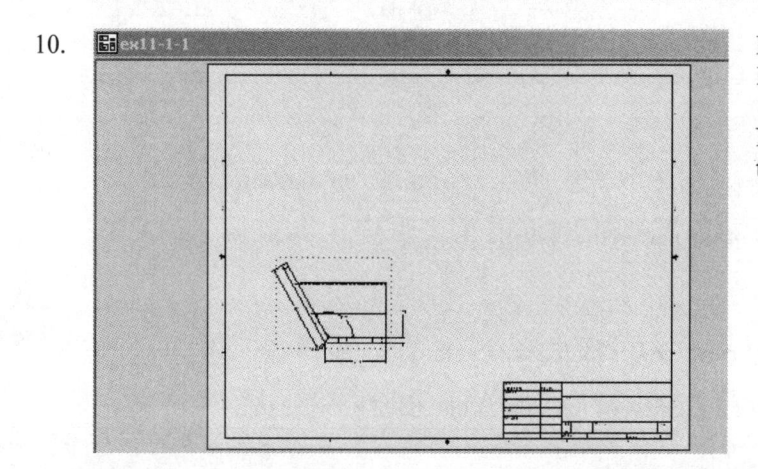

10. Place the view in the sheet in the lower left corner.

    Note that dimensions are placed with the view.

11. Some people do not like the default color of the sheet, which is a beige color.

    Go to **Tools→Document Settings**.

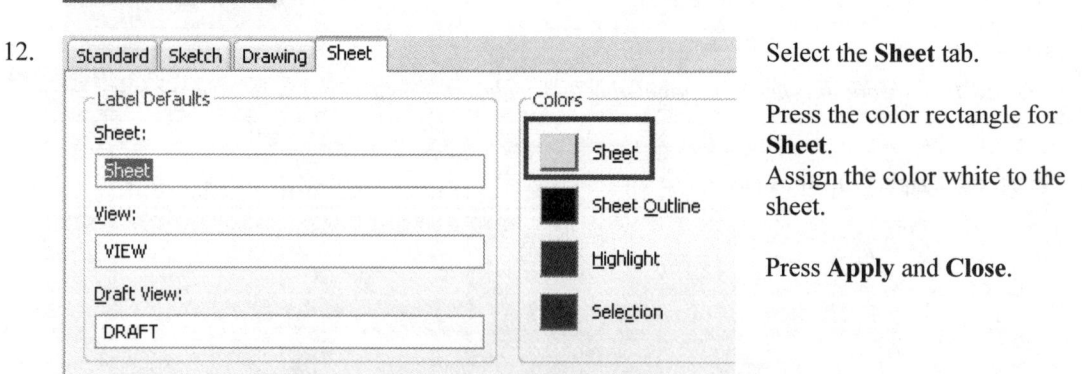

12. Select the **Sheet** tab.

    Press the color rectangle for **Sheet**.
    Assign the color white to the sheet.

    Press **Apply** and **Close**.

Note the labels; these assign the default labels that appear in your Browser and in your layout.

13. Save as *ex5-1.idw*.

Drawing Management

> Once you have set up your sheet with the format you like, create a sheet format with those settings and save it to a template.
> Some of the characters in your drawing may appear as a black rectangle. This is a graphics card issue; your plot should be OK.
> Orthographic projections are aligned to the base view and inherit its scale and display settings. Isometric projections are not aligned to the base view. They default to the scale of the base view but do not update if you change the scale of the base view.

Exercise 5-2:
Create a Projected View

File: Ex5-1.idw
Estimated Time: 15 minutes

1. Continue working with the *ex5-1.idw* file.

There is more than one way to create the projected views:

From the Base View		Select the Base View with a right pick, select **Create View→Projected** from the menu
From the Browser		Highlight the Base View in the Browser, right-click and select **Create View→ Projected**
From the Menu		**Insert→Model Views→Projected View**
From the Drawing Management toolbar		Projected View

5-5

The Power of Design: An Introduction to Autodesk Inventor

2. Highlight the base view by picking on it. (A red dotted rectangle should appear when the view is active.)

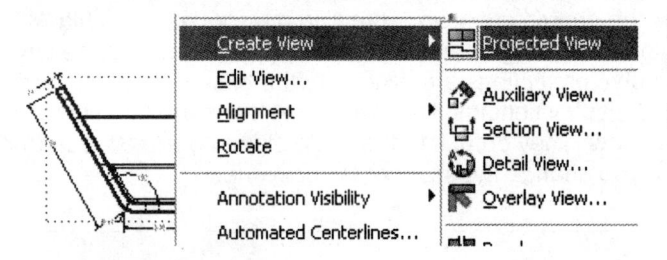

Right-click the mouse to bring up the menu.

Select **Create View→ Projected**.

3. Pick the location for the top view with the left mouse button.

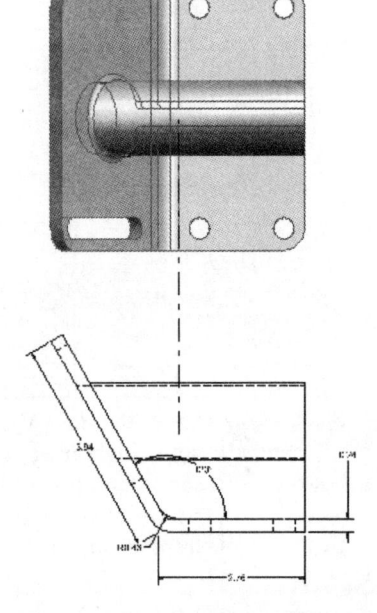

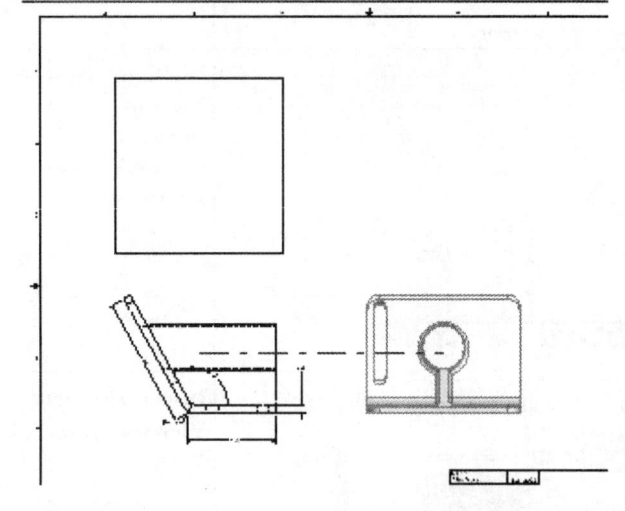

4. Move the mouse to the right of the base view.

Left pick to place the right side view.

A blank rectangle is shown to act as a placeholder for the top view, so you can keep track of which views you have already placed.

5-6

Drawing Management

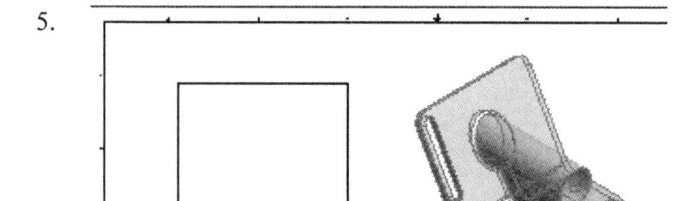

5. Move the mouse above the right side view and pick to place the isometric view.

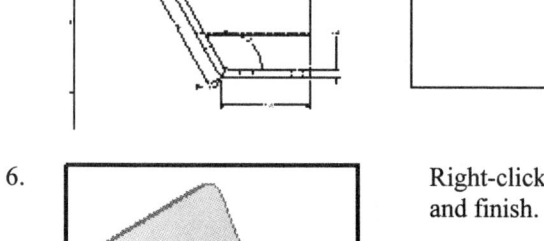

6. Right-click the mouse and select **Create** to set the views in place and finish.

7. Save the file as *ex5-2.idw*.

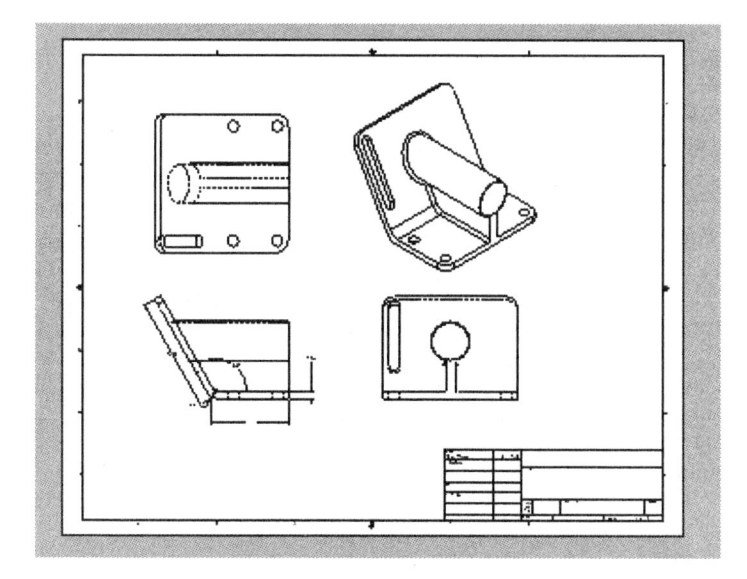

5-7

Exercise 5-3:
Adding an Auxiliary View

File: Ex5-2.idw
Estimated Time: 15 minutes

1. Continue working with the *ex5-2.idw* file.

There is more than one way to create an auxiliary view.

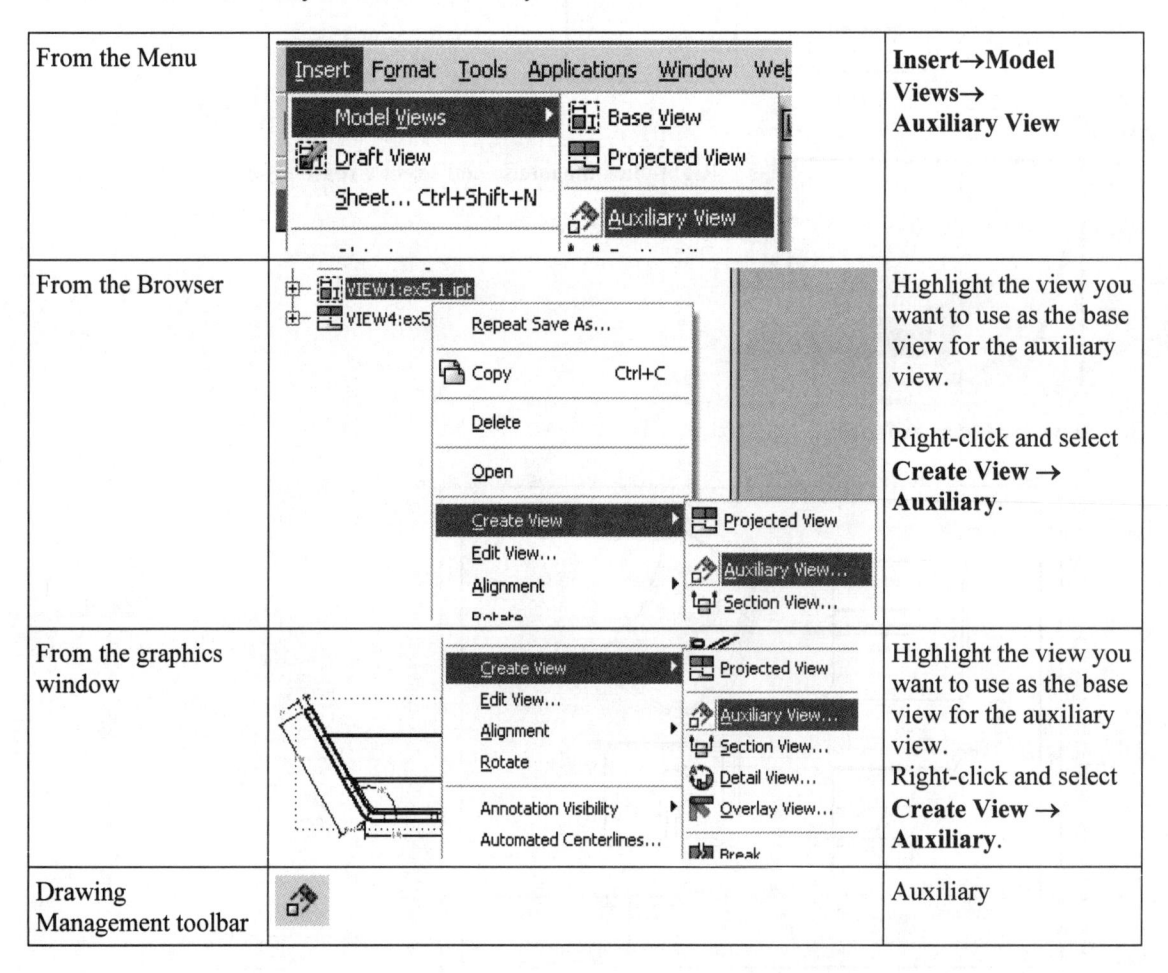

From the Menu		Insert→Model Views→ Auxiliary View
From the Browser		Highlight the view you want to use as the base view for the auxiliary view. Right-click and select **Create View → Auxiliary**.
From the graphics window		Highlight the view you want to use as the base view for the auxiliary view. Right-click and select **Create View → Auxiliary**.
Drawing Management toolbar		Auxiliary

2. Select the base view.
Right-click and select **Create View → Auxiliary**.

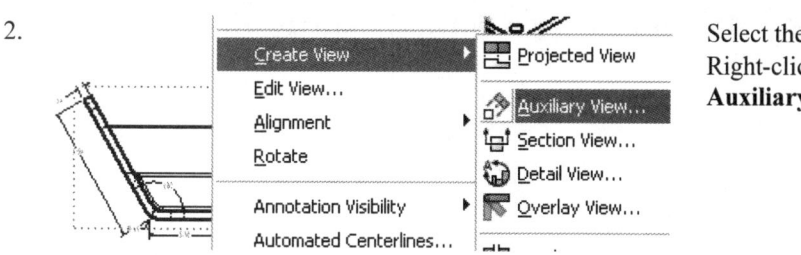

3. Drag and drop to place the auxiliary view.

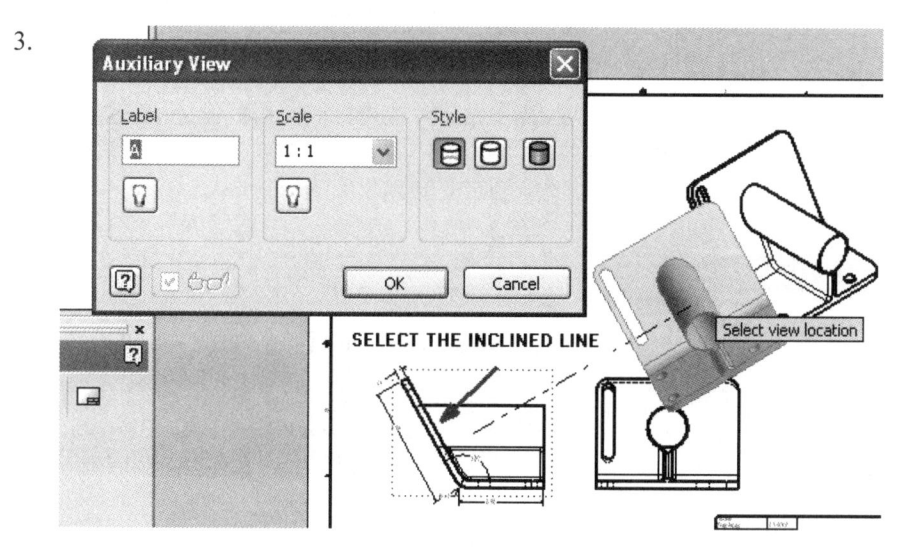

We can move the auxiliary view to a second sheet.

You can use several methods to add an additional drawing sheet.

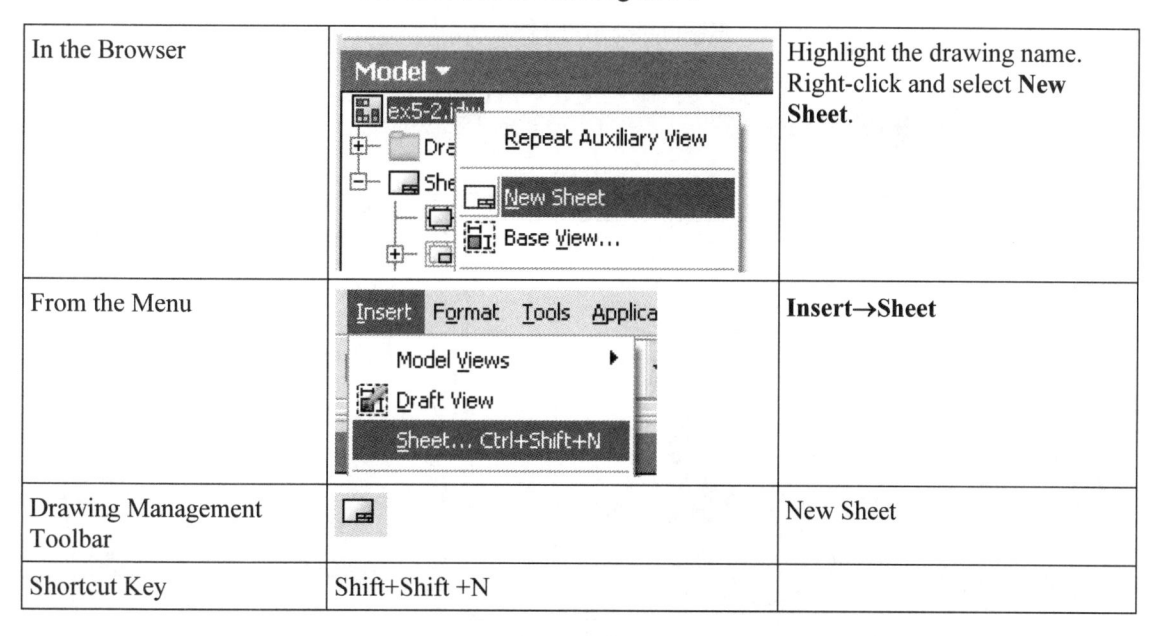

In the Browser		Highlight the drawing name. Right-click and select **New Sheet**.
From the Menu		**Insert→Sheet**
Drawing Management Toolbar		New Sheet
Shortcut Key	Shift+Shift +N	

4.  Select the **New Sheet** tool.

5. The **New Sheet** appears in the Browser and activates in the graphics window.

6. To switch back to the first sheet, highlight it in the Browser.

   Right-click and select **Activate**.

   You can also activate by double left-clicking on top of the paper sheet icon next to the sheet name.

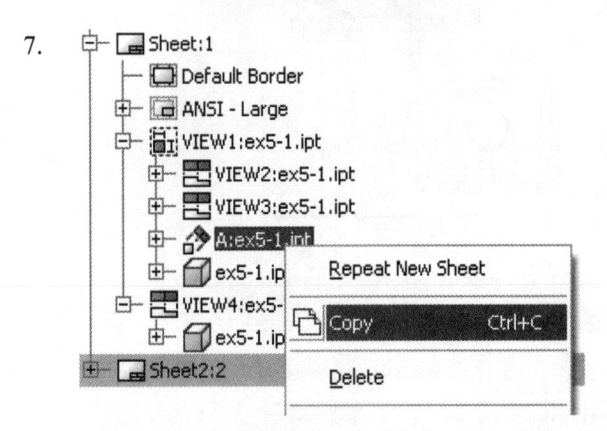

7. Expand the first sheet so you can see all the views.

   Highlight the **Auxiliary** view in the Browser. It will also highlight in the graphics window.

   Right-click and select **Copy**.

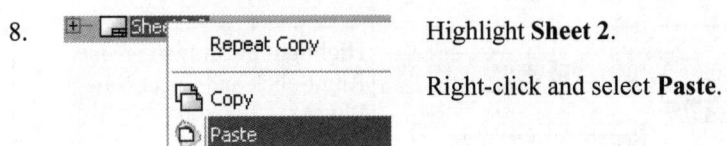

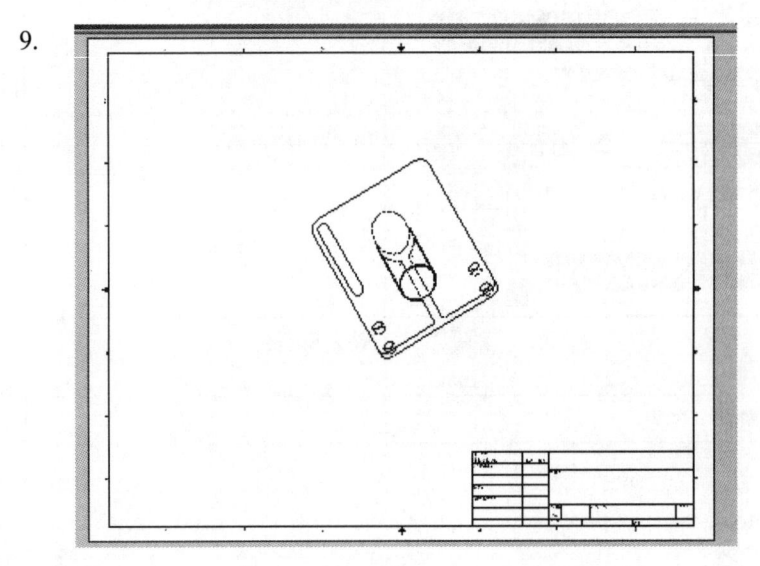

8. Highlight **Sheet 2**.

   Right-click and select **Paste**.

9. The auxiliary view is copied onto the new sheet.

   The view will appear in the same location on the sheet as its previous location.

Drawing Management

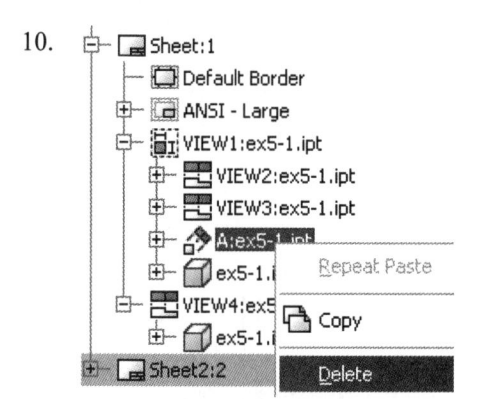

10. Activate **Sheet 1**.

Highlight the auxiliary view on Sheet 1.

Right-click and select **Delete**.

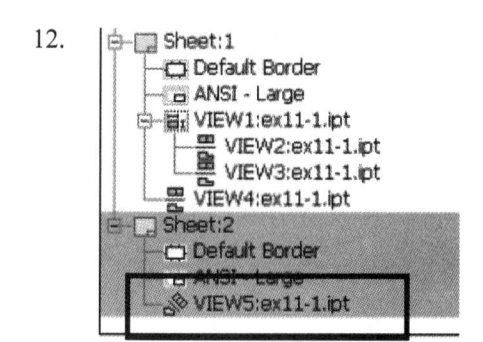

11. Press **OK**.

12. You still have the auxiliary view you copied to Sheet 2.

Save the file as *ex5-3.idw*.

5-11

Exercise 5-4:
Adding a Section View

File: New drawing using Standard
Estimated Time: 15 minutes

This exercise reviews the following:

- New Drawing
- Base View
- View Window
- Section View

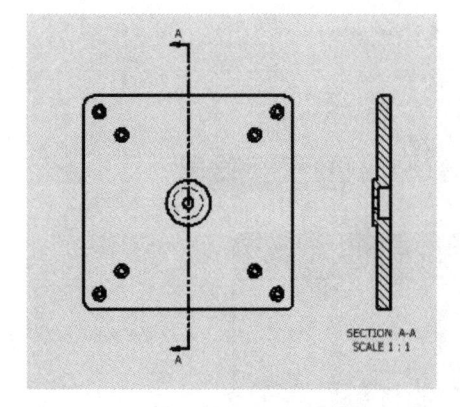

1. Start a new drawing.

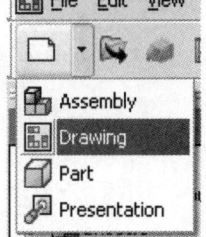

2. Select the **Base View** tool.

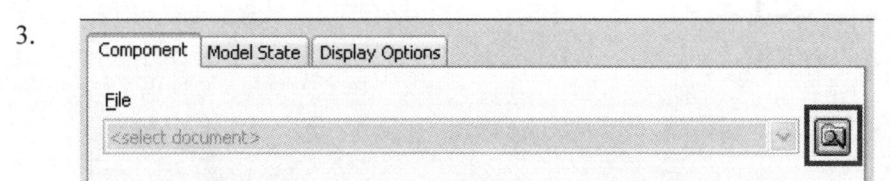

3. Select the **Browse** tool.

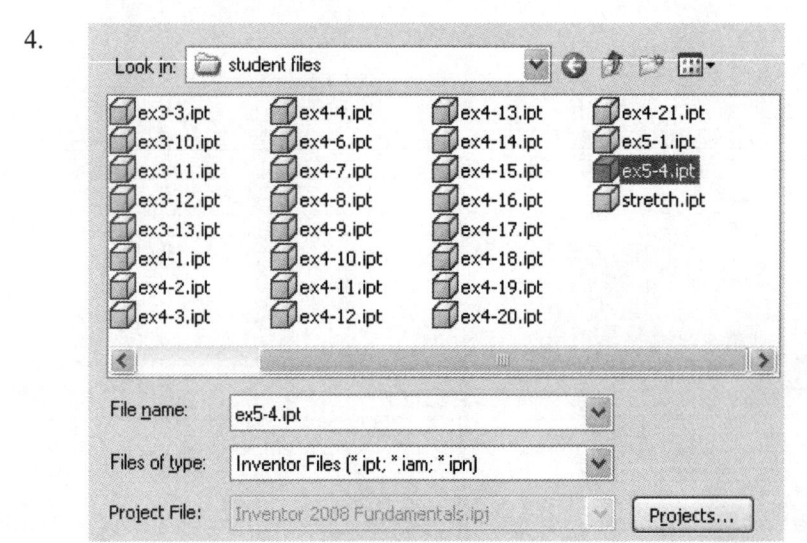

4. Locate the *ex5-4.ipt* file.

   Press **Open**.

   This file can be downloaded from the publisher's website at *www.schroff.com/resources*.

5-12

Drawing Management

5.  Select the **Change Orientation** button.

6. 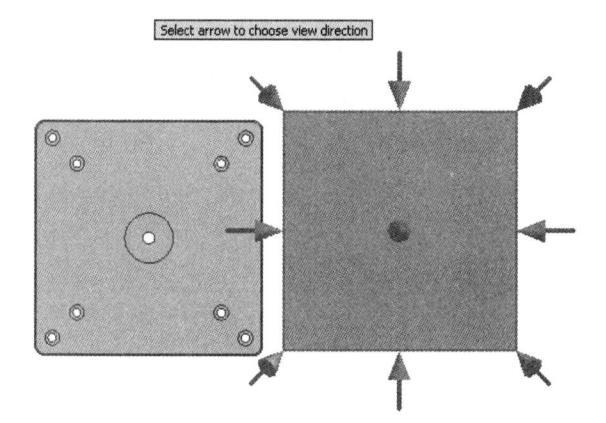 Change the view orientation so that the counter bore in the center of the plate is visible.

7.  Select the Green Check in the upper left of the screen to indicate that this is the desired view orientation.

8. 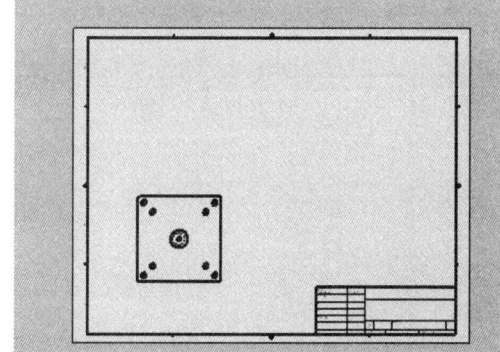 Place the base view in the sheet.

9. 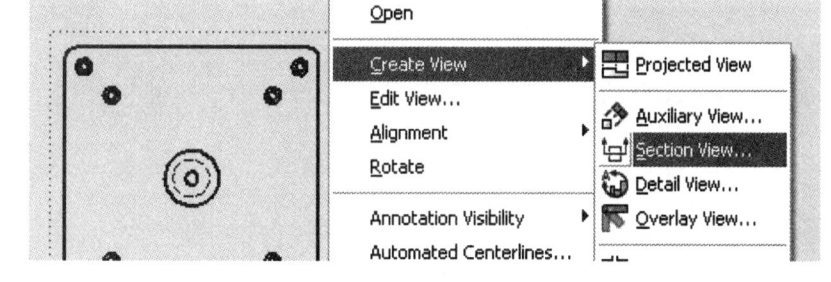 Highlight the source view in the graphics window.
Right-click and select **Create View → Section View**.

10. 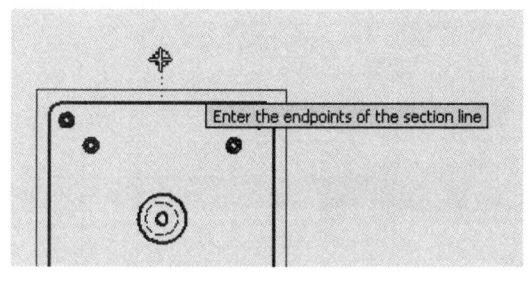 Use tracking to line up your section line with the center point of the center Counterbore hole of the part.

5-13

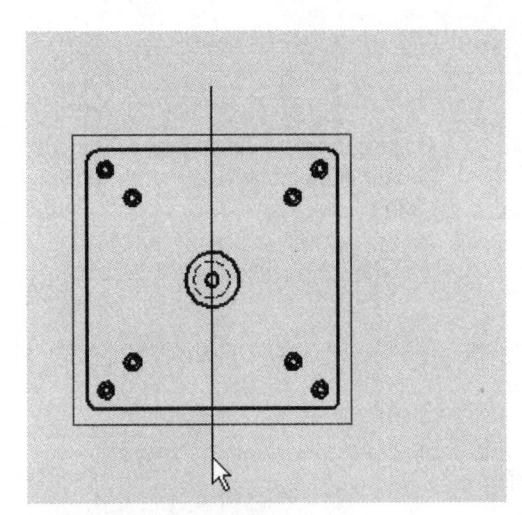

The Power of Design: An Introduction to Autodesk Inventor

11. Pick a point above the plate.
    Drag the line down in a straight line.
    Pick a point below the plate.

    If you have a problem creating a straight line, you can apply a sketch constraint; i.e., vertical/horizontal, to the line

12. Right-click and select **Continue**.

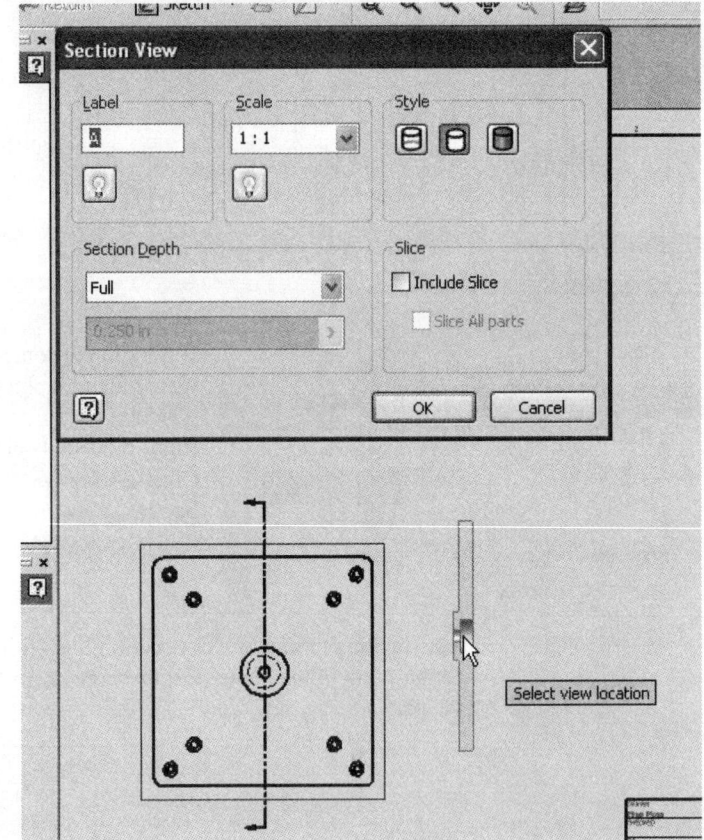

13. Drag the view to the right.
    Left-click to place.

14. Save the file as *ex5-4.idw*.

5-14

Drawing Management

> **TIP:** To get a straight section line, just draw a straight line using the object tracking. Inventor will add the shoulders, arrowheads and labels automatically. 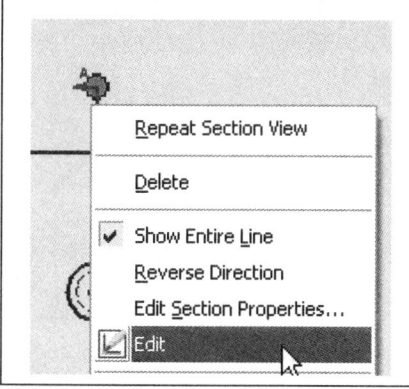
>
> To apply a sketch constraint to the section line, select the line. Right-click and select **Edit**. Place the constraint. Right-click and select **Finish Sketch.**

Exercise 5-5:
Editing a Section View

File: Ex5-4.idw
Estimated Time: 10 minutes

This exercise reviews the following:

- Section View Properties
- Modifying a Section View

1. Open *ex5-4.idw*.

2. 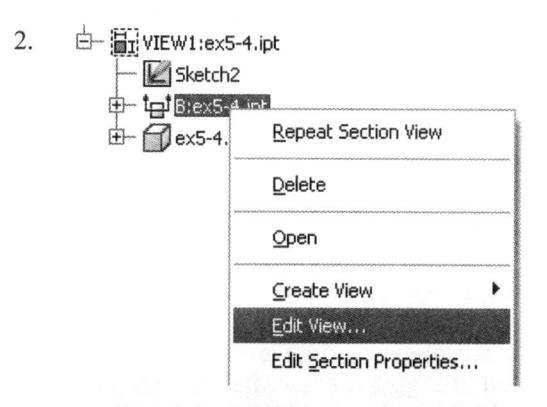 Highlight the section view.
   Right-click and select **Edit View**.

3. 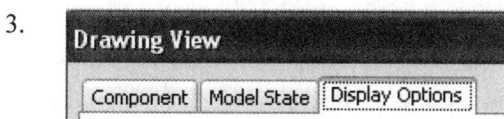 Select the **Display Options** tab.

4.  Toggle the **Visible** button for the Scale.
   Press **OK**.

5-15

5. Select the view label underneath the view.

Right-click and then select **Edit View Label**.

6. Add **FOR REFERENCE** in the edit field.

Press **OK**.

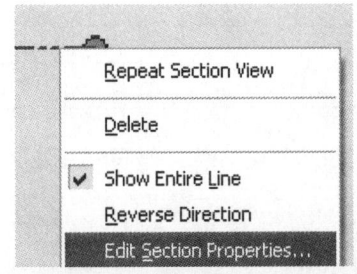

7. Select the section line.

Right-click and select **Edit Section Properties**.

8. Enable **Include Slice**.

Press **OK**.

9. Save the file as *ex5-5.idw*.

> ➢ If you get an error message stating that there is no sketch associated with your break-out view, highlight the view in the Browser and then activate sketch mode to create your sketch. This guarantees the sketch is associated with the highlighted view.
> ➢ Break-out Views require you to place an *associated sketch* with a view. To associate a sketch with a view, select the view and highlight, then select the Sketch tool..
> ➢ To set a different fence shape for your detail view, right-click and select the fence shape from the menu before clicking to indicate the outer boundary.
> ➢ You can modify the size and location of a detail view by using the grips.

Exercise 5-6:
Creating a Detail View

File: Ex5-5.idw
Estimated Time: 10 minutes

1. Open *ex5-5.idw*.

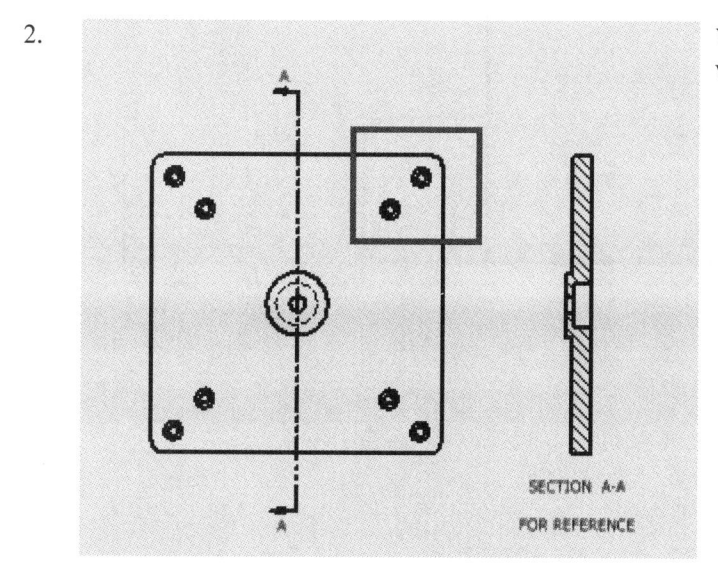

2. We'll create a detail view of the front view indicated.

3. Select the **Detail View** tool from the toolbar.

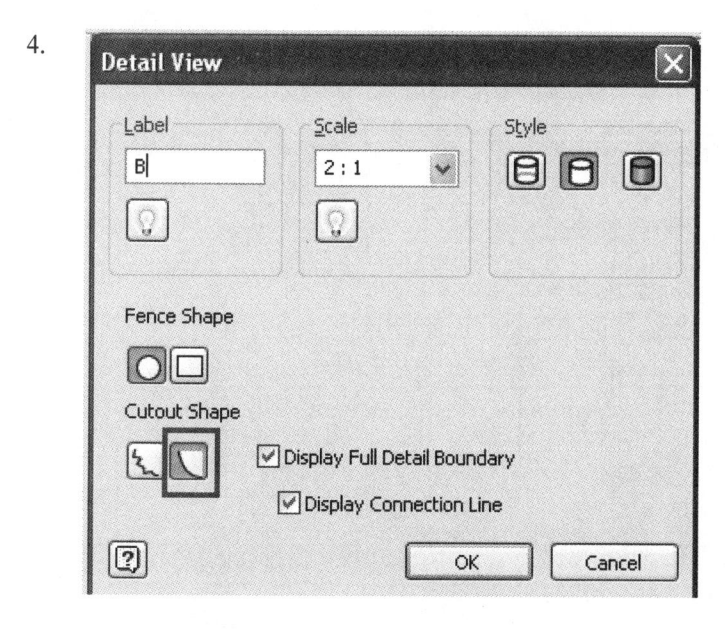

4. Select the **Front View**.

    Enable the **Cutout Shape** indicated.

    Enable **Display Full Detail Boundary**.

    Enable **Display Connection Line**.

5. [Detail View dialog shown]

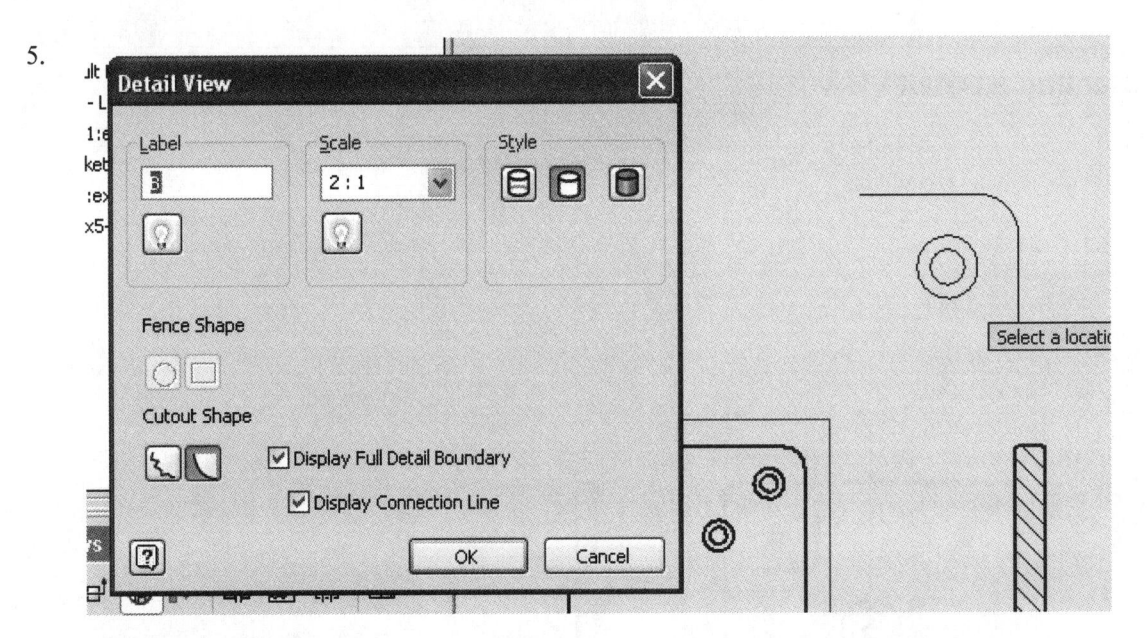

   Pick the center of the hole.

   Drag to define the circle and then pick to place.

6. Place the detail view to the right of the front view and above the section view.
   The **Scale** is set to **2** in the dialog.
   Both labels are set to be **Visible**.

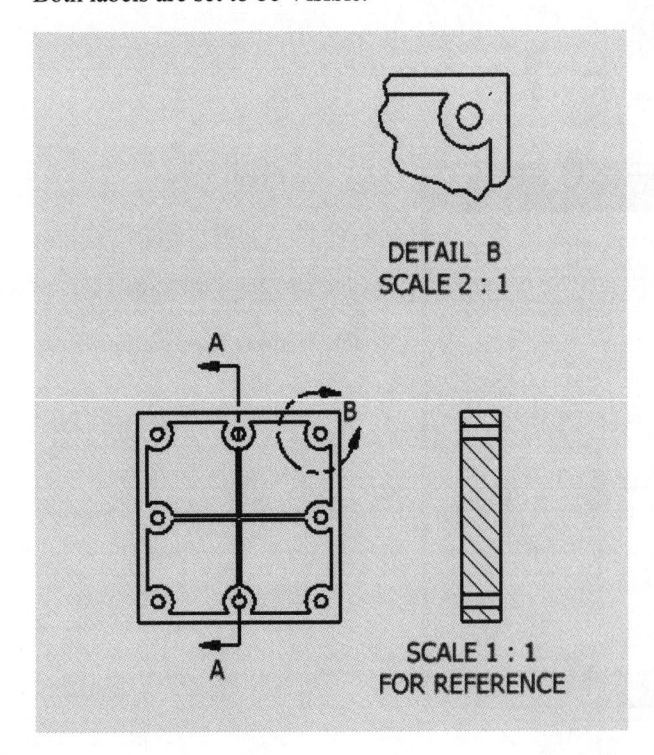

7. Save the file as *ex5-6.idw*.

Exercise 5-7:
Creating a Broken View

File: Ex5-7.ipt
Estimated Time: 15 minutes

1. Start a new drawing.

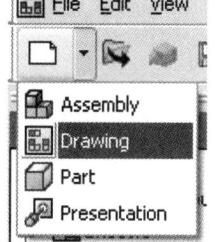

2. Right-click in the drawing window. Select **Base View**.

3. Select the **Browse** button.

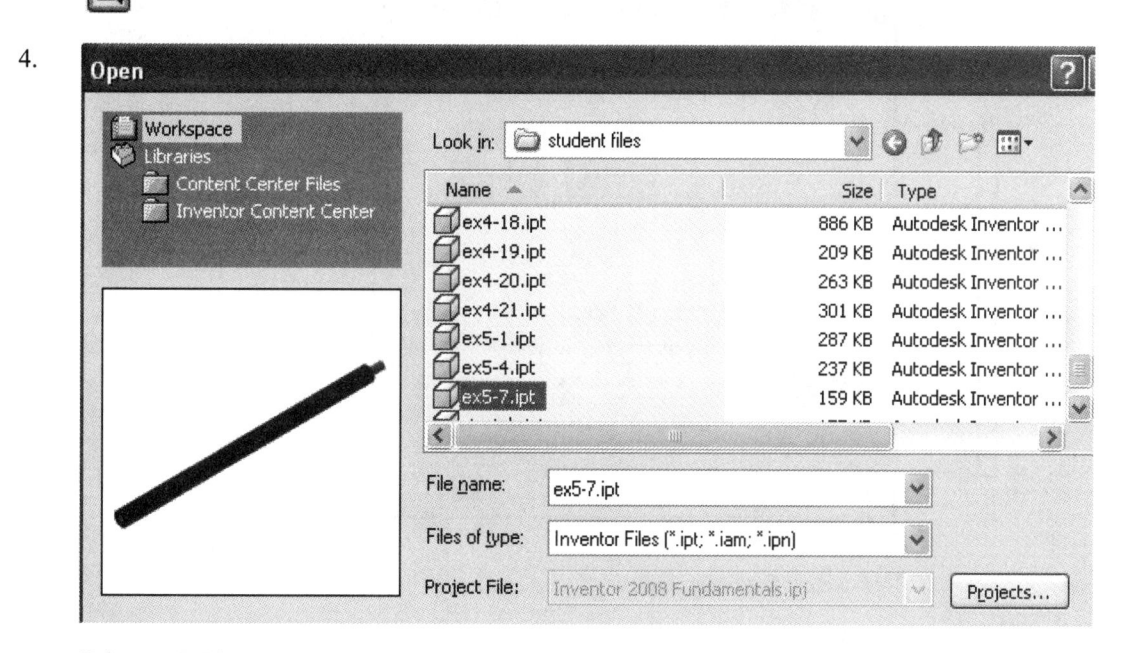

4. Select *ex5-7.ipt*.

   This file can be downloaded from the publisher's website at *www.schroff.com/resources*.

   Press **Open**.

5. Set the **Orientation** to the **Top** view.

5-19

6. Set the **Style** to **Hidden**.

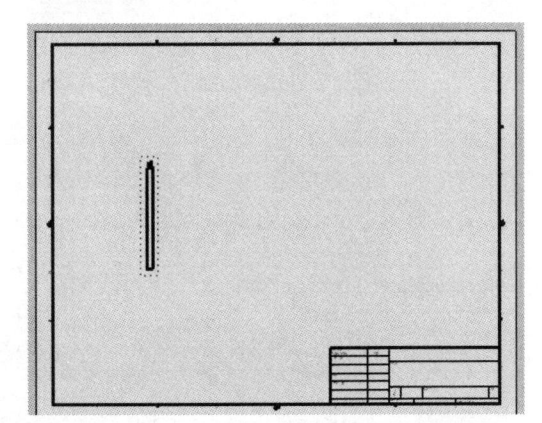

7. Place the view on the sheet.

8. Select the **Broken View** tool.

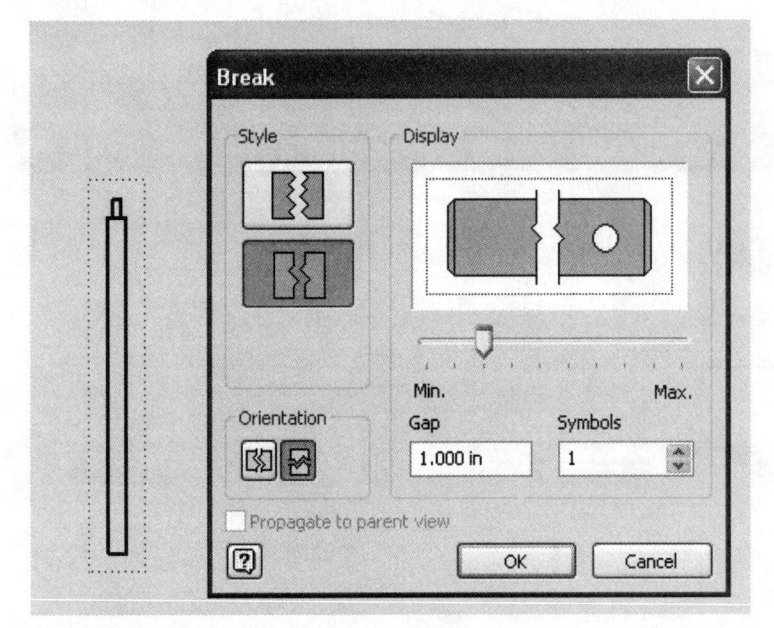

9. Select the **Rectangular** Style.

   Set the **Vertical** Orientation.

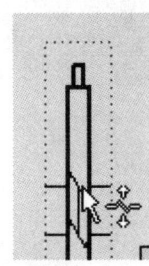

10. Select the base view.

    Pick a point on the rod.

    Drag your mouse up to establish the section you want removed.

    Pick to finish.

Drawing Management

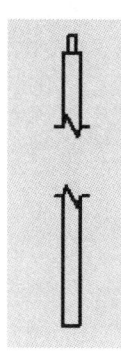

11. The view is updated to indicate the break.

    NOTE: If you have more than one view placed, all views will update to display the break.

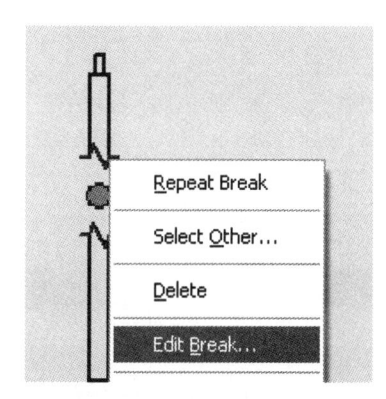

12. Select the break so it highlights.
    Right-click and select **Edit Break**.

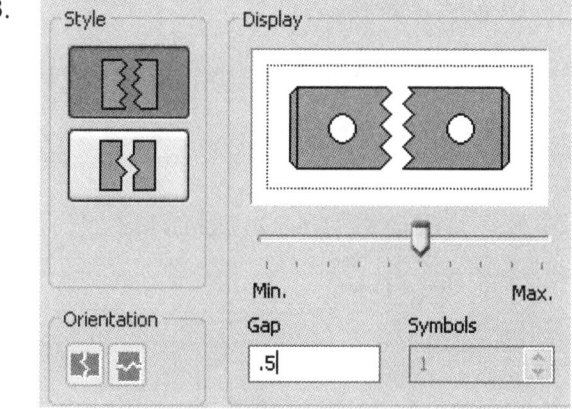

13. Change the **Style**.

    Modify the display using the slider.
    Set the **Gap** to **.5**.

    Press **OK**.

14. Save the file as *ex5-7.idw*.

5-21

Exercise 5-8:
Creating an Overlay View

File: New
Estimated Time: 10 minutes

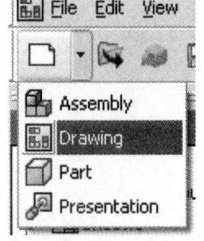

1. Start a new drawing.

2. Right-click in the drawing window.
   Select **Base View**.

3. Select the **Browse** button.

4. Browse to the *Arbor Press* folder under *Program Files\Autodesk\ Inventor <Version>\Samples\Models\Assemblies*.

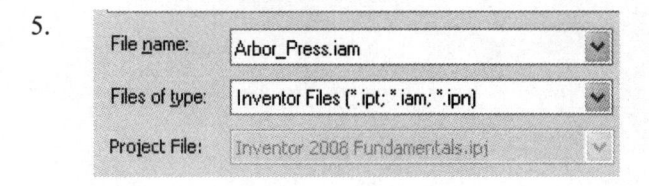

5. Locate the ***Arbor Press.iam*** file.

   Press **Open**.

6.

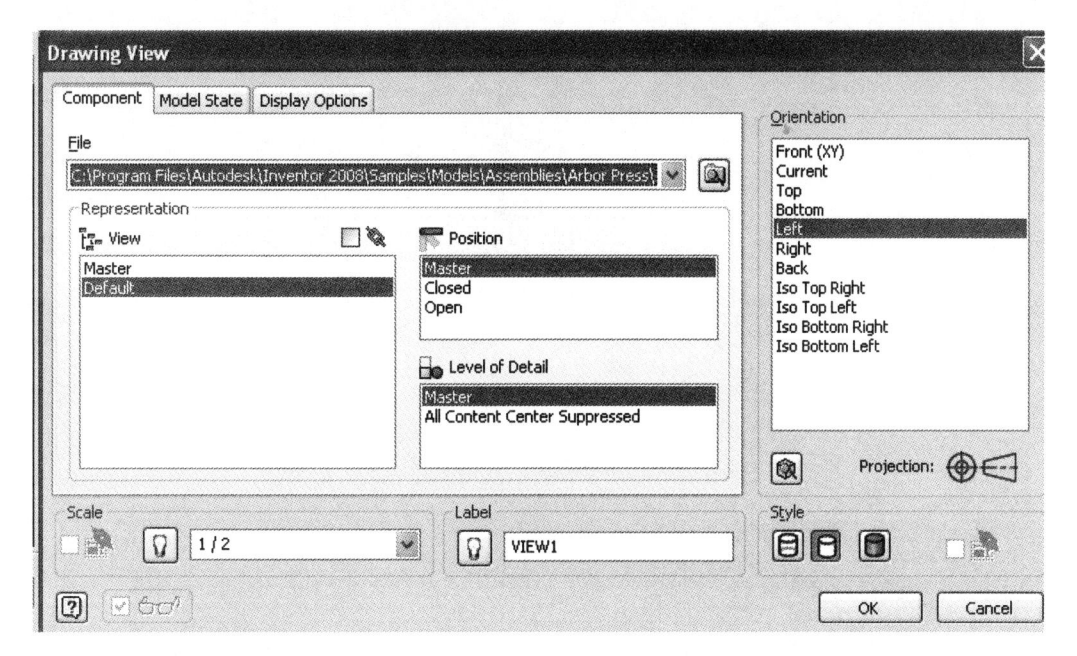

Set the **View** to **Default**.
Set the **Position** to **Master**.
Set the **Level of Detail** to **Master**.
Set the **Orientation** to **Left.**
Set the **Style** to **Hidden**.
Set the **Scale** to ½.
Place the view on the sheet.

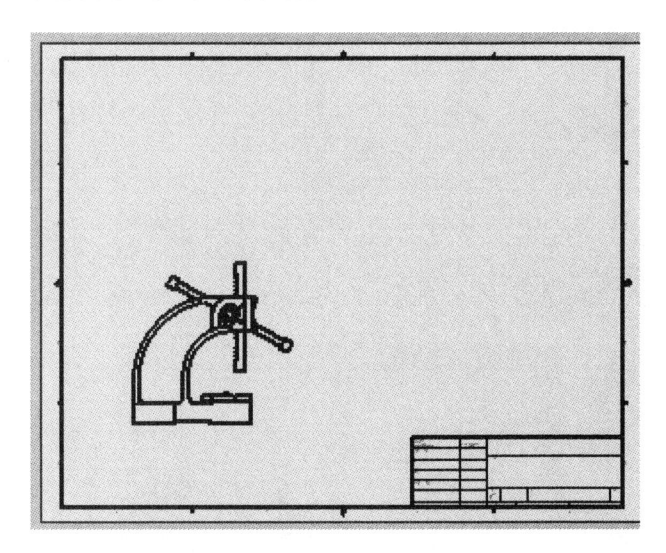

7. Select the **Overlay** tool.
Select the Base View.

8.

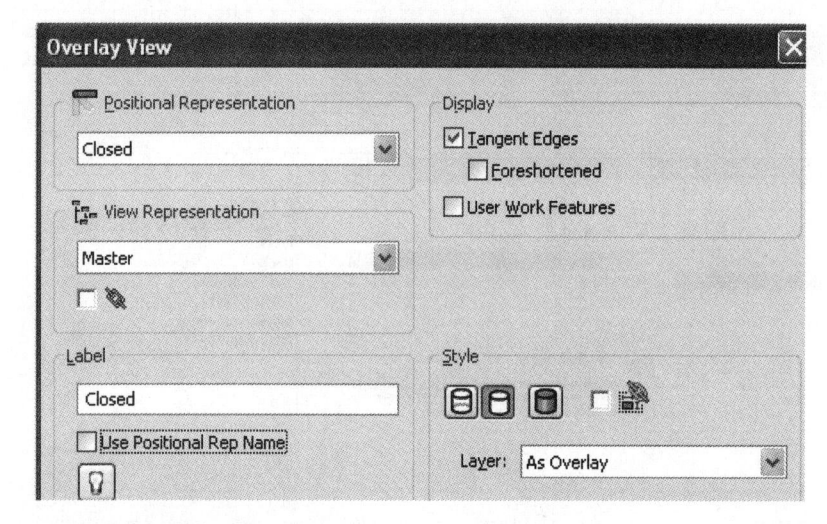

Select **Closed**.
Enable **Tangent Edges**.
Disable **Foreshortened**.
Set **View Representation** to **Master**.
Disable **Use Positional Rep Name**.
Enable **Hidden Style**.
Set **Layer** to **As Overlay**.
Press **OK**.

9.   Zoom in to see the first overlay.

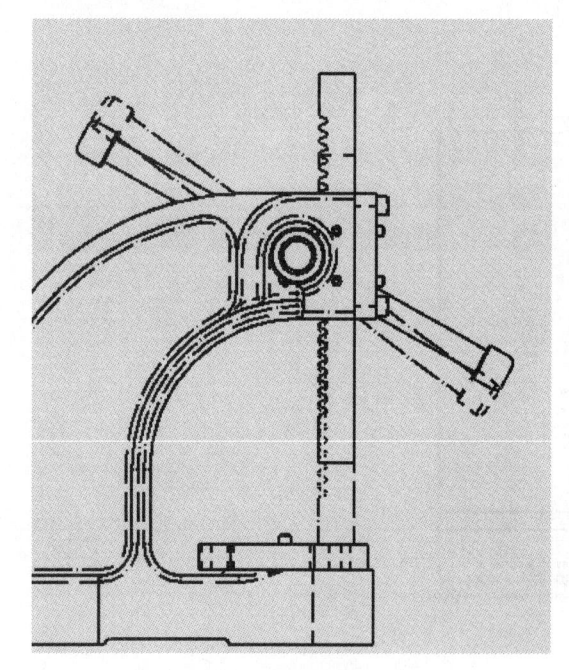

10. Select the **Overlay** tool.
     Select the Base View.

Drawing Management

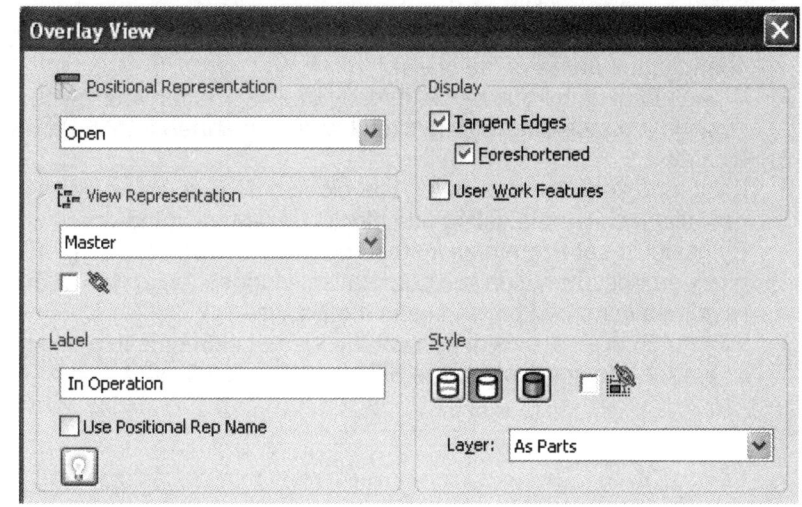

11.

Select **Open**.
Enable **Tangent Edges**.
Enable **Foreshortened**.
Set **View Representation** to **Master**.
Disable **Use Positional Rep Name**.
Type **In Operation** in the **Label** name field.
Enable the Label **Visible**.
Enable **Style to Hidden**.
Set **Layer** to **As Parts**.
Press **OK**.

12.

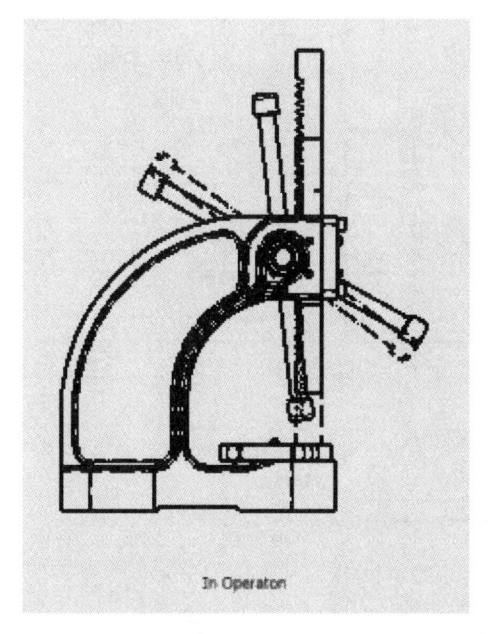

The next overlay is placed.

Save the file as *ex5-8.idw*.

5-25

The Power of Design: An Introduction to Autodesk Inventor

> Clicking in the Sheet Name edit box and typing in a new name changes the name of the sheet.
> When defining the title block, it really doesn't matter where you position it on the sheet. The sketch will be inserted into the proper location.
> If you don't delete the existing title block before you start creating a new title block, the existing title block will remain in the drawing.
> By using Sheet Properties instead of using a Prompted Entry for title block entries, the value will automatically update when the user redefines the sheet properties in the Browser.
> When you use dimensions to set the size of elements in a title block or border, the dimensions are hidden when you finish editing.

Exercise 5-9:
Creating a Slice View

File: ex5-9.idw (download from the publisher's website at *www.schroff.com/resources*)
Estimated Time: 15 minutes

1. Open a drawing.
   Locate *ex5-9.idw*.

2. Select the **Slice** tool.

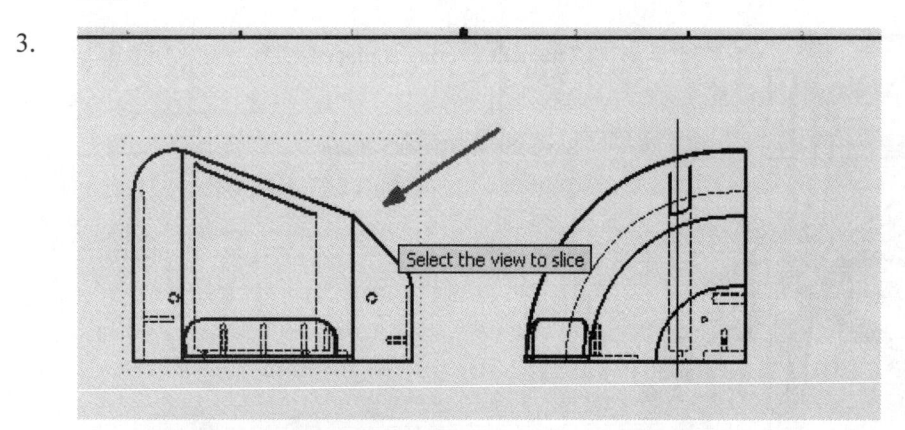

3. Select the view on the left.

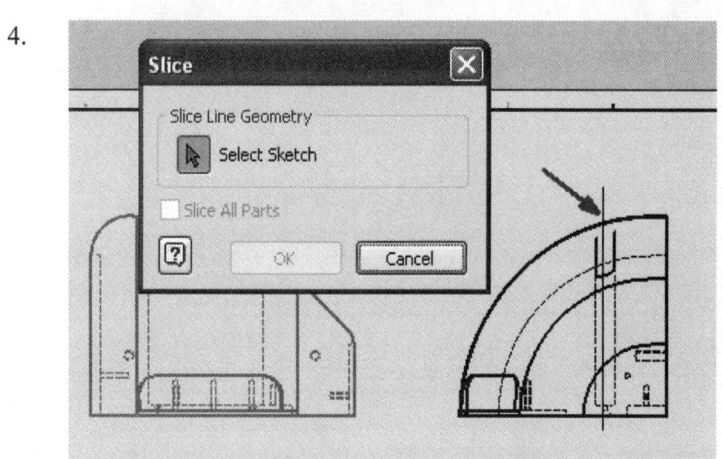

4. Select the vertical line on the right view.

   Press **OK**.

5-26

Drawing Management

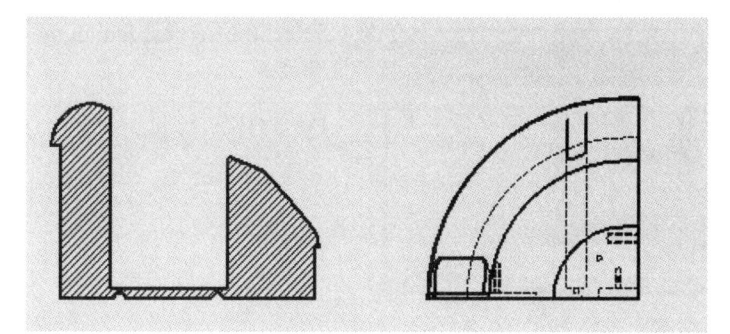

5. You see the result.

6. Press the **UNDO** button.

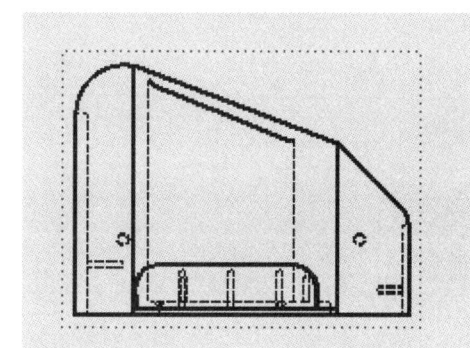

Select the left view.

7. Sketch — Select the **Sketch** tool.

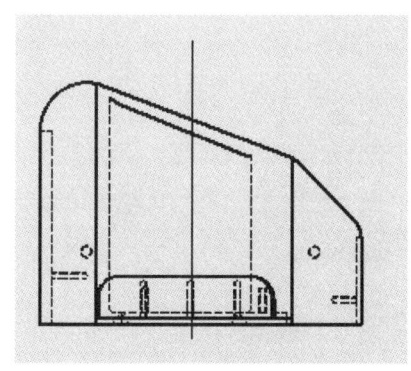

8. Draw a vertical line midway through the part.

The Slice tool only works with vertical lines.

Right-click and select **Finish Sketch**.

9. Select the **Slice** tool.

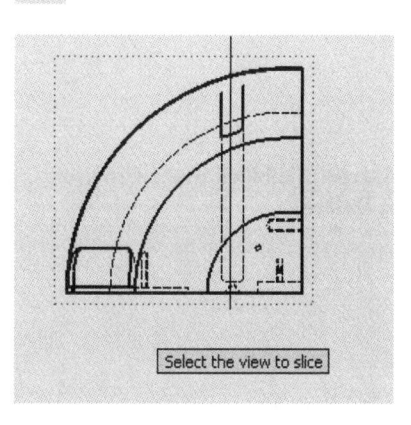

10. Select the right view.

5-27

11. Select the vertical line in the left view.

Press **OK**.

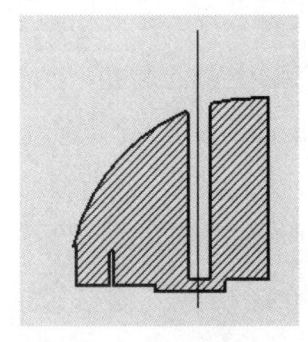

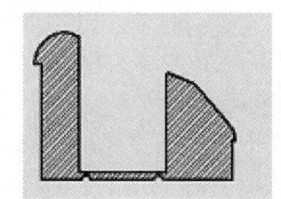

12. Compare with the first result.

13. Close without saving.

Exercise 5-10:
Creating a Custom Title Block

File: New using Standard (inches)
Estimated Time: 60 minutes

This exercise will review the sketch tools used to create a custom title block.

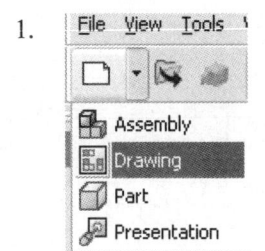

1. Open a new Drawing file.
   Save as *ex5-10.idw*.

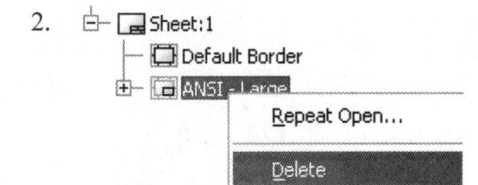

2. Highlight the **ANSI-Large** title block in the Browser.
   Right-click and select **Delete**.

Drawing Management

3. Go to **Format→Define New Title Block**.

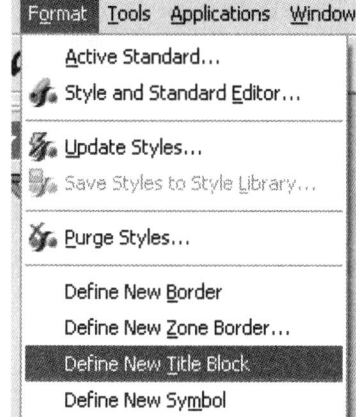

4. Use **Zoom Window** to zoom into the title block area.

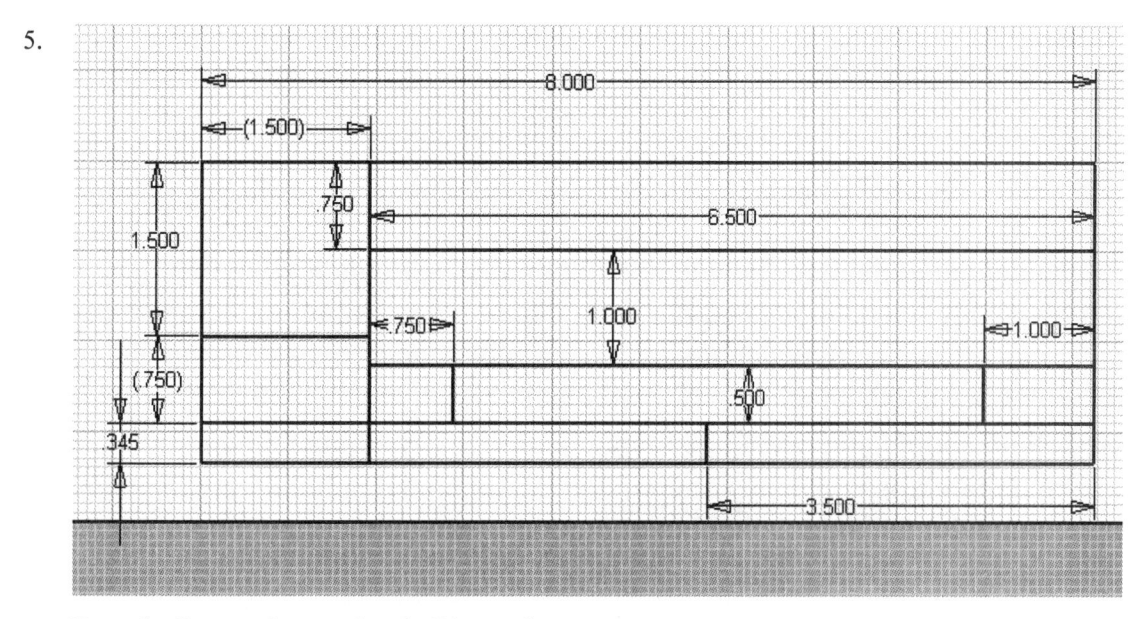

5. Draw the lines as shown using the **Line** tool.

6. To insert a company logo, use **Insert→Insert Image**.

   Next, indicate the area the picture is to be placed by picking two points to form a rectangle.

7. To see a preview of the image, set your view to **Thumbnails**.

5-29

The Power of Design: An Introduction to Autodesk Inventor

8. Local Disk (C:)
    Program Files
    Autodesk
    Inventor 2008
    Web

    There are several bmp files in the Web folder under Inventor you can use if you don't have a bmp file available.

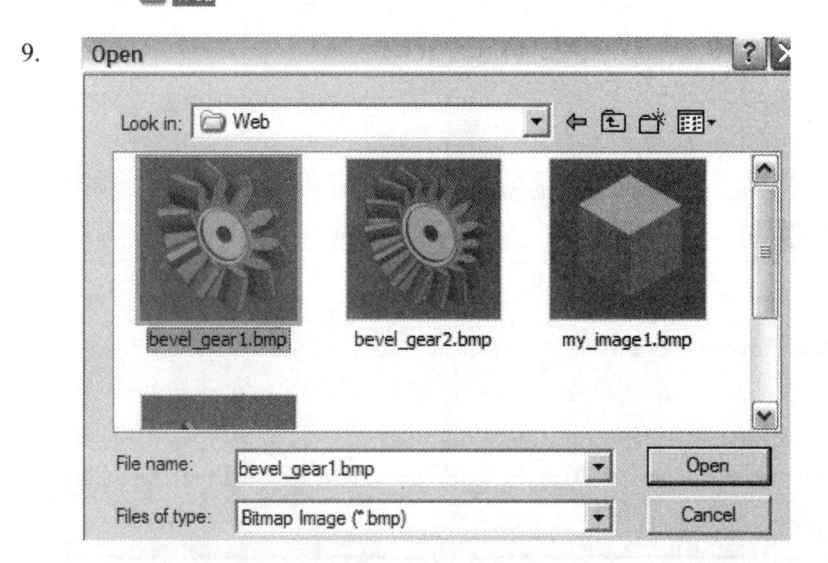

9. Browse to the Web folder. Select the *bevel_gear1.bmp*. Press **Open**.

10. Right-click and select **Done**.

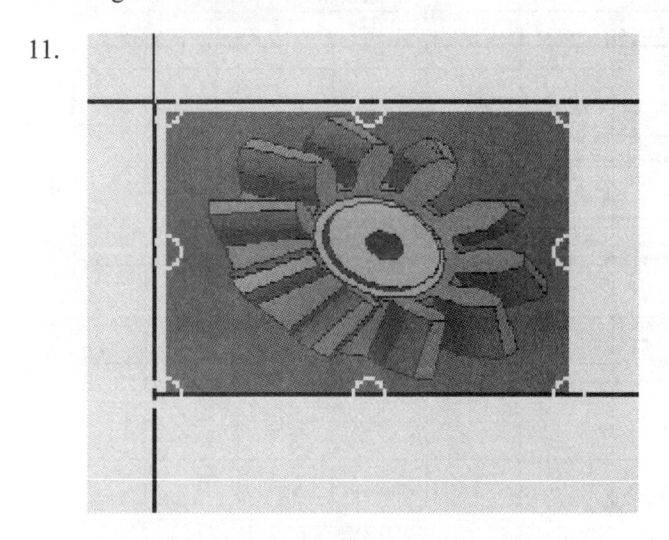

11.

    You can use the corner grips on the bitmap to reposition and resize so it fits properly.

12. Select the **Text** tool.

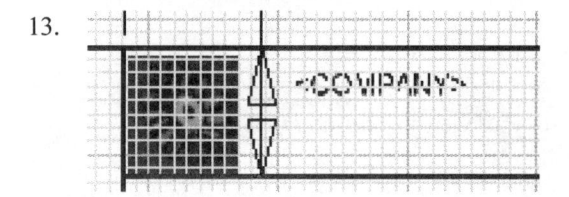

13. Draw a rectangle next to the logo to indicate the location for company information.

5-30

14. Select **Properties – Model** under **Type**.

15. Select **COMPANY** under **Property**.

    The Company name that is placed in the File Properties will automatically fill in the title block.

16. Set the text height to **.218**.

17. Press the **Add Text Parameter** button.

18. The property will appear in the text window.
    Press **OK**.

19. Note that you are still in Text mode.
    Draw a rectangle next to the Company field.

20. Leave Type and Property blank.

    Enter in an address and website URL.

    This is a static value which can not be changed.

    Press **OK**.

21. Right-click and select **Done**.

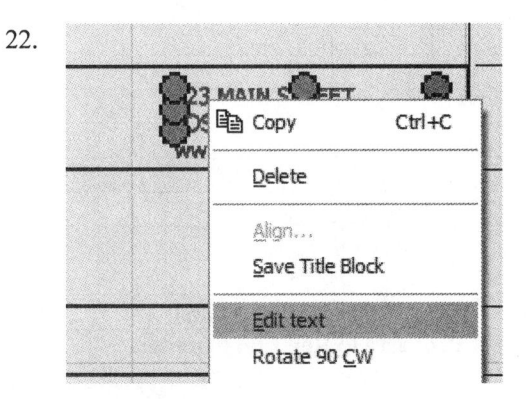

22. With **Text Box** enabled, right-click and select **Edit Text**.

5-31

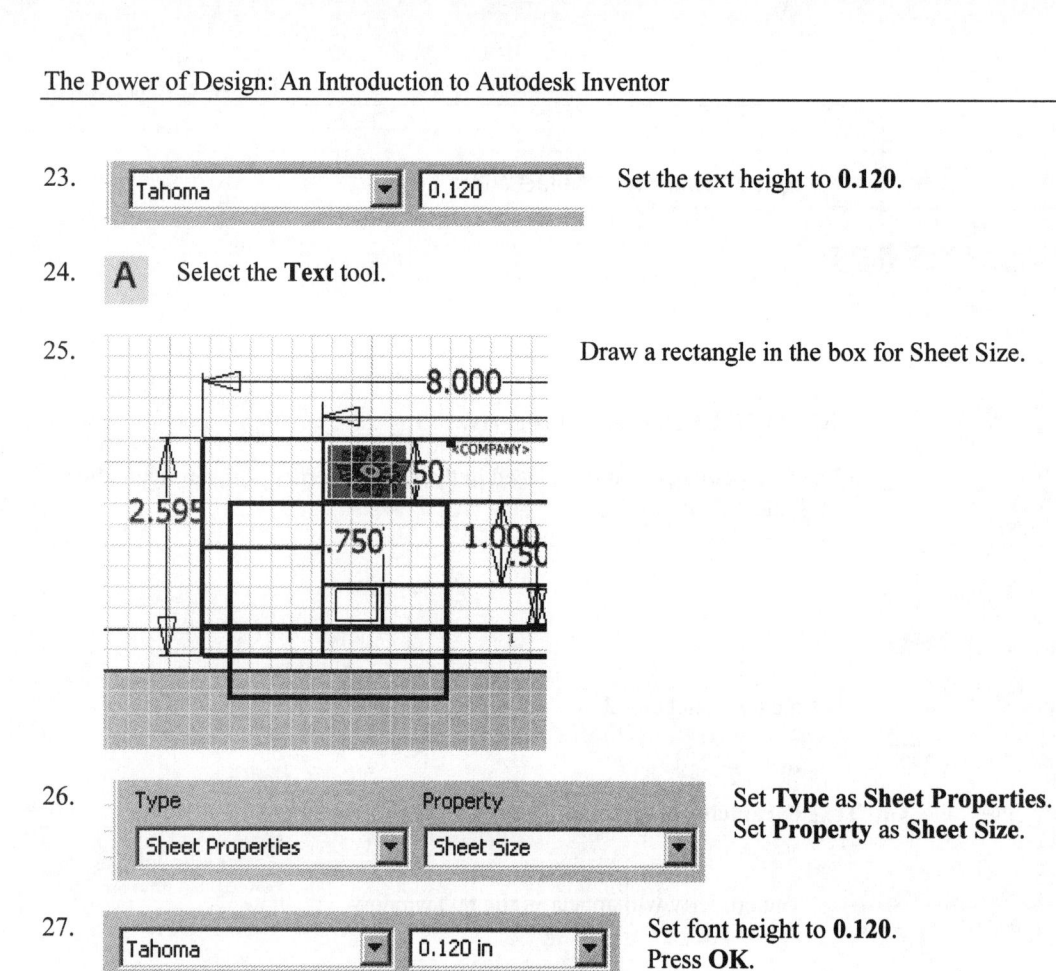

23. [Tahoma | 0.120] Set the text height to **0.120**.

24. **A** Select the **Text** tool.

25. Draw a rectangle in the box for Sheet Size.

26. [Type: Sheet Properties | Property: Sheet Size] Set **Type** as **Sheet Properties**.
    Set **Property** as **Sheet Size**.

27. [Tahoma | 0.120 in] Set font height to **0.120**.
    Press **OK**.

28.  Press the **Add Text Parameter** button.

29.  <Sheet Size> The property will appear in the text window.
    Press **OK**.

30.  **A** Select the **Text** tool.

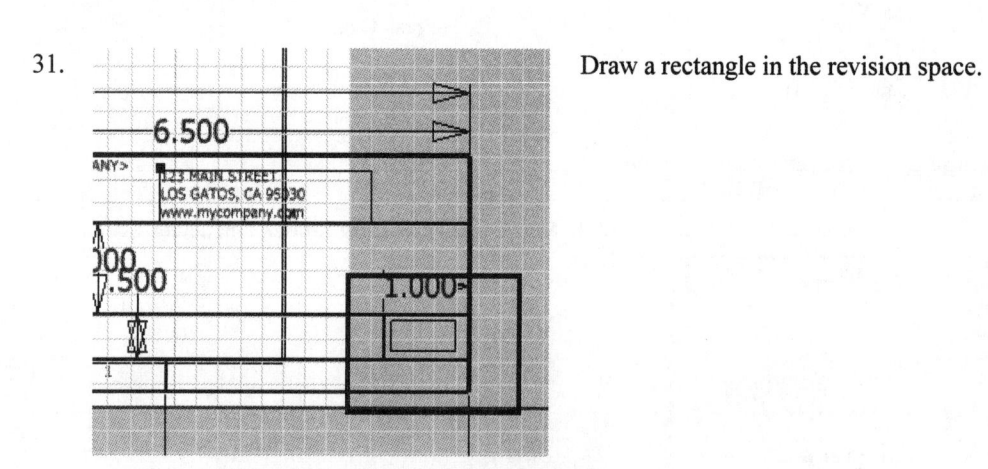

31. Draw a rectangle in the revision space.

32.  [Type: Properties - Model | Property: REVISION NUMBER] Set **Type** to **Properties -Model**.
    Set **Property** to **Revision Number**.

5-32

33. [Tahoma | 0.120 in] Set font height to **0.120**.
    Press **OK**.

34. [icon] Press the **Add Text Parameter** button.

35. [<REVISION NUMBER>] The property will appear in the text window.
    Press **OK**.

36. [image of title block with rectangle] Draw a rectangle in the title space.

37. [Tahoma | 0.240 in | B I U; Type: Properties - Drawing; Property: TITLE]

    <TITLE>

    Set **Type** to **Properties - Drawing**.
    Set **Property** to **TITLE**.
    Set the text height to **0.240**.

38. [icon] Press the **Add Text Parameter** button.
    The property will appear in the text window.
    Press **OK**.

39. [image of title block with rectangle] Draw a rectangle in the part number space.

40. [Tahoma | 0.218 | B I U; Type: Properties - Model; Property: PART NUMBER]
    Set **Type** to **Model Properties**.
    Set **Property** to **Part Number**.
    Set text height to **0.218**.

41. [icon] Press the **Add Text Parameter** button.
    The property will appear in the text window.
    Press **OK**.

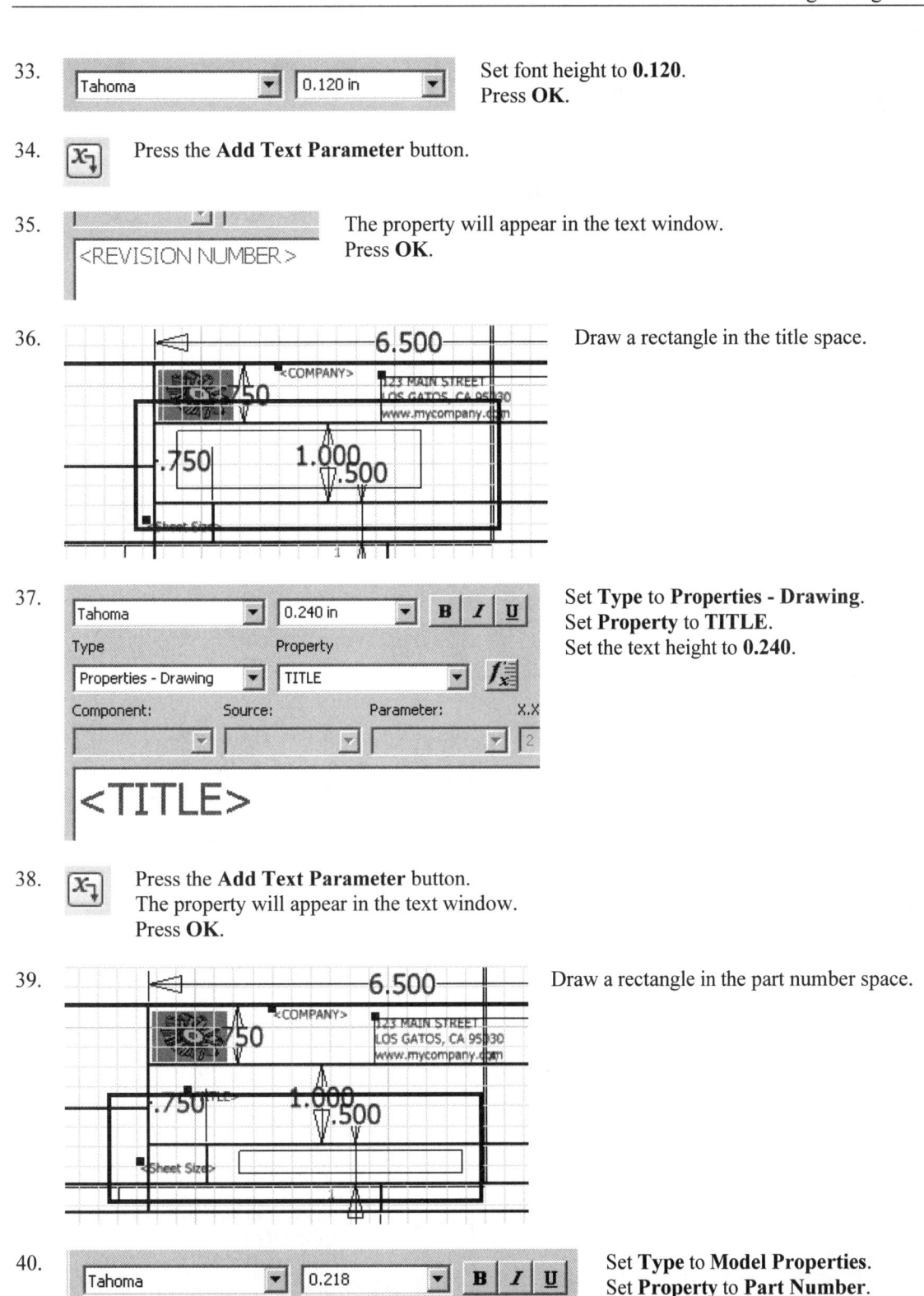

The Power of Design: An Introduction to Autodesk Inventor

42. Draw a rectangle in the sheet number space.

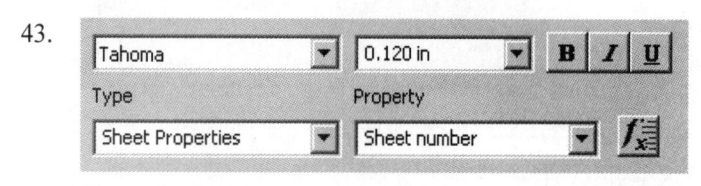

43. Set **Type** to **Sheet Properties**.
Set **Property** to **Sheet Number**.
Set **Height** to **0.120**.

44. Press the **Add Text Parameter** button.
The property will appear in the text window.
Press **OK**.

45. Draw a rectangle in the file name space.

46. Set **Type** to **Properties - Drawing**.
Set **Property** to **FILENAME AND PATH**.
Set **Height** to **0.120**.

47. Press the **Add Text Parameter** button.
The property will appear in the text window.
Press **OK**.

48. Draw a rectangle in the drafter name space.

49. Set **Type** to **Properties - Drawing**.
Set **Property** to **AUTHOR**.
Set **Height** to **0.156**.

50. Press the **Add Text Parameter** button.
The property will appear in the text window.
Press **OK**.

Drawing Management

51. Draw a rectangle in the Date space.

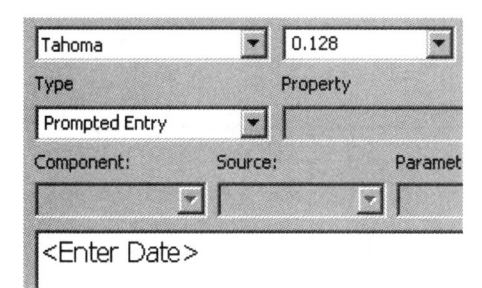

52. To add a prompted entry (similar to an attribute):
    Set **Type** to **Prompted Entry**.
    Set **Prompt** to '**Enter Date**'.
    Set **Height** to **0.128**.

    Press **OK**.

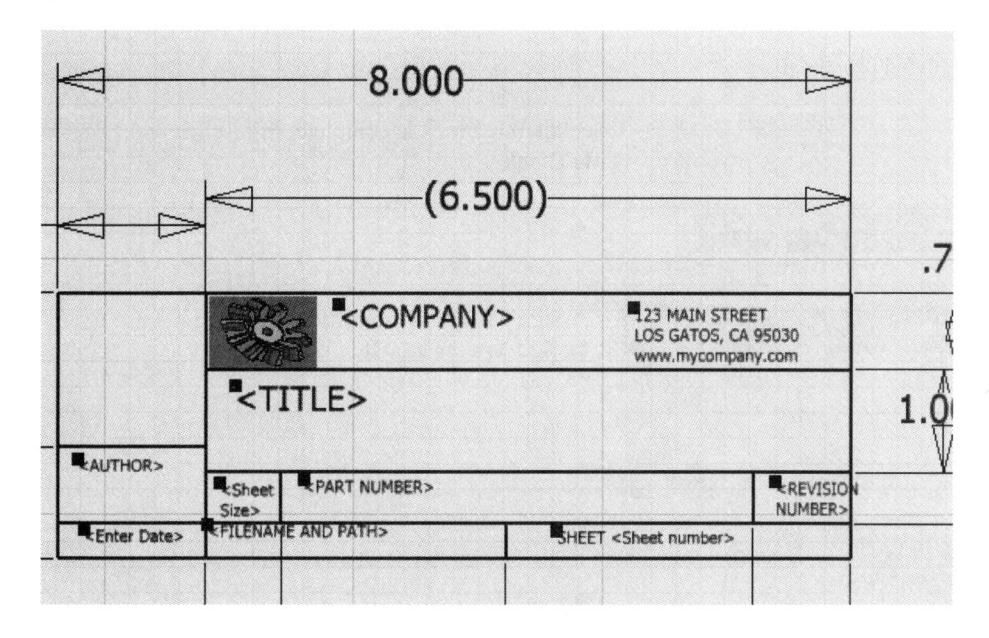

The title block automatically inserts values into the Model Properties fields using the values stored in the Properties dialog box for each model.

TIP: Inventor will only accept bitmaps for insertion, so your file must have a *.bmp extension.

5-35

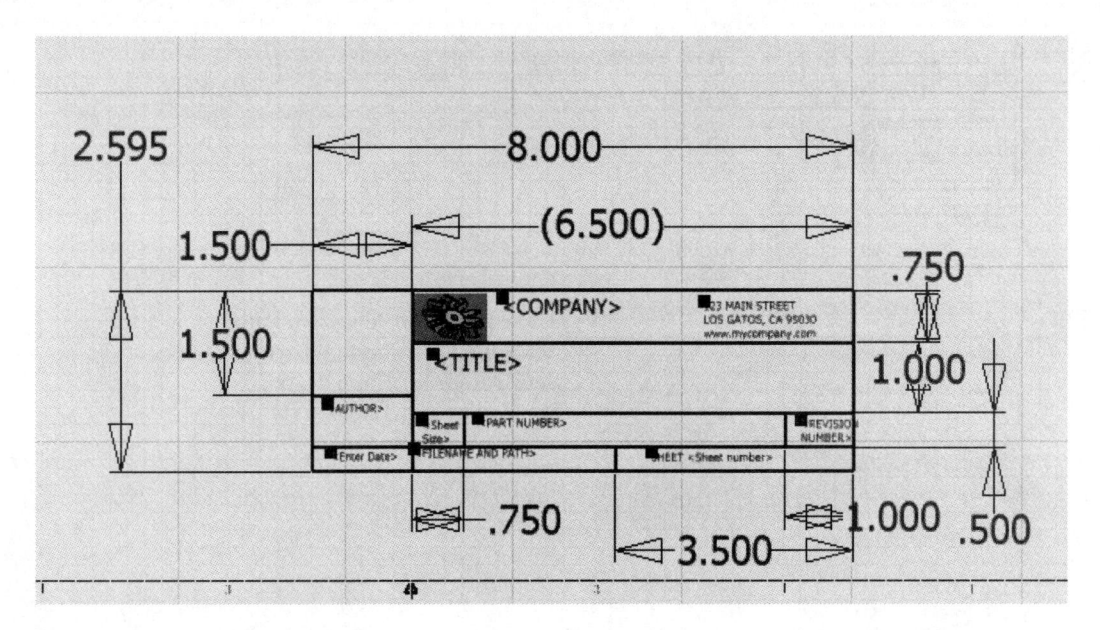

The completed title block.

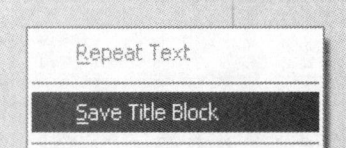

53. Once the title block is complete, right-click and select **Save Title Block**.

54. Assign a name to your title block. Do not use punctuation marks. Spaces are OK.

55. Your new title block name automatically appears in the Browser.

    To insert, select and double click.

56. You will then be prompted for any prompted entry fields that were defined.

Drawing Management

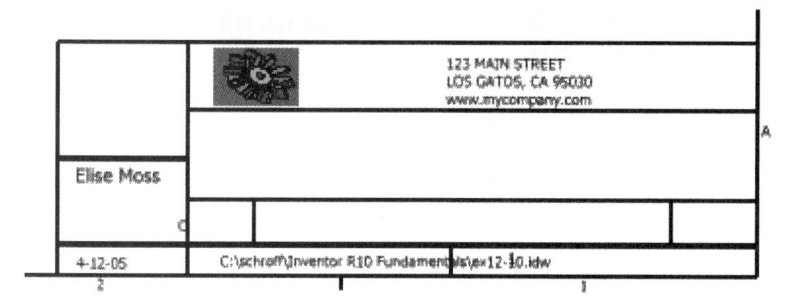

When you first place the title block, most of the fields will show blank. The only ones filled in are those linked to **Sheet Properties**.

Remember, most of your fields are set by **Model Properties** and will vary depending on the model you use for your drawing.

Once you place a view in the drawing with the associated model properties defined, the title block should fill in correctly.

57. Save the file as *ex5-10.idw*.

If you would like to add fields to be used in your title block, you can add custom fields under the **Custom** tab in the **iProperties** dialog box.

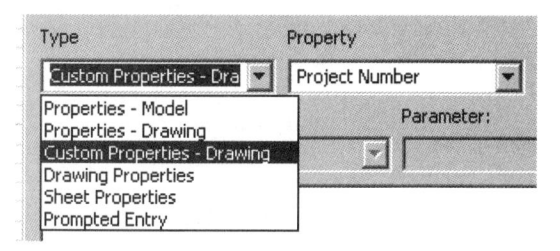

The **Custom** fields will then appear in the **Text** dialog box.

Exercise 5-11:
Adding Custom properties

File: Standard (inches).idw
Estimated Time: 30 minutes

This exercise will demonstrate how to add custom properties to a drawing template. If you add custom properties to the template, the properties will be available to you for any new drawing using that template.

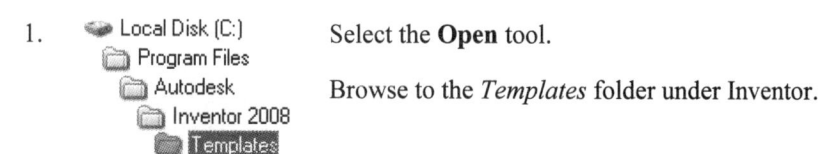

1. Select the **Open** tool.

   Browse to the *Templates* folder under Inventor.

2. Open the *Standard.idw* template file. This is the file that is used every time you start a new drawing.

3. Go to **File→iProperties**.

5-37

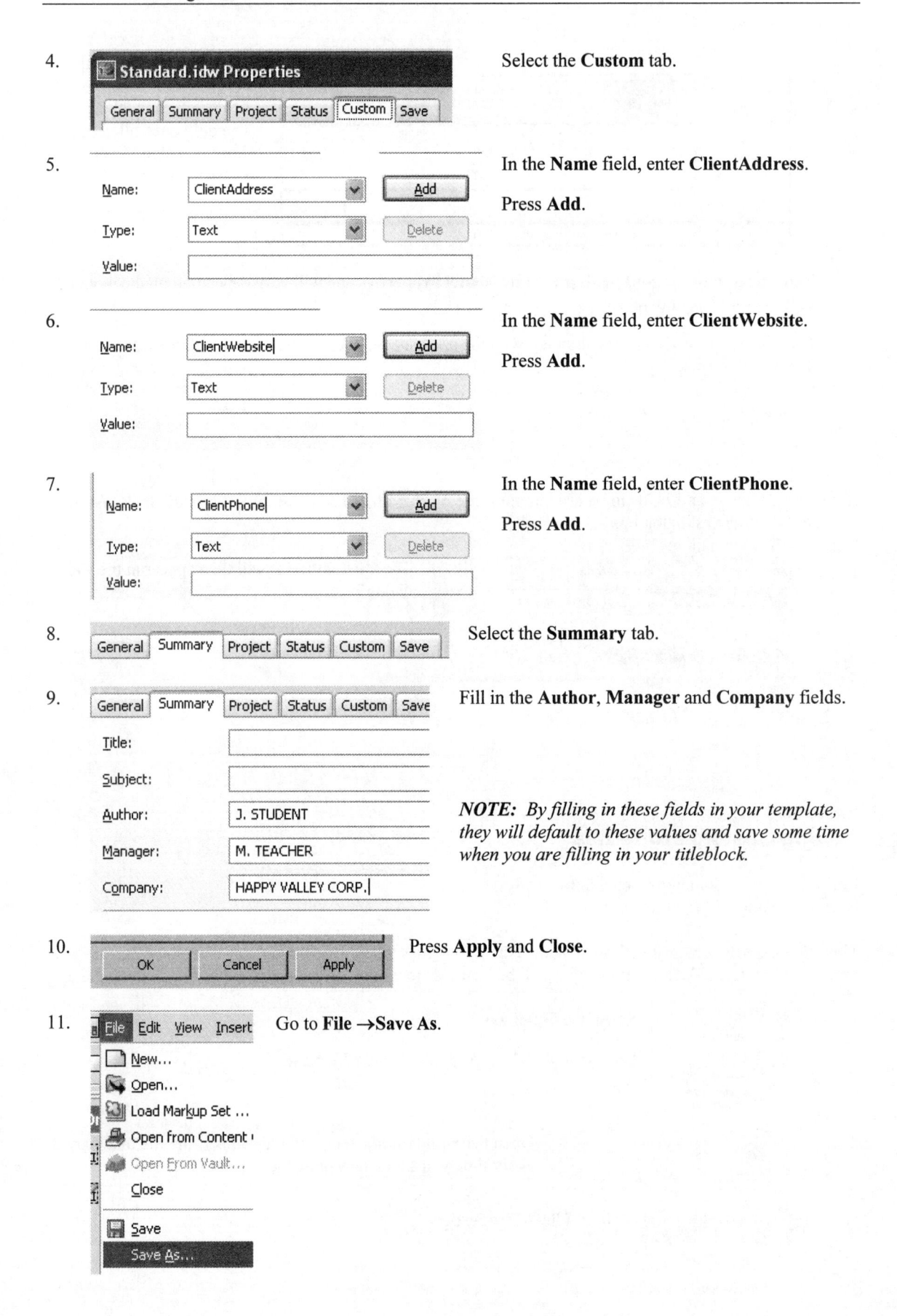

4. Select the **Custom** tab.

5. In the **Name** field, enter **ClientAddress**.
   Press **Add**.

6. In the **Name** field, enter **ClientWebsite**.
   Press **Add**.

7. In the **Name** field, enter **ClientPhone**.
   Press **Add**.

8. Select the **Summary** tab.

9. Fill in the **Author**, **Manager** and **Company** fields.

   NOTE: By filling in these fields in your template, they will default to these values and save some time when you are filling in your titleblock.

10. Press **Apply** and **Close**.

11. Go to **File → Save As**.

12. Name your file *custom.idw*.

   This file needs to be saved in the Templates folder or you can store your templates on the file server and re-direct your templates path in your options.

13. Press **Yes**.

    Note that you are now working in the saved drawing file.

14. Go to **Tools→Document Settings**.

15. Select the **Drawing** tab.

16. Under **Properties in Drawing**, select **Browse**.

17. Browse to the *Templates* folder where you stored your custom properties.

18. Select the *custom.idw* file.
    Press **Open**.

5-39

19. [Autodesk Inventor 2008 dialog: "The location of the selected file is not in the active project. To ensure that the file can be found when you open files that reference it, add the location to the project or move the file to a location specified in the project."]   Press **OK**.

20. Press **Apply** and **Close**.

21. [iProperties icon]   Go to **File→iProperties**.

22. [iProperties Summary tab showing Title, Subject, Author: J. STUDENT, Manager: M. TEACHER, Company: HAPPY VALLEY CORP.]   Verify that the properties you created in the *custom.idw* are now available in this file.

23. Save and close the file.

Exercise 5-12:
Copy a Titleblock

File: Custom.idw
Estimated Time: 10 minutes

Learn how to copy a titleblock from one drawing into another.

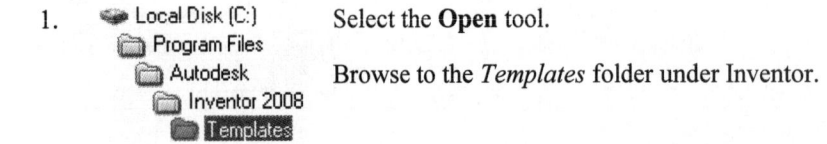

1. [Folder tree: Local Disk (C:) → Program Files → Autodesk → Inventor 2008 → Templates]   Select the **Open** tool.

   Browse to the *Templates* folder under Inventor.

2. [File name: Custom.idw]   Open the ***Standard.idw*** template file. This is the file that is used every time you start a new drawing.

3. [Recent files list: 1 Custom.idw, 2 Standard.idw, 3 C:\Schroff\...\ex5-10.idw, 4 C:\Schroff\...\ex5-11.idw, 5 C:\Schroff\...\ex12-10.idw, 6 C:\Schroff\...\ex5-9.idw, 7 C:\Schroff\...\ex5-9.ipt]   Under the File menu, there is a list of recently opened files.

   Locate and open ***ex5-10.idw***... the file where you created your custom title block.

Drawing Management

4. Expand the *Title Blocks* folder under Drawing Resources.
   Highlight the **SDC-Block**.
   Right-click and select **Copy**.

5. Go to **Window**.
   Select the custom.idw file you have open.

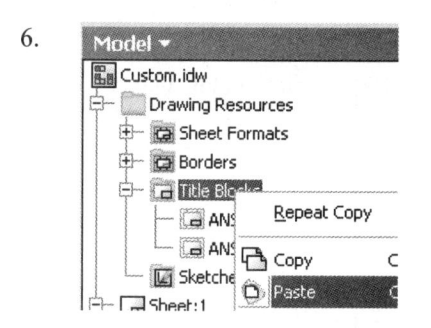

6. Expand the *Title Blocks* folder under Drawing Resources.

   Highlight the Title Blocks category in the Browser. Right-click and select **Paste**.

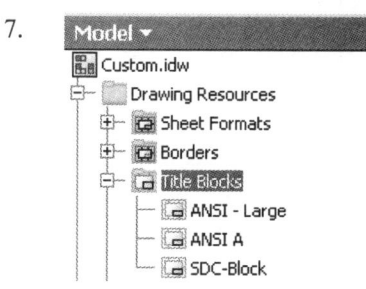

7. The block is now available to be inserted.

   NOTE: You must delete any existing title block in the sheet before you can insert a new title block.

8. Save and close both drawings.

> - A single model dimension cannot be used in multiple views on the same sheet.
> - To edit a completed sketch, select a sketch element in the graphics window or Browser, right-click, and select edit to reactivate the sketch.
> - You can use a property field in a sketched symbol to create a block with attributes.
> - You cannot edit the default border after it is placed. To change the border, delete it and insert a new border with the desired properties.

5-41

The Power of Design: An Introduction to Autodesk Inventor

Exercise 5-13
Define New Symbol

File Name: New (Standard using inches) idw
Estimated Time: 10 minutes

1. Start a New Drawing file.

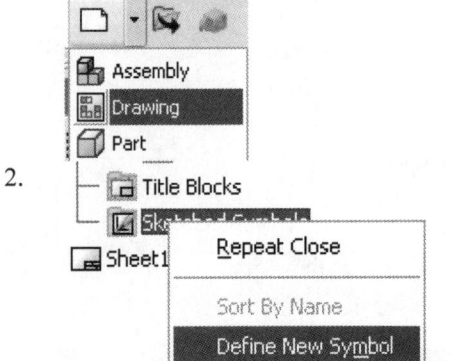

2. In the Browser, under Drawing Resources, go to **Sketched Symbols**, right-click and select **Define New Symbol**.

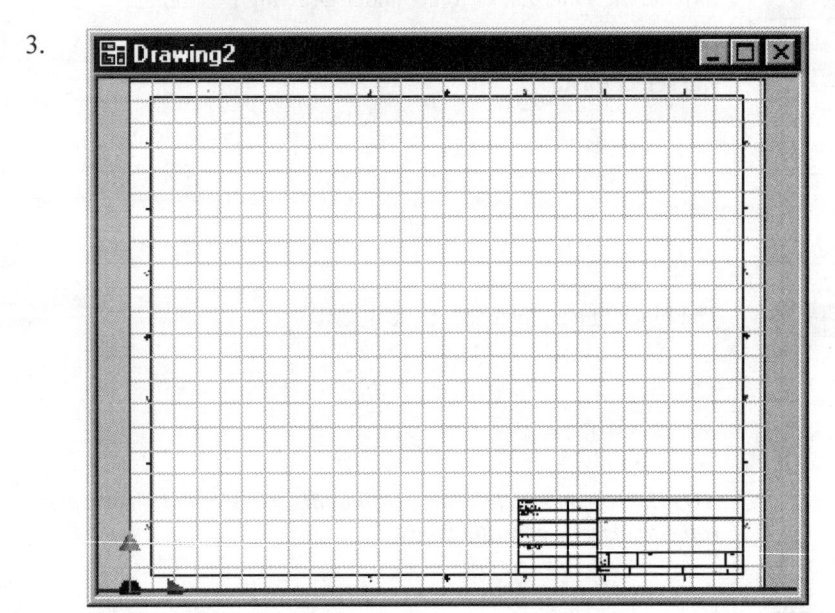

3. Your window will change to Sketch Mode.

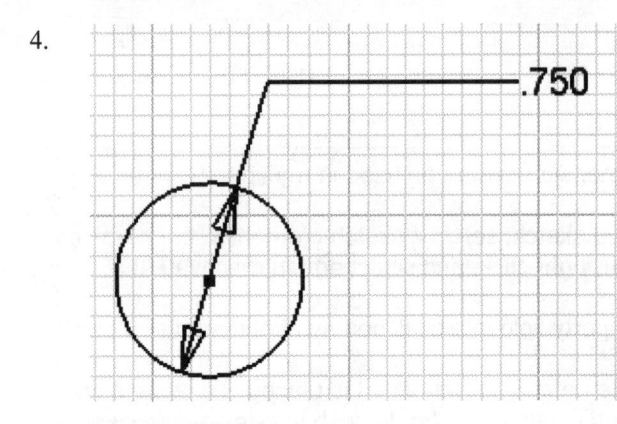

4. Draw a circle with **0.750** diameter.

5-42

Drawing Management

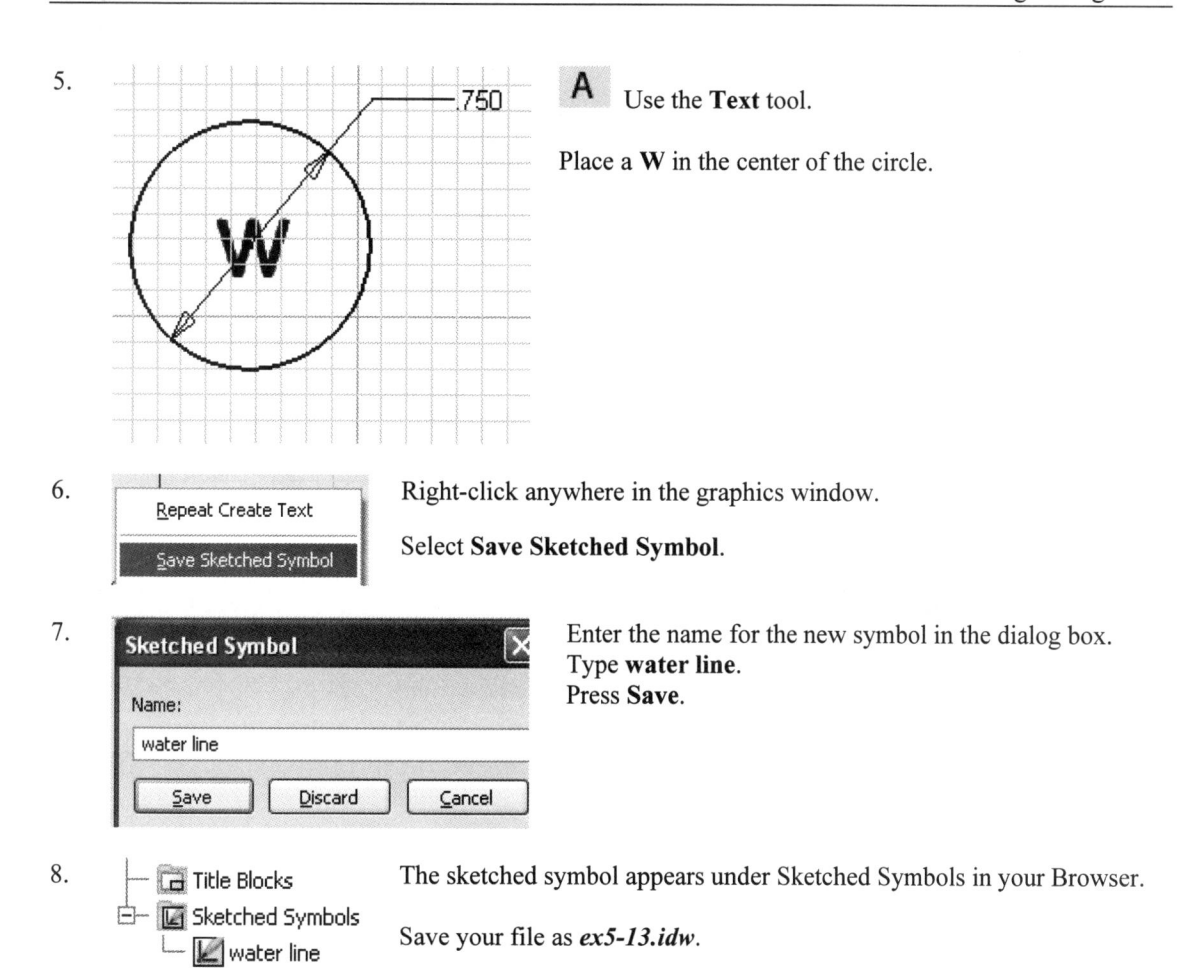

5. Use the **Text** tool.

   Place a **W** in the center of the circle.

6. Right-click anywhere in the graphics window.

   Select **Save Sketched Symbol**.

7. Enter the name for the new symbol in the dialog box.
   Type **water line**.
   Press **Save**.

8. The sketched symbol appears under Sketched Symbols in your Browser.

   Save your file as *ex5-13.idw*.

To add a sketched symbol to a drawing, double-click the symbol name in the Browser.

Sketched symbols are either associated with a sheet or with a view. If a sketched symbol is associated with a sheet, it is considered a symbol. If a sketched symbol is associated with a view, it is considered a callout.

You can add a sketched symbol to a drawing view as a callout, where a leader is automatically added. The symbol is associated to the view. If you delete the view, the sketched symbol is deleted. If you copy the view, the sketched symbol is copied.

The Power of Design: An Introduction to Autodesk Inventor

Exercise 5-14
Inserting a Symbol

File: Ex5-13.idw
Estimated Time: 15 minutes

1. Open *ex5-13.idw*.

2. Highlight the sketched symbol called **water line**.

   Right-click and select **Insert**.

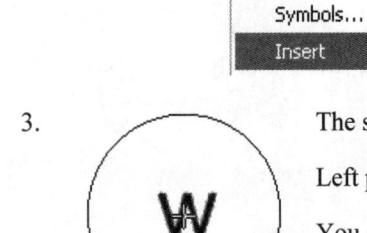

3. The symbol appears at the end of your cursor.

   Left pick to place.

   You can continue placing symbols as a multiple action.

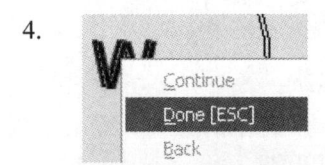

4. Right-click and select **Done** when you are finished placing callouts.

5. Left-click the arrow on the **Drawing Views Panel**.
   Select **Drawing Annotation Panel**.

6. Select the **Symbols** tool.

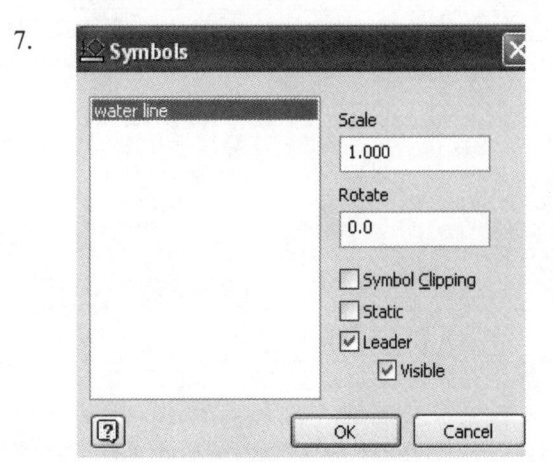

7. A dialog appears.

   A list of all the sketch symbols available in the active drawing file appears on the left.

   Symbol Clipping is used to trim existing dimensions, leaders, and extension lines that cross over the leader applied with the symbol.

   Leader is enabled to automatically add a leader to the symbol.

   Press **OK**.

5-44

Drawing Management

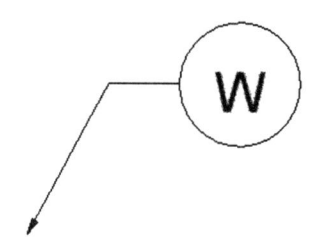

8. Select a point for the start of the leader (where the arrowhead is to be placed).
Select a second point for the start of the shoulder.
Select a third point for the end of the shoulder (and to place the symbol).

9. Right-click and select **Continue**.

   (Select **Done** if you do not want to place a second call-out.)

   The callout symbol is placed.

   You can place another or right-click and select **Done**.

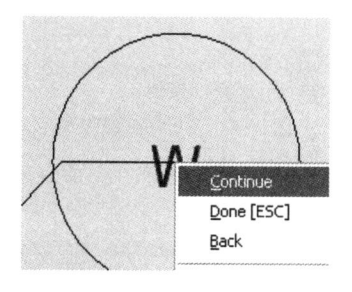

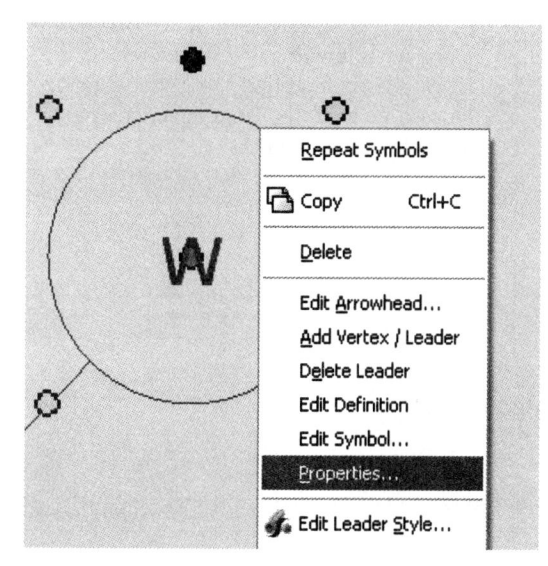

10. Select the callout.

    Right-click.
    Notice that in the menu you can edit the **Arrowhead, Add Vertex/Leader/Delete Leader.**

11. Select **Delete Leader**.

12. Save the file as *ex5-14*.

5-45

The Power of Design: An Introduction to Autodesk Inventor

Exercise 5-15
Creating a Symbol with Attributes

File: Ex5-13.idw
Estimated Time: 30 minutes

1. Open *ex5-13.idw*.

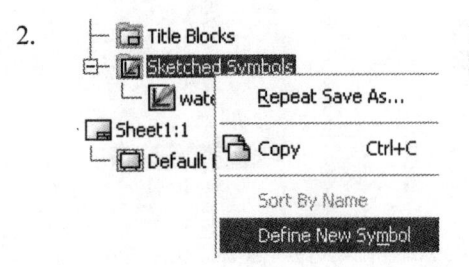

2. Highlight **Sketched Symbols** in the Browser.
   Right-click and select **Define New Symbol**.

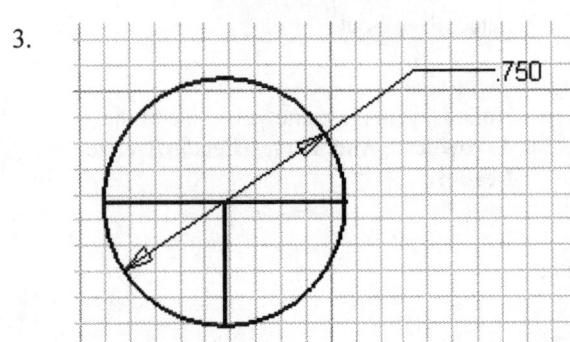

3. Draw a circle with a **0.750** diameter.
   Draw a horizontal line across the diameter.
   Draw a vertical line from the center point down to the lower quadrant.

4. Use the **Text** tool to create three **Prompted Entries**.

5. Set each field so it will be centered in its area.

5-46

Drawing Management

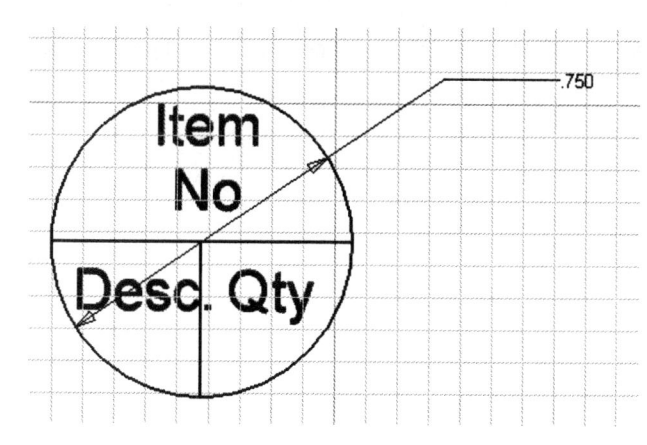

6. Create a prompted entry for **Item No**, **Qty**, and **Description**.

7. Right-click and select **Save Sketched Symbol**.

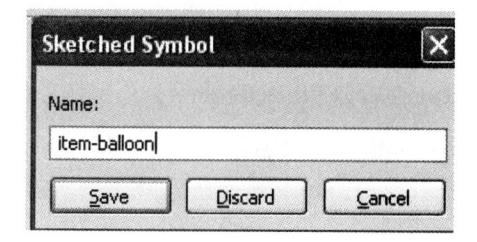

8. Edit the **Name** to **item-balloon**.

9. Save as *ex5-15.idw*.

Exercise 5-16
Editing Symbols

File: Ex5-15.idw
Estimated Time: 15 minutes

1. Open the *ex5-15.idw* file.

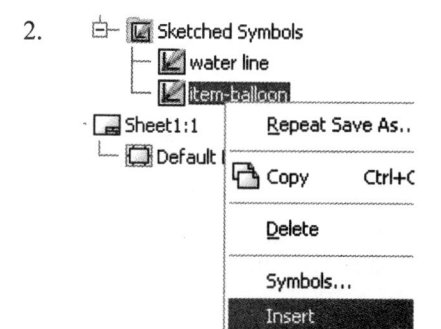

2. Highlight **item-balloon** in the Browser. Right-click and select **Insert**.

5-47

3. A field entry dialog will pop up for each prompted entry.

Type in **1** for the **Item No** field.
Type **Item** for the **Description** field.
Type **1** for the **Qty** field.

Press **OK**.

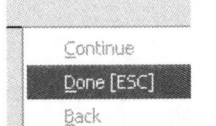

Right-click and select **Done**.

4. Select the symbol.

Right-click.

To modify the symbol, you can select **Edit Definition**.

To modify the attributes, select **Edit Field Text**.

Select **Edit Field Text**.

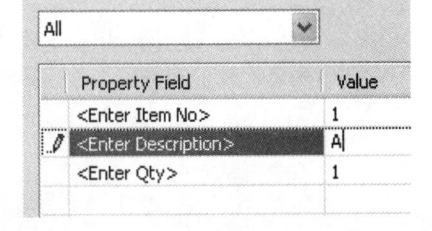

5. A dialog box will appear where it is easy to change the values for each field.

Change the value of the **Description** to **A**.

Press **OK**.

6. The field is not justified properly.

Select the symbol.

Right-click and select **Edit Definition**.

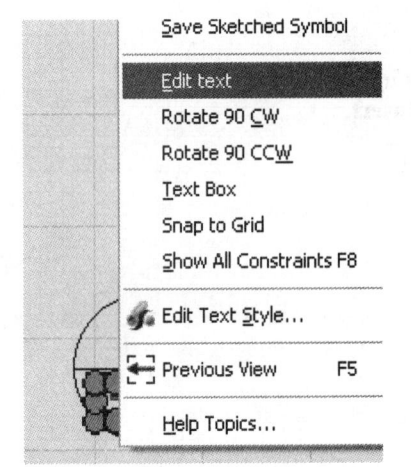

7. Highlight the field that is not justified properly.

Right-click and select **Edit Text**.

8. Set the **Justification** to **Top Center**.
   Press **OK**.

9. Shift the text so it will be located properly.

10. Right-click and select **Save Sketched Symbol**.

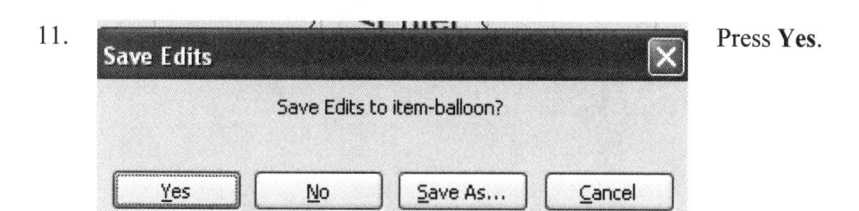

11. Press **Yes**.

12. Enter **A** for the **Description** value.

    Press **OK**.

13. The symbol updates and is corrected.

    Save the file as *ex5-16.idw*.

Review Questions

1. Deleting a base view automatically deletes all dependent views.

    A. True
    B. False

2. A single model dimension cannot be used in multiple views on the same sheet.

    A. True
    B. False

3. Sketch overlays are used to clip or edit drawing views.

    A. True
    B. False

4. Drawing views cannot be copied from one sheet to another.

    A. True
    B. False

5. You can only create a view using the default orientations, i.e. Front, Top, Right.

    A. True
    B. False

6. 'Show Contents' is used to:

    A. List the views in a drawing
    B. List the features in the part
    C. List the format of a title block
    D. List the format of a sheet

7. The 'Fill Sketch Region' tool is used to:

    A. Add color to a profile in a title block
    B. Add color to a sketch overlay
    C. Add color to a view
    D. Add hatching to a section view

8. If you change the scale of the base view from 1:2 to 1:1, the scale of the isometric view:

    A. will change to 1:1
    B. will change to 3:4
    C. will remain 1:2
    D. will change to 4:3

9. The Section View tool creates all the section view types listed EXCEPT:

    A. Full
    B. Half
    C. Offset
    D. Revolved

ANSWERS: 1) B; 2) A; 3) B; 4) B; 5) B; 6) B; 7) A; 8) C; 9) D

Lesson 6
Drawing Annotation Toolbar

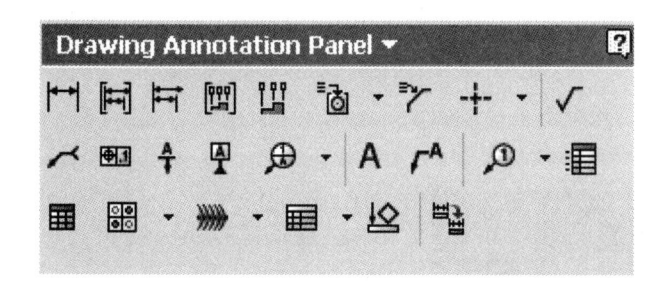

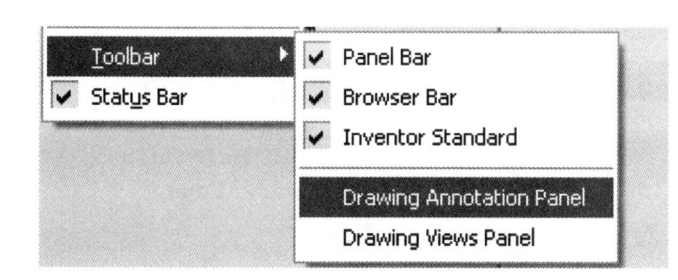

The Drawing Annotation toolbar can be accessed through the **View→Toolbar** menu.

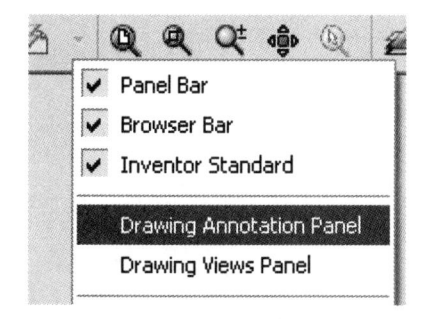

Right-clicking on the Standard toolbar also allows you to enable available toolbars.

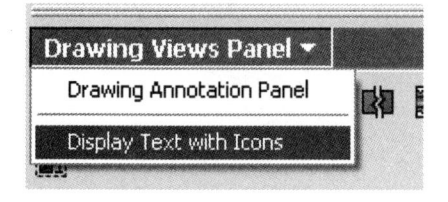

Left-clicking on the arrow at the top of the Panel Bar can also activate the Drawing Annotation toolbar. Using this method will retire the Drawing Management Panel bar.

Exercise 6-1
Using the Style Manager Library

File: None
Estimated Time: 60 minutes

1. Close Inventor.
2. Using Windows Explorer, create a new folder called Styles. This will be the location of your new Style Library.

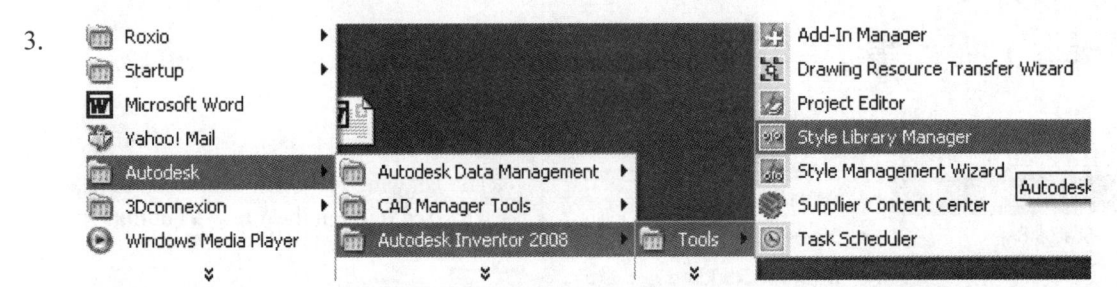

3. Go to **Programs→Autodesk→Autodesk Inventor 2008→Tools→Style Library Manager**.

4. Select **Create New Style Library** (located in the upper right corner of the dialog).

5. Select **Create Empty Style Library** from the drop-down list.

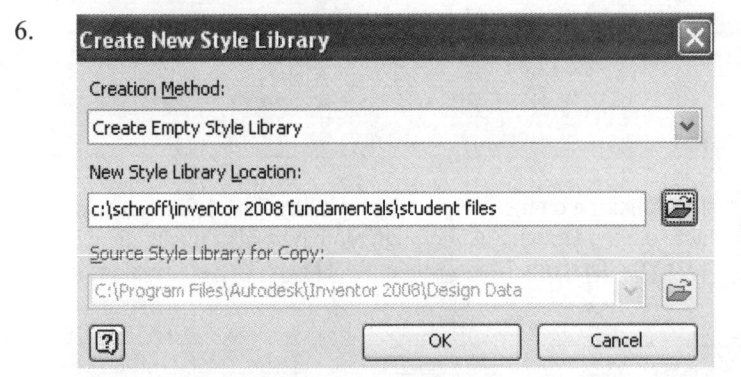

6. Browse for a folder to save the style library file.

   Press **OK**.

7. Press **Exit** to close the Style Library Manager.

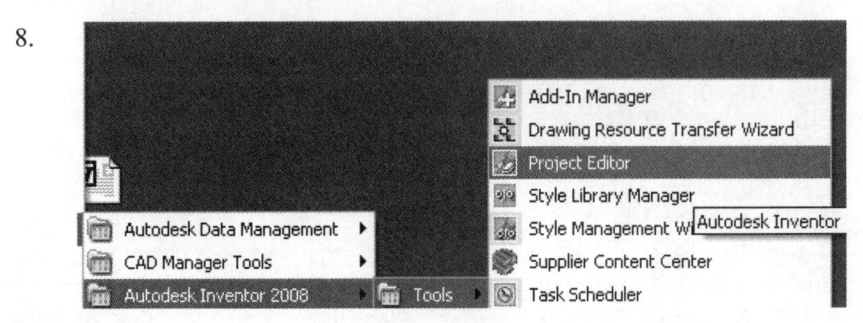

8. Go to **Programs→Autodesk→Autodesk Inventor <Version>→Tools→Project Editor**.

Drawing Annotation

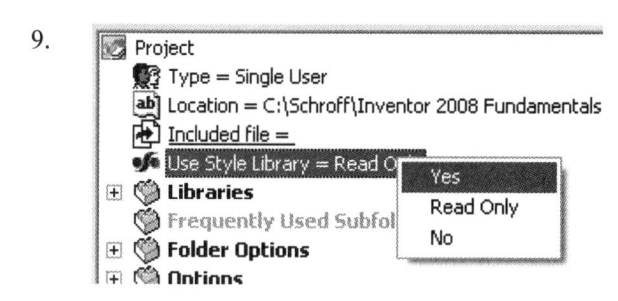

9. Set **Use Styles Library** to **Yes**.

   This allows you to modify the Style Library.

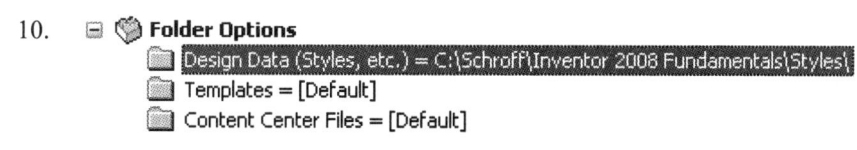

10. Set the folder options for styles to the path where you saved the new style.

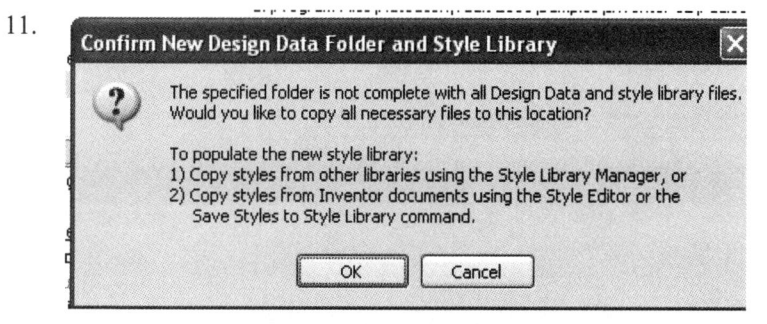

11. Press **OK**.

    It may take a few minutes to copy all the styles over.

12. Save the changes to the project and close the dialog.

13. Launch Inventor.

14. Close the Project Editor.

15. Close the dialog.

16. Create an empty idw file by holding down Ctl+Shift and selecting the **New** drawing icon under **File**.

    A drawing sheet will open with no border or titleblock.

6-3

17. Go to **Tools→Application Options**.

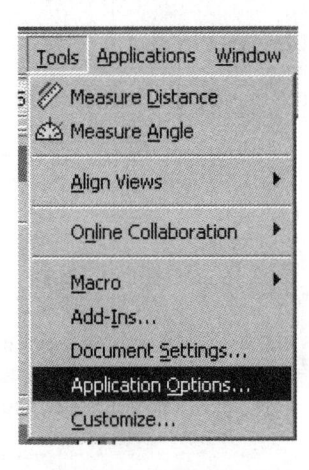

18. Verify that the Design Data (and Default Style Library is set to the Styles path where you saved your styles.

    If it is not, reset it to the correct folder.

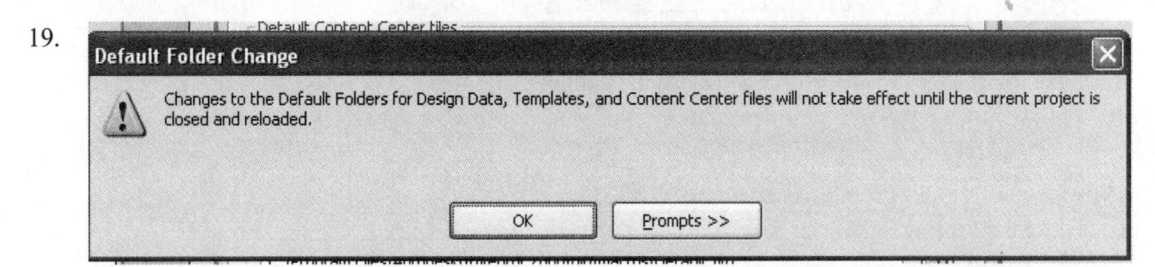

19. Press **OK**.

20. Close Inventor without saving the file and re-start it.

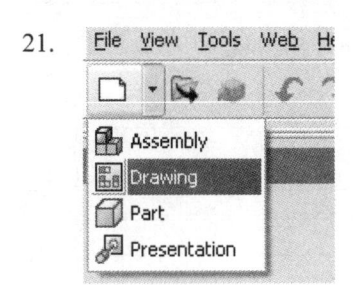

21. Create an empty idw file by holding down Ctl+Shift and selecting the **New** drawing icon under **File**.

    A drawing sheet will open with no border or titleblock.

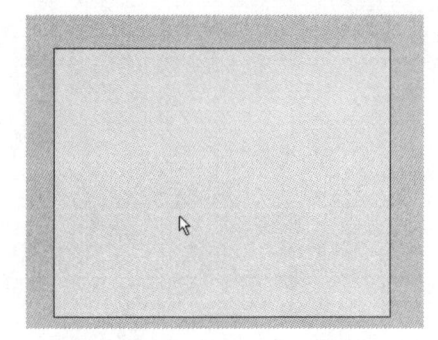

Drawing Annotation

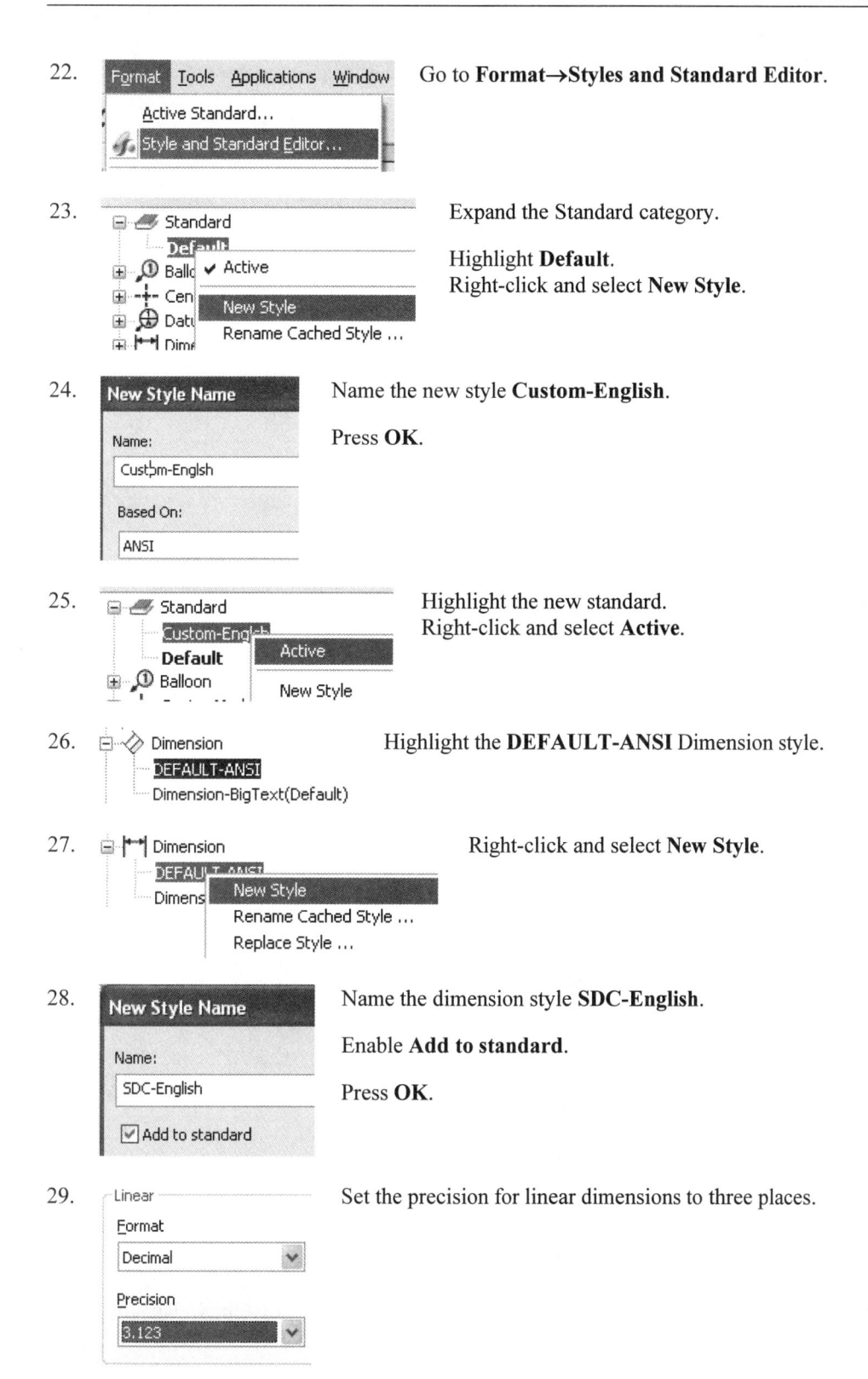

22. Go to **Format→Styles and Standard Editor**.

23. Expand the Standard category.

    Highlight **Default**.
    Right-click and select **New Style**.

24. Name the new style **Custom-English**.

    Press **OK**.

25. Highlight the new standard.
    Right-click and select **Active**.

26. Highlight the **DEFAULT-ANSI** Dimension style.

27. Right-click and select **New Style**.

28. Name the dimension style **SDC-English**.

    Enable **Add to standard**.

    Press **OK**.

29. Set the precision for linear dimensions to three places.

30. 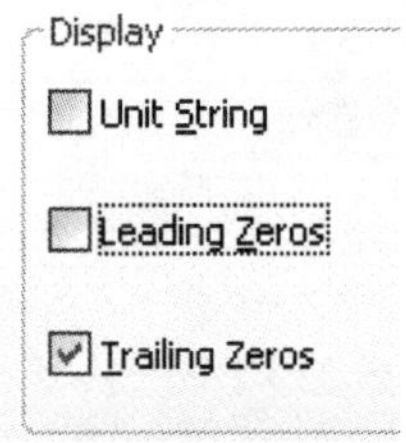 Disable the **Leading Zero** for linear dimensions to conform with ANSI standards.

31. Select the **Display** tab.

32. 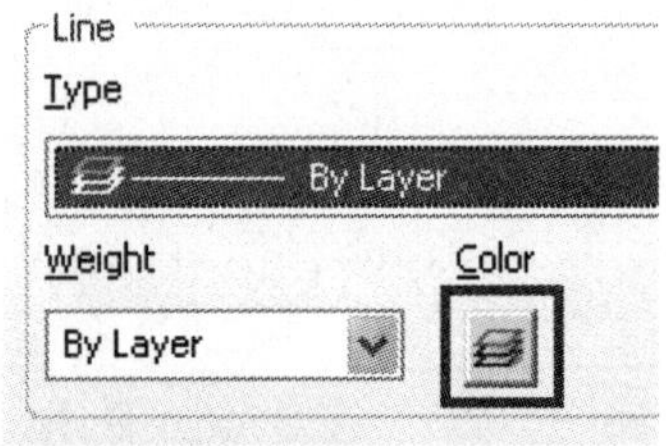 Press the **Color** button.

33.  Set the **Color** to **Red**.

34. Set the **Gap** to **.125**.

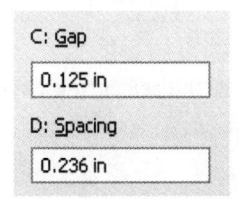

35. 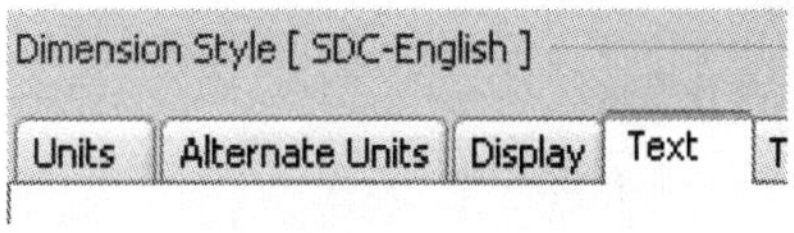 Select the **Text** tab.

36. 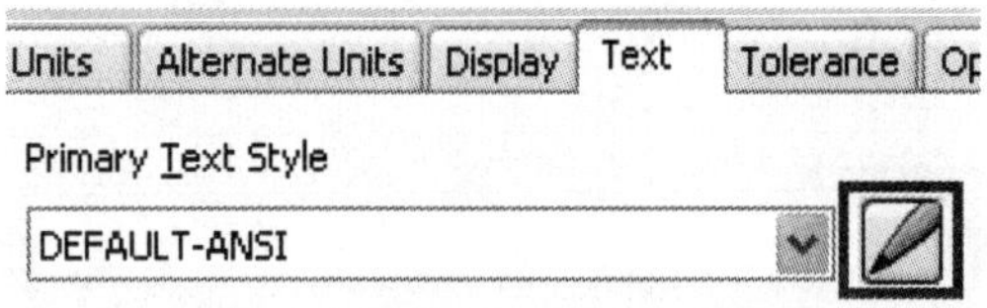 Select the **Edit** button next to **Primary Text Style**.

37. If you are prompted to save the style, press **Yes**.

38. 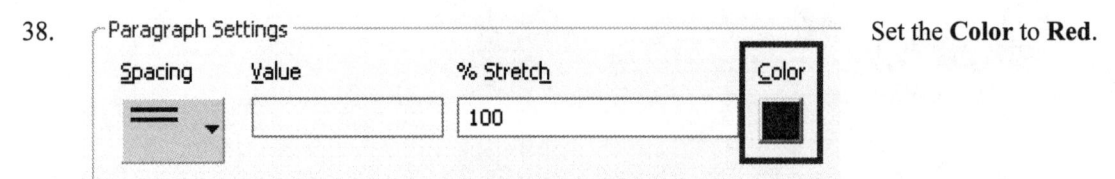 Set the **Color** to **Red**.

39.  Press **Save**.

6-6

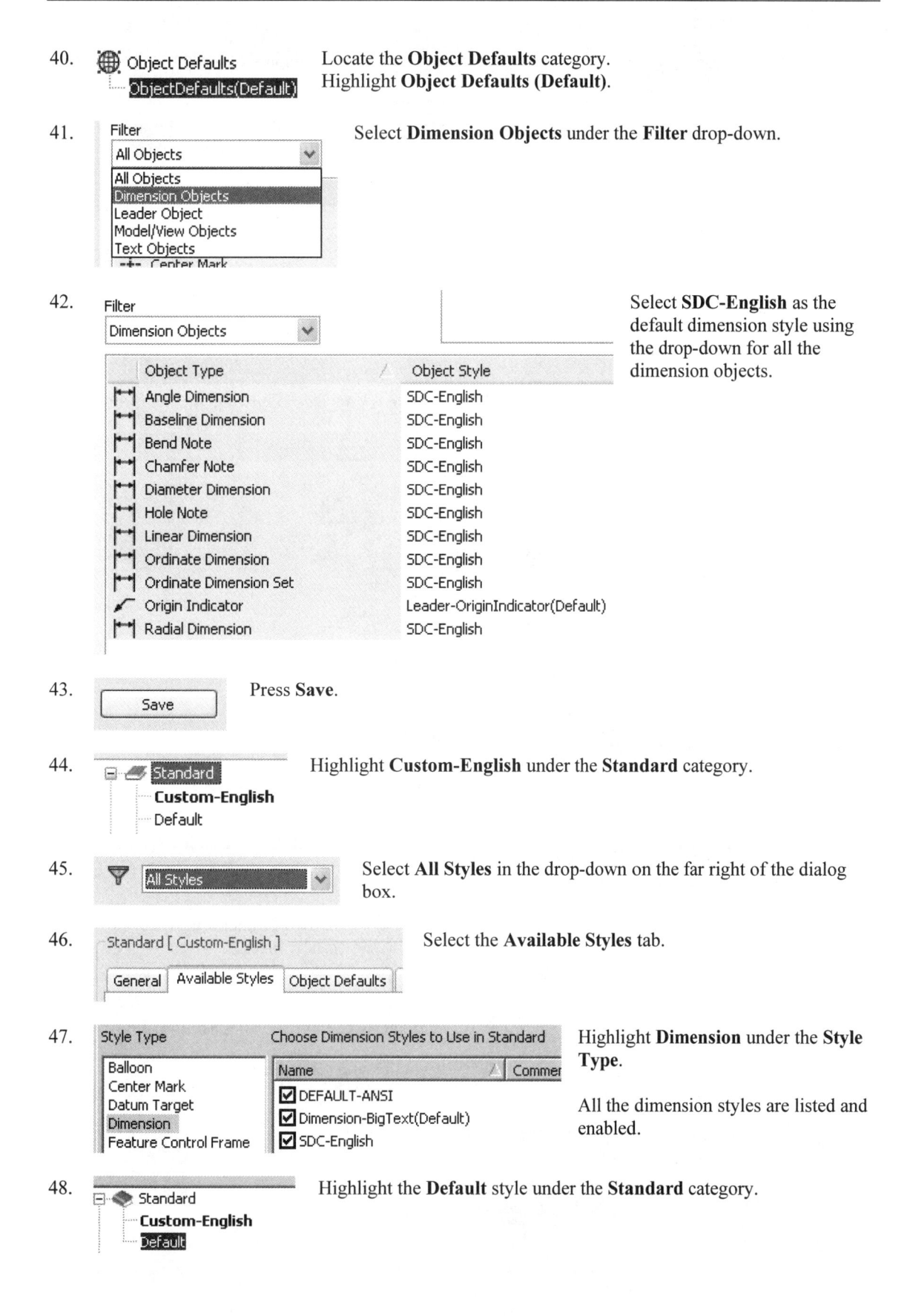

Drawing Annotation

40. Locate the **Object Defaults** category.
Highlight **Object Defaults (Default)**.

41. Select **Dimension Objects** under the **Filter** drop-down.

42. Select **SDC-English** as the default dimension style using the drop-down for all the dimension objects.

43. Press **Save**.

44. Highlight **Custom-English** under the **Standard** category.

45. Select **All Styles** in the drop-down on the far right of the dialog box.

46. Select the **Available Styles** tab.

47. Highlight **Dimension** under the **Style Type**.

All the dimension styles are listed and enabled.

48. Highlight the **Default** style under the **Standard** category.

6-7

49. Highlight **Dimension** under the **Style Type**.

    All the dimension styles are listed, but they are not all enabled.

50. Set the filter to only show the **Active Standard**.

51. Only the **Custom-English** style is now listed under the Standard category.

52. Select the **Available Styles** tab in the main dialog.
    Highlight **Dimension**.
    Disable all the dimension styles *except* for **SDC-English**.

53. Select **All Styles** from the drop-down list.
    This will refresh the list.
    If prompted to save, press **Yes**.
    Select **Active Standard** from the drop-down list.

54. Select the **Available Styles** tab.
    Highlight **Dimension**.
    Only the dimension style you created is now listed.

55. Highlight **SDC-English** under **Standard**.
    Right-click and select **Export**.

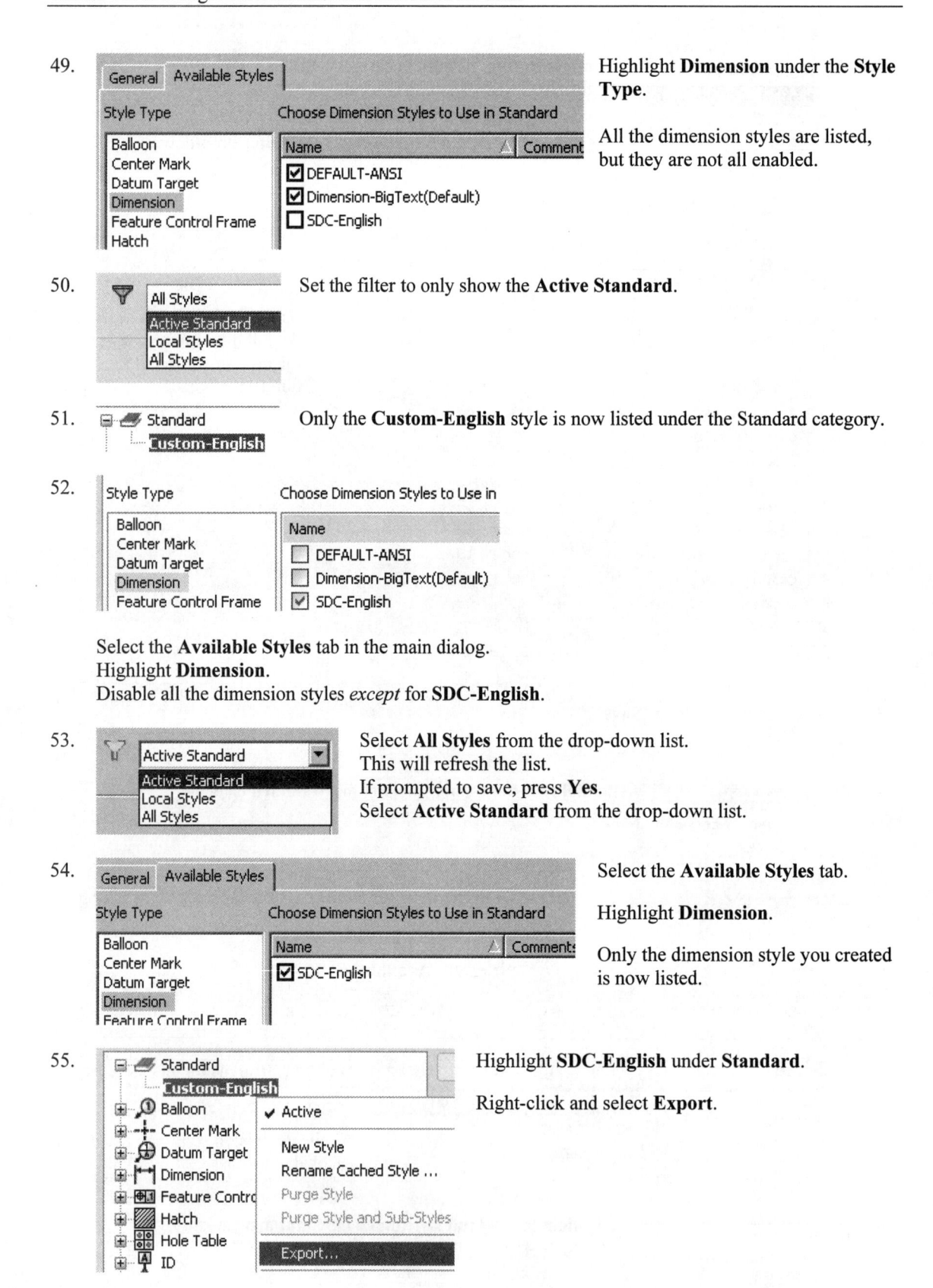

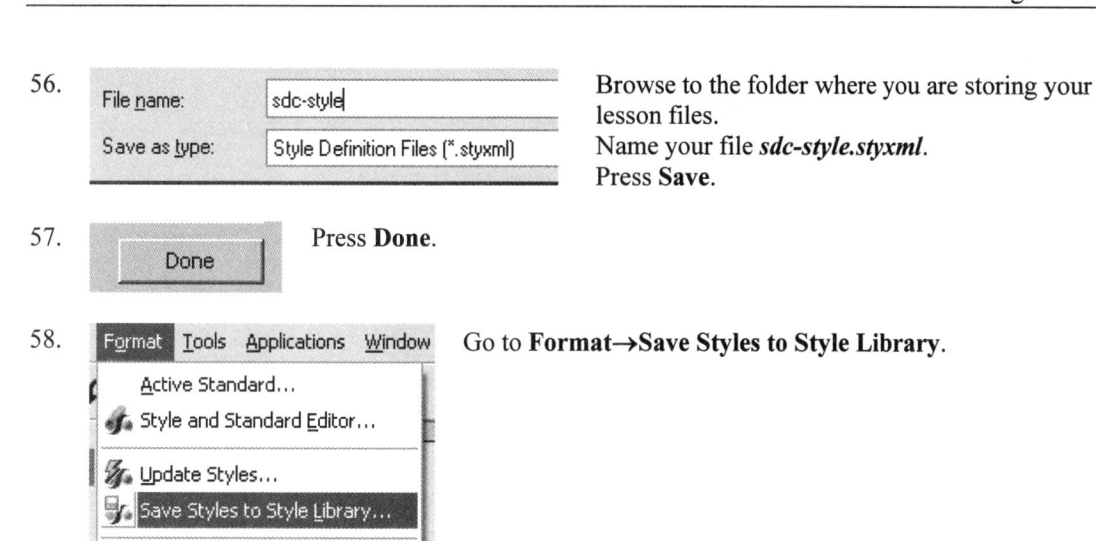

56. File name: sdc-style
    Save as type: Style Definition Files (*.styxml)

    Browse to the folder where you are storing your lesson files.
    Name your file *sdc-style.styxml*.
    Press **Save**.

57. Press **Done**.

58. Go to **Format→Save Styles to Style Library**.

59. 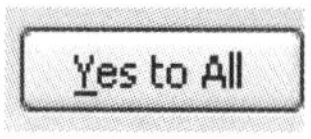 Press **Yes to All**.

60. Press **OK**.

61. 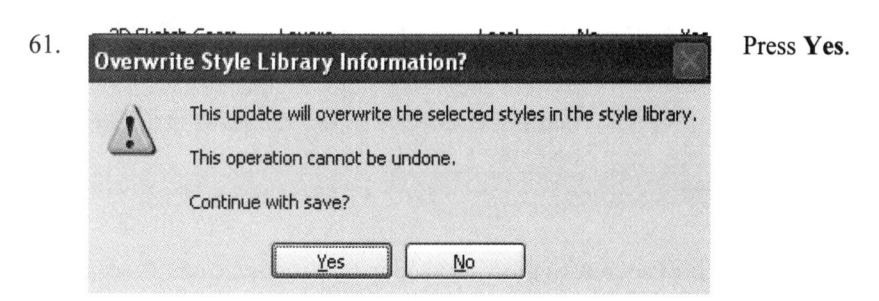 Press **Yes**.

62. Close the file without saving.

> - The reason it requires a change to the project before you can save styles is so that a CAD Manager in a department can control who has control over company styles and standards.
> - When you set a Style Library to Yes, you have Read/Write functionality. When you set a Style Library to Read Only, styles are available to other users who are working on the project, but they can't modify the style. If you set a Style Library to No, then the document will not be accessing the style library; instead you will set up styles inside each document.
> - The Thread.xls and Clearance.xls files should be copied into the Styles folder that is assigned to the project or you will not be able to use data from those files.
> - To ensure that you have no problems exporting, importing, copying, and deleting styles make sure that your project has the Style Library pointed to the path where you store your styles. The default path is \Program Files\Autodesk\Inventor 10\Design Data.

The Power of Design: An Introduction to Autodesk Inventor

Exercise 6-2
Applying a General Dimension

File: Ex5-2.idw
Estimated Time: 20 minutes

You can add a General Dimension to a drawing view using the following methods:

From the Drawing Annotations Panel	General Dimension +D
Shortcut Key	D
Right-click in the Graphics Window	

1. Locate *ex5-2.idw* and press **Open**.

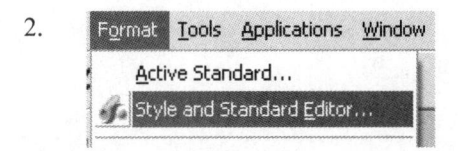

2. Go to **Format→Style and Standard Editor**.

3. Select the **Import** button.

4. Locate the *sdc-style.styxml* file saved in the previous exercise.

   Press **Open**.

   If you don't see the **Custom-English** standard in the browser, check to see if your filter is set to **All Styles**.

5. Set the **Custom-English** standard as the active style.

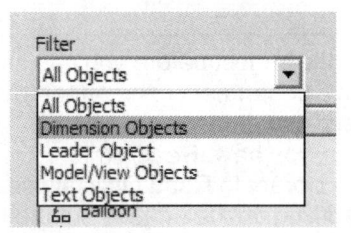

   Set the pane to filter for **Dimension Objects**.

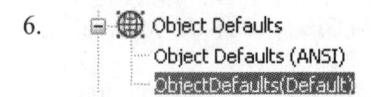

6. Locate the **Object Defaults** category.
   Highlight **Object Defaults (Default)**.

7. Note that **SDC-English** is already set as the default dimension style.

   Press **Done**.

6-10

Drawing Annotation

8. Left-click the arrow on the **Drawing Views Panel**.
   Activate the **Drawing Annotation Panel**.

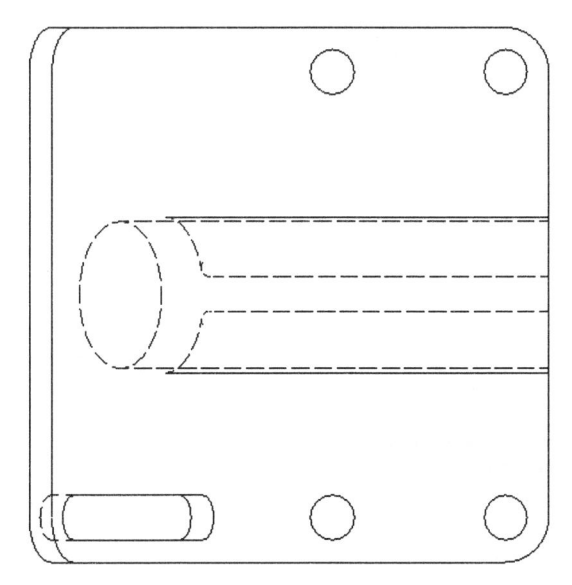

9. Use **Zoom Window** to zoom into the top view.

10. Select the **General Dimension** tool.

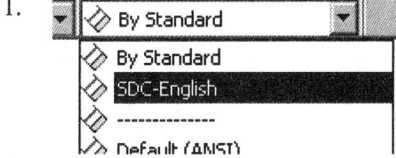

11. Select **SDC-English** from the style drop-down to use that style to apply dimensions.

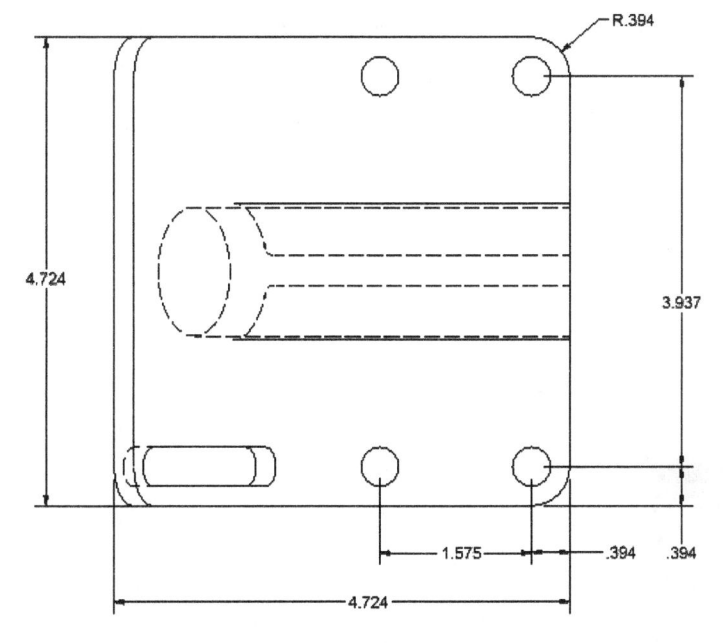

12. Using the **General Dimension** tool, apply the dimensions shown. The dimensions should appear with the style changes you applied.

13. Save the file as *ex6-2.idw*.

6-11

The Power of Design: An Introduction to Autodesk Inventor

Exercise 6-3
Applying a Baseline Dimension

File: Ex6-2.idw
Estimated Time: 15 minutes

1. Locate *ex6-2.idw* and press **Open**.

2. Zoom into the right side view.

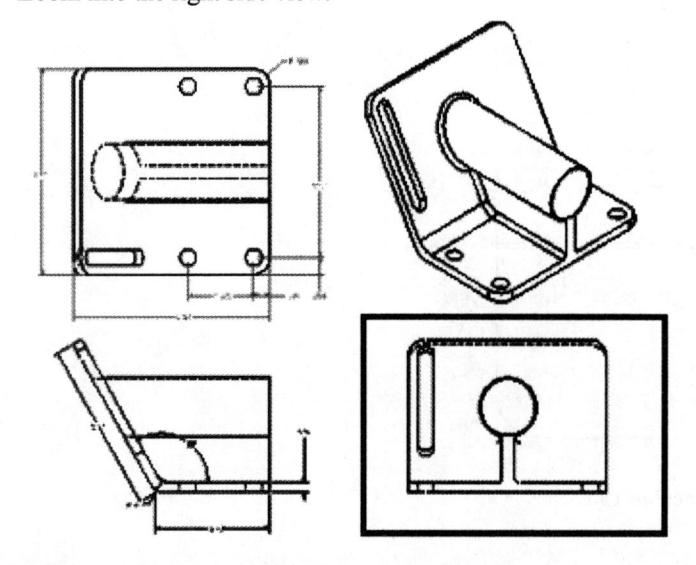

3. Select the **Baseline Dimension Set** tool.

   Note that the active standard is **SDC-English**.

4. You'll be prompted to select an edge to be included in the baseline dimension.

   Select the edges and arcs indicated by the arrows.

   Right-click and select **Continue**.

6-12

5. 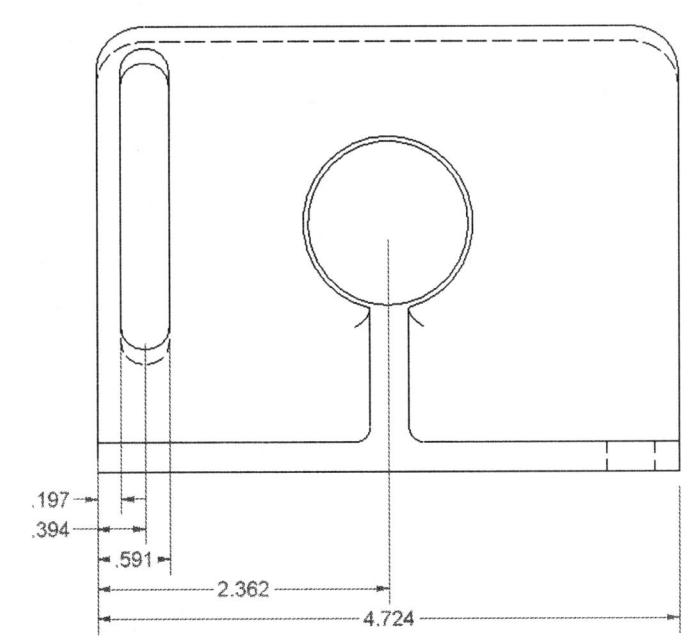	Pick a point below the view.

	This will place the baseline dimensions.

6. 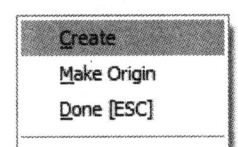	Right-click and select **Create**.

7. 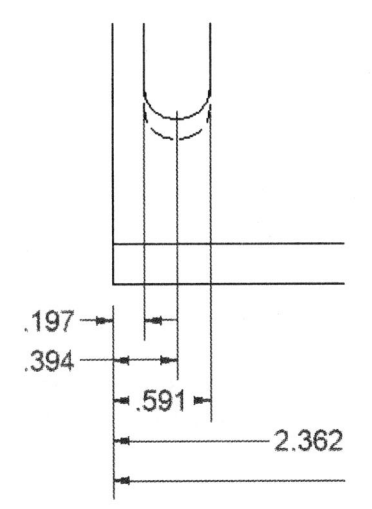	Normally when dimensioning a slot, we would only have the dimension to the center of the slot. This means that we should delete the .591 dimension and the .197 dimension.

8.	Select the .591 dimension and right-click.

6-13

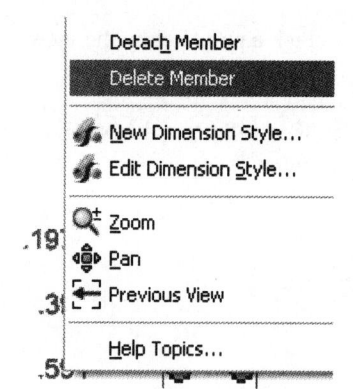

9. If you select '**Delete**' you will delete the entire set of dimensions.

Instead, select the '**Delete Member**' option.

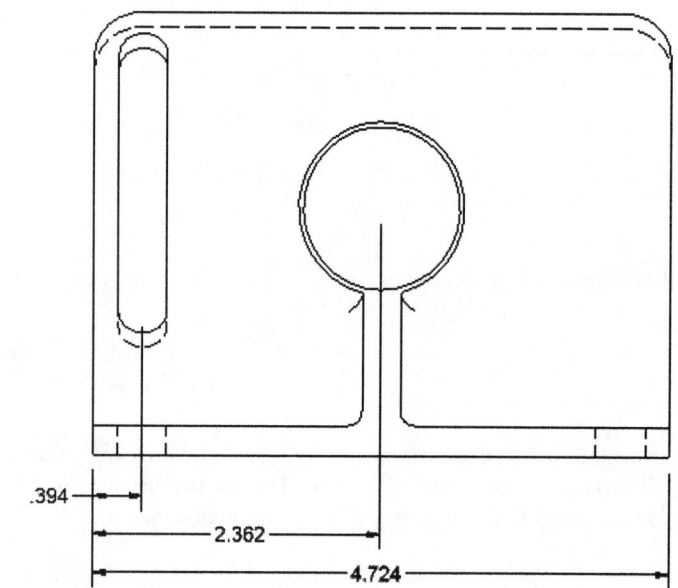

10. Repeat to delete the .197 dimension.

You can reposition the dimensions so they are spaced properly by simply picking to activate the grips and moving them into the correct location.

11. Save as *ex6-3.idw*.

> ➤ You can also type 'A' as a shortcut to activate the Baseline Dimension Set command.
> ➤ Once you select the first point to place the first ordinate dimension, all ordinate dimensions placed following will be placed automatically based on the selection.
> ➤ Hole notes can be added only to hole features created using the Hole feature tool in parts.
> ➤ The first point picked when placing an ordinate dimension is automatically assigned the 0 value. Place all the dimensions desired for that axis (whether horizontal or vertical), then right-click to select Create. Then place the dimensions for the other axis.
> ➤ Pressing **O** on the keyboard will also initiate the Ordinate dimension command.
> ➤ When you place ordinate dimensions using the ordinate dimension set command, the dimensions are linked as a group. To add or delete members of the set, you need to right-click and use the right-click shortcut menu.
> ➤ You can automatically calculate the number of holes in a view using the <QTY> parameter. Simply type <QTY> below the hole note.

Exercise 6-4
Applying Ordinate Dimensions

File: Ex6-2.idw
Estimated Time: 15 minutes

1. Open the *ex6-2.idw* file.

2. Zoom into the right side view.

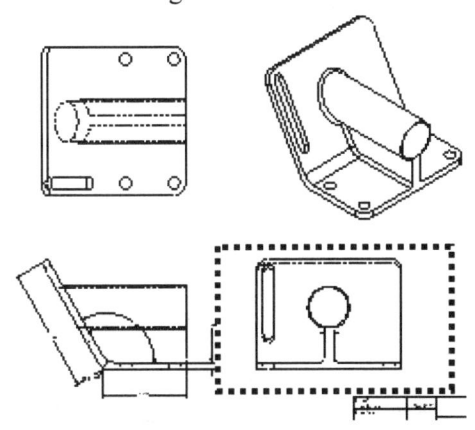

3. Select the **Ordinate Dimension Set** tool.

4. Note the style is already set to **SDC-English**.

5. Pick the lower left corner of the bracket.
   Pick to select the center of the arc.
   Pick to select the center of the cylinder.
   Pick the far right corner of the bracket.
   Right-click and select **Make Origin**.
   Select the lower left corner of the bracket.
   Right-click and select **Create**.

   Your ordinate dimensions are placed.

6. Select the **Ordinate Dimension** tool.

7. Note the style is already set to **SDC-English**.

8. You'll be prompted to select a view to dimension.
   Select the right side view.

9.

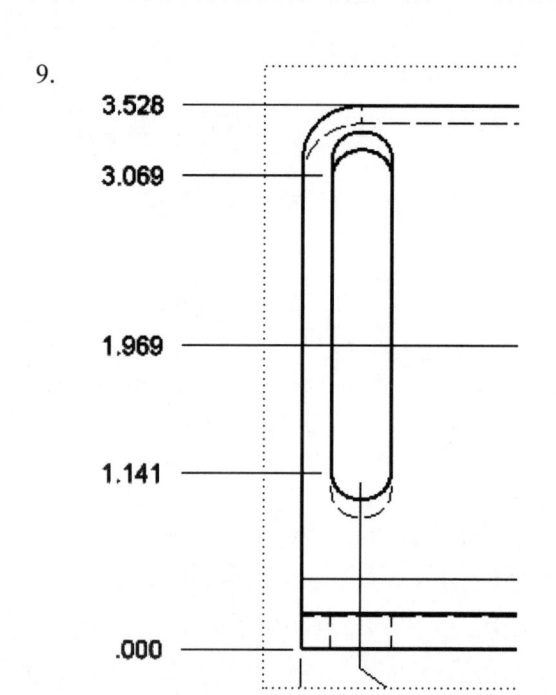

Select the left lower corner of the bracket.

You are still in Ordinate Dimension mode.

NOTE: *Verify that you are in the correct dimension style.*

If you look in the panel, the **Ordinate Dimension** button is still depressed.

Select the center of the lower arc to start the ordinate dimension.

Select a point to the left of the view.

10.

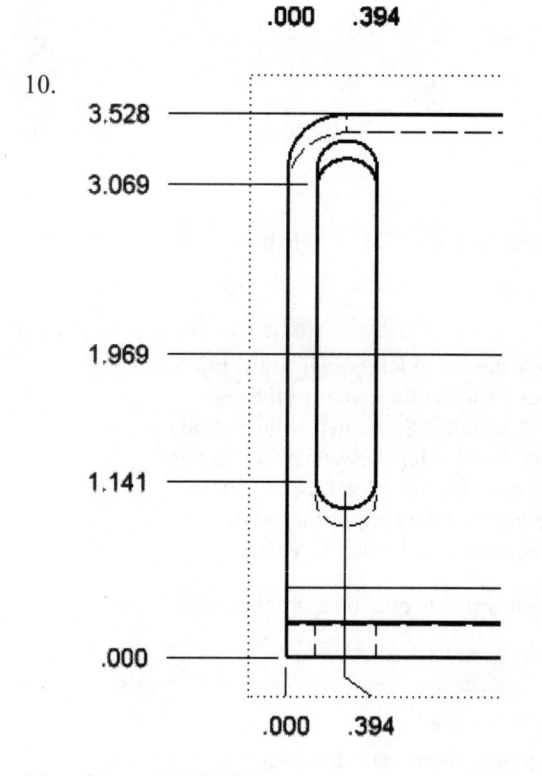

Select the top arc to start a second ordinate dimension.

Select a point to the left of the view.

Select the top of the bracket.

Select a point to the left of the view.

Right-click and select **Make Origin**.

Select the lower left corner of the bracket.

Right-click and select **Create**.

11. Save as *ex6-4.idw*.

Exercise 6-5
Adding a Hole Note

File: Ex6-4.idw
Estimated Time: 15 minutes

1. Open the *ex6-4.idw* file.

2. Zoom into the top view.

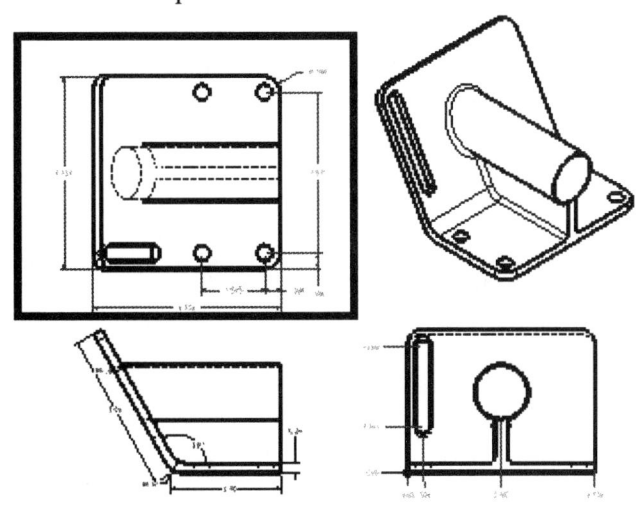

3. Select the **Hole/Thread Notes** tool.

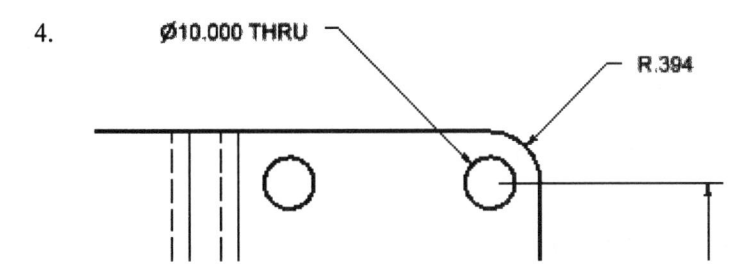

4. Select the top right circle in the view.

   Place the dimension as shown.

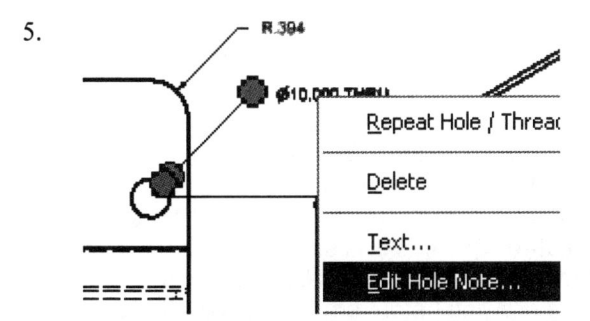

5. Select the Hole Note.

   Right-click and select **Edit Hole Note**.

6. Disable **Use Default**.

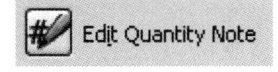

   Select the **Edit Quantity Note**.

7. Enable **Number of holes in view (normal)**.

Select the **Qty** button.
<QTY> appears in the edit box.
Highlight this and press CTL-C to copy it.

Press **OK**.

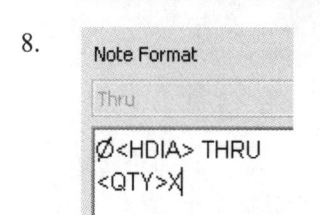

8. Paste the clipped text beneath the hole note.
Type an **X** to indicate a multiple.

Press **OK**.

9. Ø10.000 THRU
4X

The hole note updates.

10. Save the file as *ex6-5.idw*.

Tips & Tricks

- If the features form a circular pattern, the center mark for the pattern is automatically placed when you have selected all of the members.
- The units and appearance of hole notes are controlled using the Dimension Style.
- Pressing 'C' on the keyboard will also activate the Center Mark command.
- There is more than one type of center mark available.
- You can add dimensions to isometric views.
- The drawing annotation tools can be selected before or after you select the view. You can select the tool and then the view or vice versa.
- No centerline is added between the final feature in a centered pattern and the beginning feature. You can manually extend the line if needed.
- If you attach a note leader line to a view or to geometry within a view, the note is moved or deleted when the view is moved or deleted.
- The active drafting standard controls the default text format. To change the default text for a drawing or template, modify the drafting standard. Go to Format→Standards and select the Common tab.
- A first-level component can be either a subassembly or a part.
- If you add balloons to a drawing before creating a parts list, the balloons will show the item numbers of first-level components. You can select a balloon and change it to show the item number of the part to which it is attached.
- Typing 'B' will initiate the Balloon command.
- To change the default heading location and title text for parts lists, use Format>Standards>Parts List to change the attributes of the active drafting standard.

Drawing Annotation

Exercise 6-6
Adding Center Marks

File: Ex6-5.idw
Estimated Time: 15 minutes

1. Open the *ex6-5.idw* file or continue working in the open drawing.

2. Select the top view.

   Right-click and select **Automated Centerlines**.

3. A dialog appears.

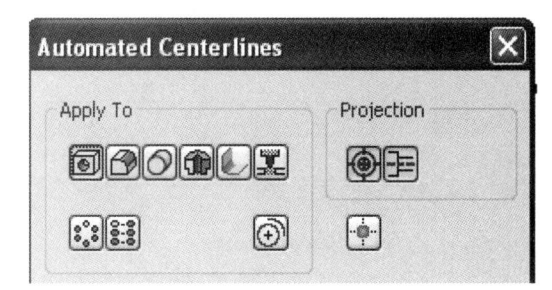

   You can apply center marks to holes, fillets, cylinders, revolves, patterns, bends, punches or sketches. You can set the projection for a top or projected view.

   Under **Apply To**: enable **Holes**.
   Under **Projection**: enable **Top View**.

   Press **OK**.

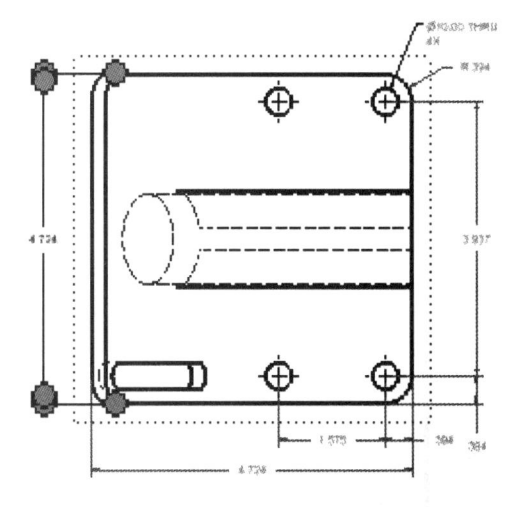

4. Center marks are added to the four holes in the view.

5. Select the Front view.

   Right-click and select **Automated Cenerlines**.

   Enable **Holes**.
   Enable **Projected**.

   Press **OK**.

6-19

The Power of Design: An Introduction to Autodesk Inventor

6.

Center lines are added to the holes.

7. Save the file as *ex6-6.idw*.

Exercise 6-7
Adding Center Lines

File: Ex6-6.idw
Estimated Time: 10 minutes

1. Open the *ex6-6.idw* file or continue working in the open drawing.

2. Zoom into the right side view.

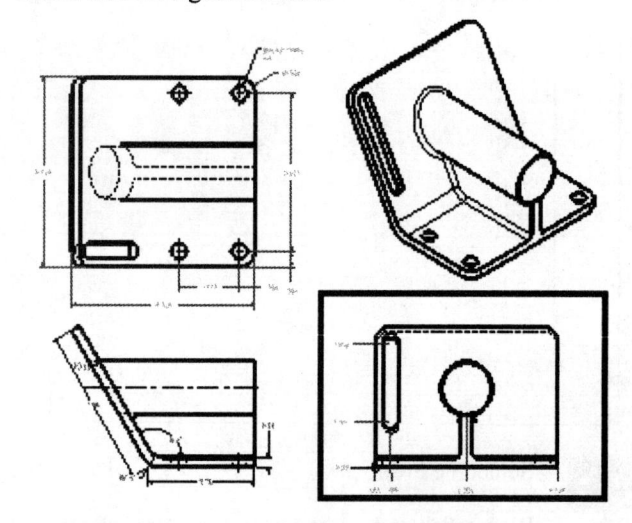

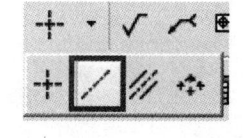

3.   Select the **Centerline** tool.
     (This is part of the Center Mark flyout.)

6-20

Drawing Annotation

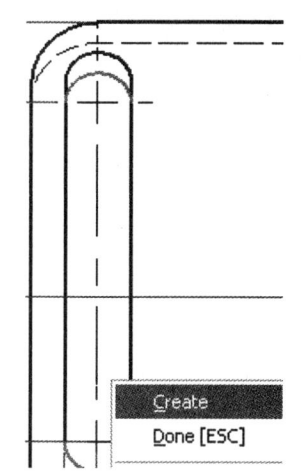

4. Select the center point of the top arc.

   Drag the line down vertically and select the center point of the bottom arc.

   Right-click and select **Create**.
   Right-click and select **Done**.

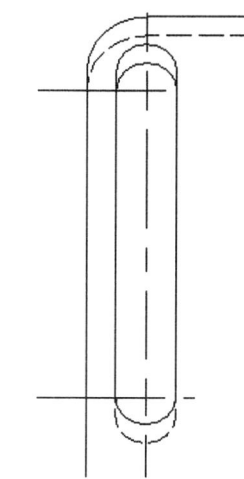

5. A center line appears in the slot.

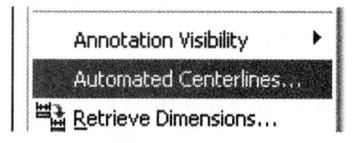

6. Select the **Right Side** view.
   Right-click and select **Automated Centerlines**.

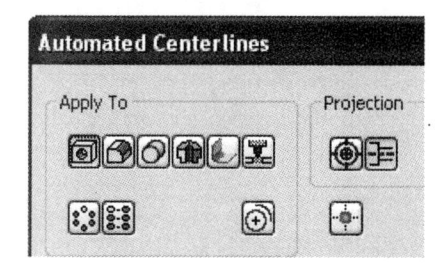

7. In this view, we want to place center marks at the cylinder and for the holes indicated by the arrows.

   Enable '**Hole**'.
   For **Projection**, enable both top and side projections.

   Press **OK**.

6-21

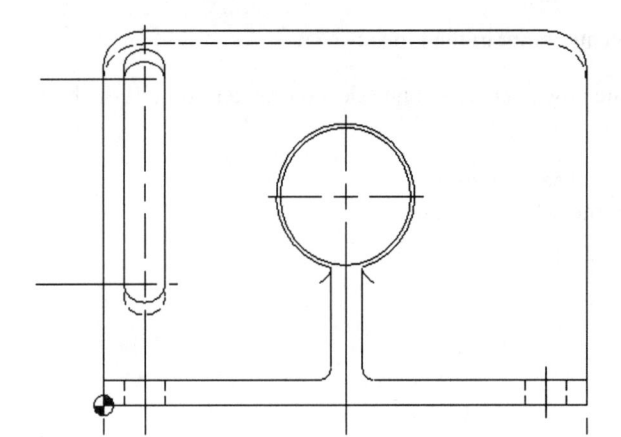

8. Center lines are added to the holes.

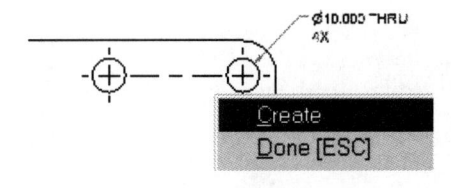

9. Zoom into the **Top View**.

   Place a centerline between the two top holes.

   Right-click and select **Create**.

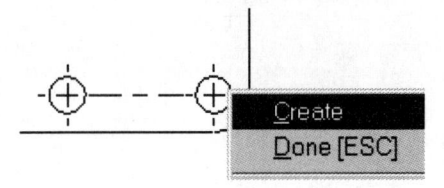

10. Place a centerline between the two bottom holes.

    Right-click and select **Create**.
    Right-click and select **Done**.

11. Save as *ex6-7.idw*.

Exercise 6-8
Adding a Centered Pattern

File: Views-6.ipt (This file is available in the Tutorial Files folder under Inventor or as a free download from the publisher's website.)
Estimated Time: 15 minutes

1. Start a **New** drawing.

2. Select the **Base View** tool.

3. Locate the *views-6.ipt* file.

   It is located under the Tutorial Files folder or can be downloaded from the publisher's website.

6-22

Drawing Annotation

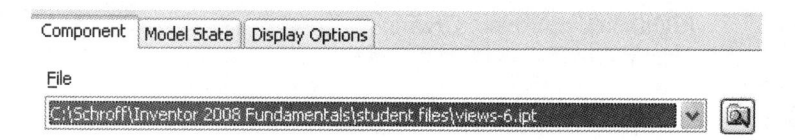

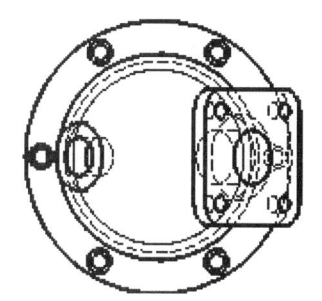

4. Select the **Top** view and place it in your sheet.

5. Switch to the **Drawing Annotation Panel**.

6. Select the **Centered Pattern** tool.

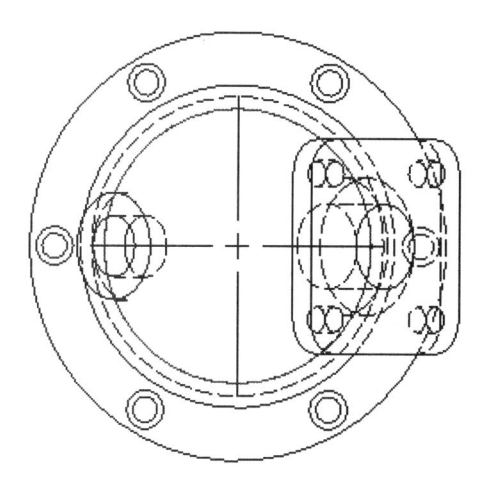

7. Select the center hole to indicate the center point for your centered pattern.

8. Inventor will add a center mark to the pattern center.

    Next select the center for each hole in the pattern.
    Move around the circle clockwise selecting each circle's center point in order.

    Be sure you select the circle you started with to close the pattern or you will have a gap.

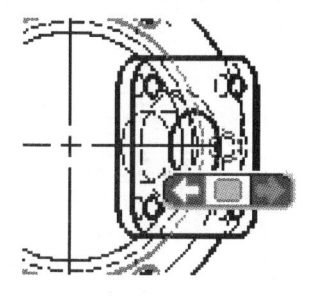

    You can use the **Select Other** tool to tab through the circles until the correct circle is highlighted.

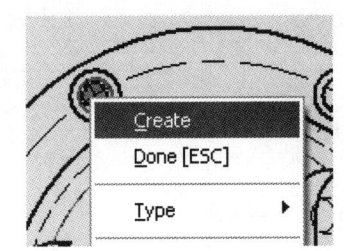

The Power of Design: An Introduction to Autodesk Inventor

9. Right-click and select **Create**.
Then right-click and select **Done**.

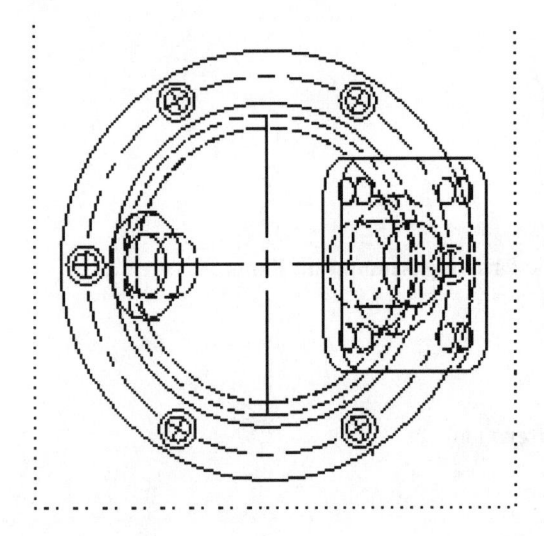

10. Save as *ex6-8.idw*.

Exercise 6-9
Adding Balloons

File: Ex8-2.iam
Estimated Time: 15 minutes

1. Start a **New** drawing.

2. Select the **Base View** tool.

3. Locate and select the *NewLinkRod.iam* file located in the Tutorial files under Inventor or from the publisher's website.

6-24

Drawing Annotation

4. Place a view as shown.

   Hint: Use the Front view.

5. Activate the **Drawing Annotation Panel**.

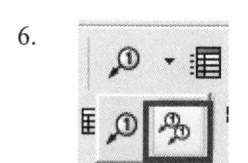

6. Select the **Auto Balloon** tool.

   This is located under the **Balloon** flyout.

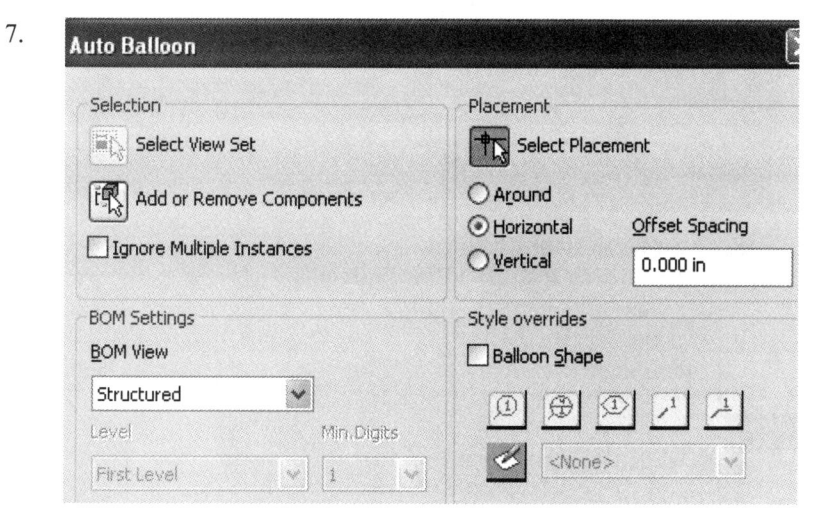

7. The first enabled button prompts you to select a view.
   Select the base view you just placed.

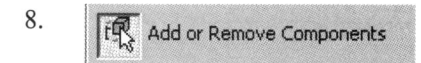

8. Window around the view.
   Set the **Placement** to **Horizontal**.

9. Disable **Ignore Multiple Instances**.

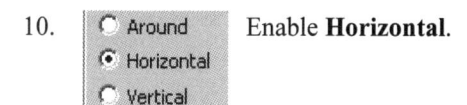

10. Enable **Horizontal**.

11. Press the **Select Placement** button to place the balloons.

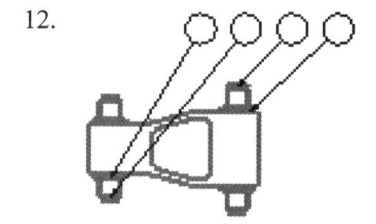

12. Pick a point above the view to place the balloons.

    Press **Apply** and close the dialog.

6-25

13. Balloons are added to each part.

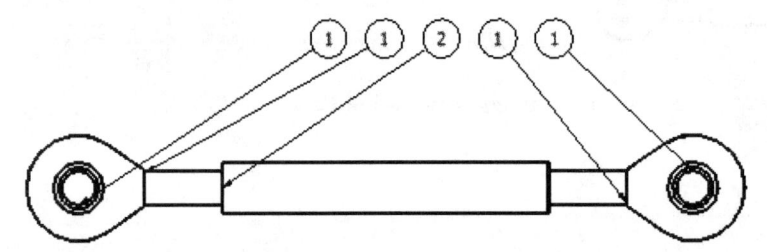

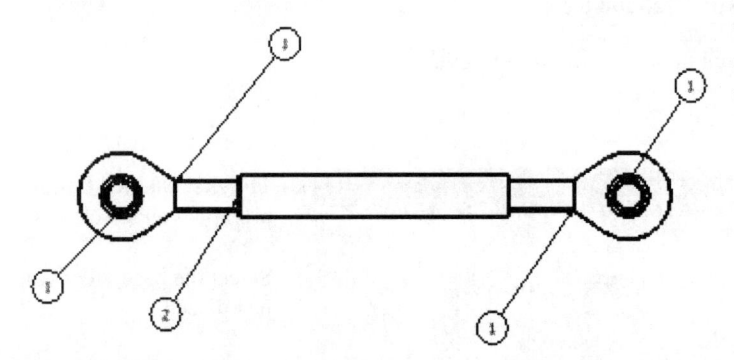

14. To move the balloons, select to activate the grips and move to the desired location.

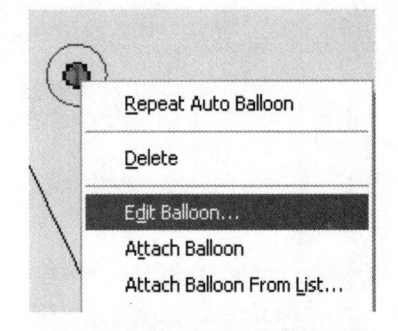

15. Select an **Item 1** balloon.
    Right-click and select **Edit Balloon**.

16. Under **Balloon Value**:
    In the **Override** column, enter **1 (4X)**.

    Press **OK**.

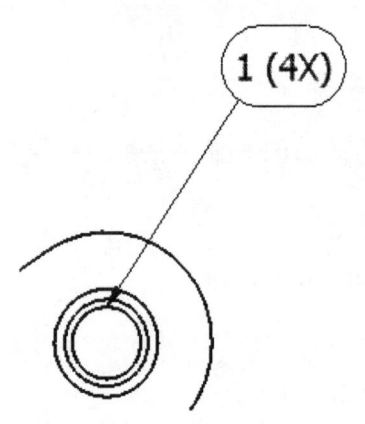

17. The balloon expands for the additional note.

Drawing Annotation

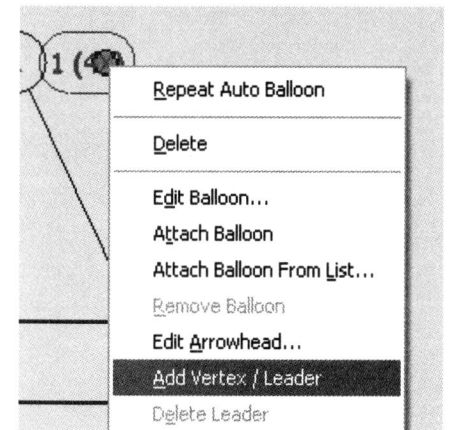

18. Select the **Item 1** balloon.

    Right-click and select **Add Vertex/Leader**.

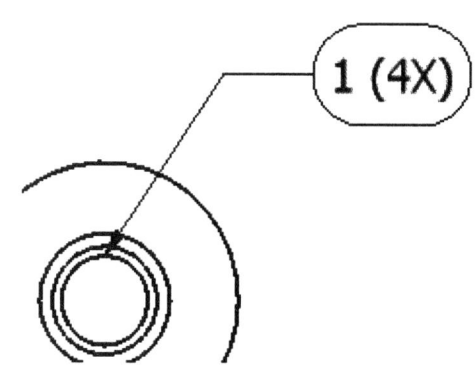

19. Pick the leader line to add a point to the line.
    A point is added to create a shoulder.

    Activate the balloon grips, then move the balloon to position the shoulder.

20. To delete a balloon, simply select, right-click and select **Delete**.

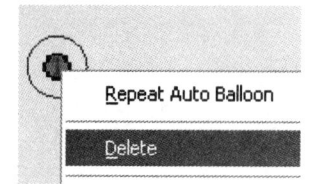

21. Delete the remaining Item 1 balloons.

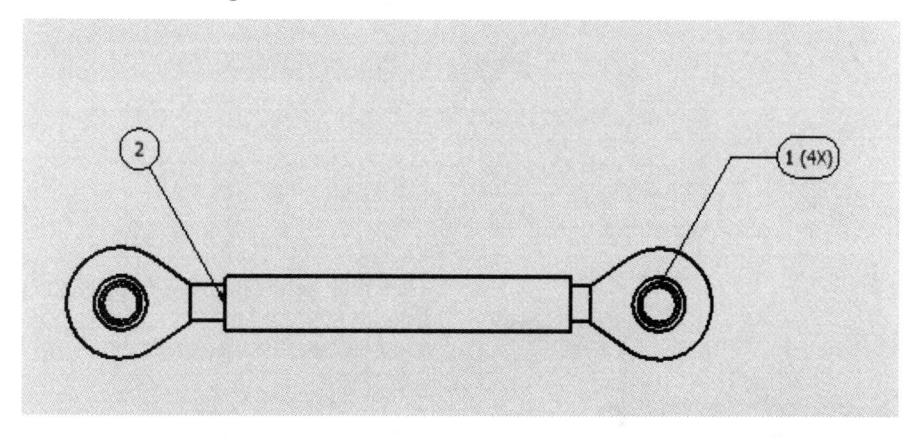

22. Save the file as *ex6-9.idw*.

Exercise 6-10
Adding a Parts List

File: Ex6-9.idw
Estimated Time: 30 minutes

1. Open *ex6-9.idw* or continue working in the previous file.

2. Select the **Parts List** tool.

3. Select the view.

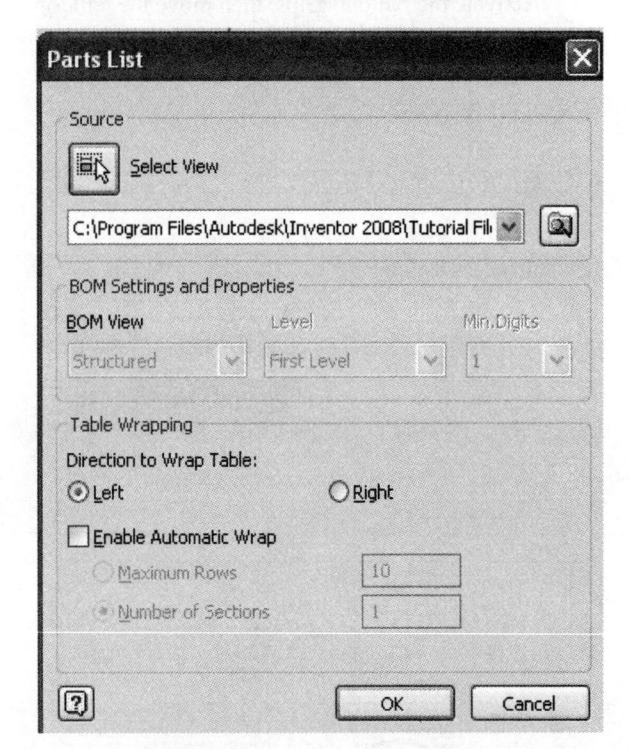

   Press **OK**.

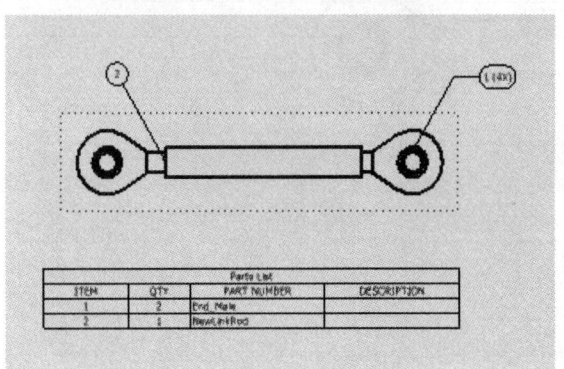

   Then pick below the view to place the parts list.

6-28

Drawing Annotation

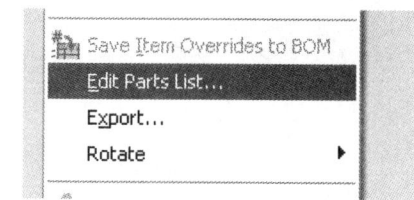

4. Select the **Parts List**.

   Right-click and select **Edit Parts List**.

5. Right-click on the **+/-** button in the table.

   Select **Table Layout**.

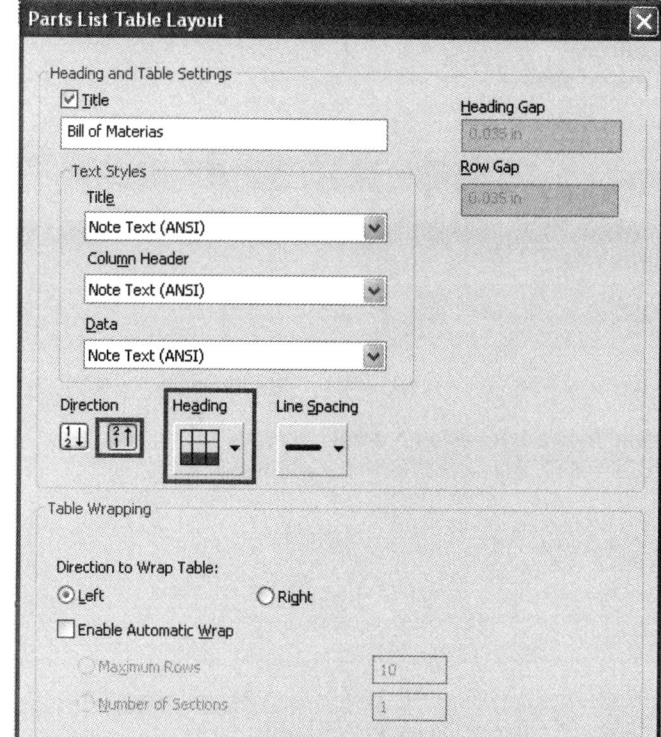

6. Change the title to **Bill of Materials**.

   Enable **Bottom** under **Heading**.

   Set the **Table Direction** to **Descending**.

   Press **OK**.

7. Press **OK** to exit the 'Edit Parts List' dialog.

8.

   The Parts List changes to show the header on the bottom and the title is updated.

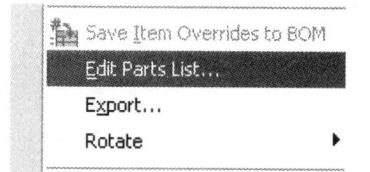

9. Select the parts list again.
   Right-click and select **Edit Parts List**.

10. Select the **Column Chooser** tool.

6-29

The Power of Design: An Introduction to Autodesk Inventor

11.

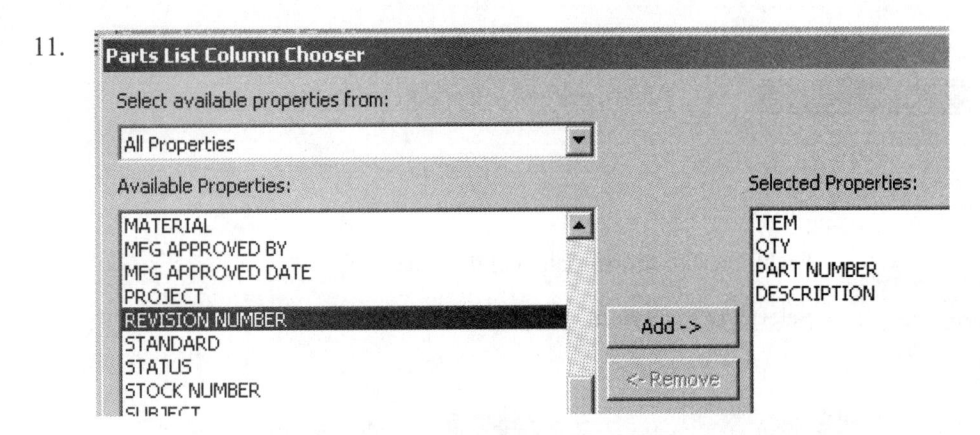

Highlight **Description** in the right column.
Highlight **Revision Number** in the left column.

Select the **Add** button.

12. Highlight **Revision Number**.
Use the **Move Up** button to position **Revision Number** above **Description**.

Press **OK**.

13. Highlight the **Revision Number** column.
Right-click and select **Format Column**.

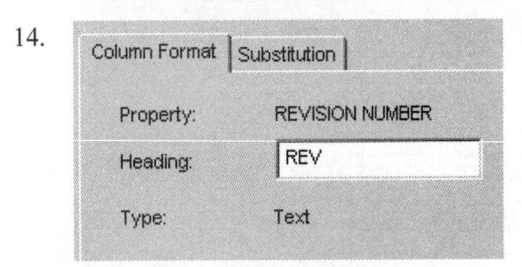

14. Change the name to **REV**.

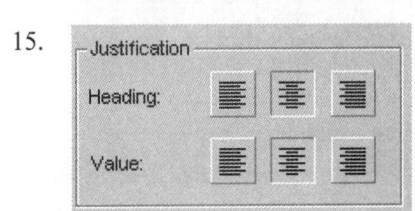

15. Change the **Justification** for the **Value** to **Centered**.

6-30

Drawing Annotation

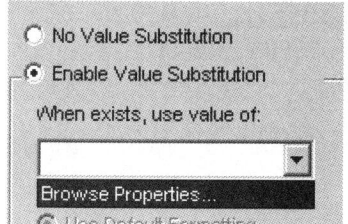

16. Select the **Substitution** tab.
    Select **Browse Properties** from the drop-down.

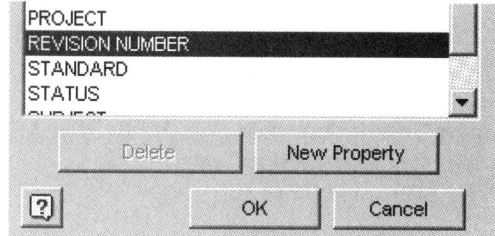

17. Select **Revision Number** from the drawing properties list.

    Press **OK** twice.

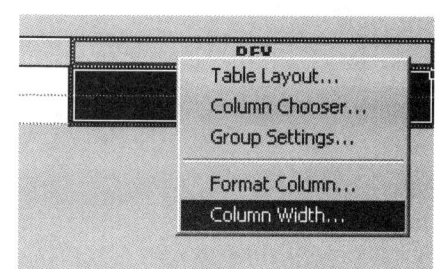

18. Highlight the **REV** column.
    Right-click and select **Column Width**.

    Press **OK**.

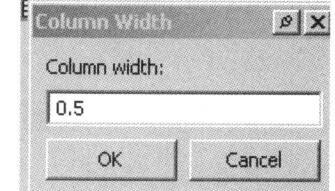

19. Set the **Column Width** to **0.5**.

    Press **OK** twice to close the dialog.

20.

ITEM	QTY	PART NUMBER	REV	DESCRIPTION
2	1	NewLinkRod		
1	2	End_Male		

    Bill of Materials

    The parts list updates to reflect the changes.

21. Go to **File→Design Assistant**.

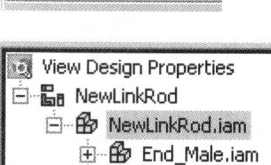

22. Expand the Parts Browser.

6-31

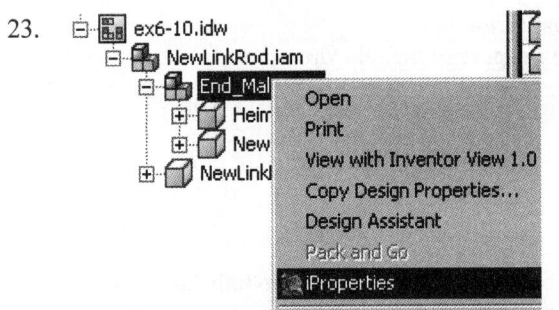

The Power of Design: An Introduction to Autodesk Inventor

23. Highlight the *End-Male.iam* part.
Right-click and select **iProperties**.

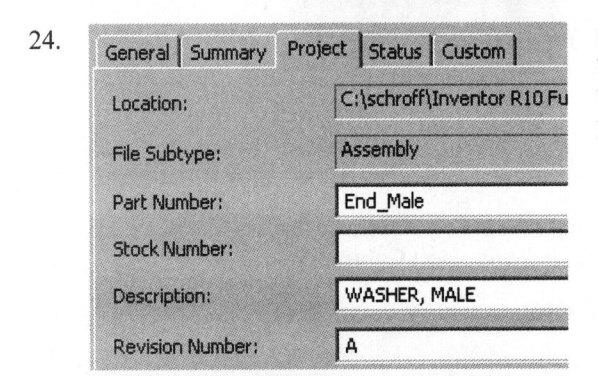

24. Select the **Project** tab.
Under **Description**, enter **WASHER, MALE**.
Under **Revision Number**, enter **A**.
Press **Apply** and **Close**.

25. Highlight the *NewLinkRod.ipt*.
Right-click and select **iProperties**.

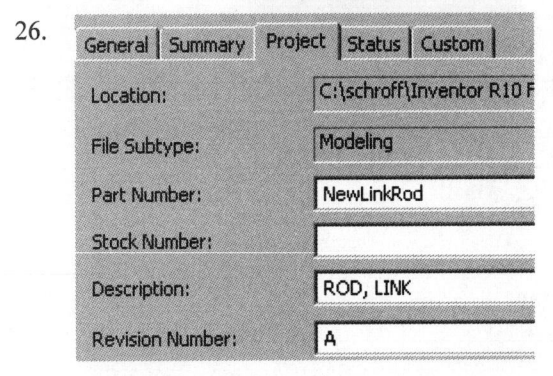

26. Select the **Project** tab.
Under **Description**, enter **ROD, LINK**.
Under **Revision Number**, enter **A**.
Press **Apply** and **Close**.

27. Highlight the assembly.

28.
Name	Part Number	Sta...	Revision Number
NewLinkRod.ipt	NewLinkRod		A
End_Male.iam	End_Male		A

The table updates.

6-32

Drawing Annotation

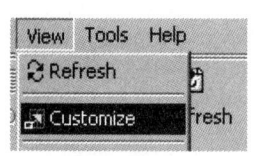

29. Go to **View→Customize**.

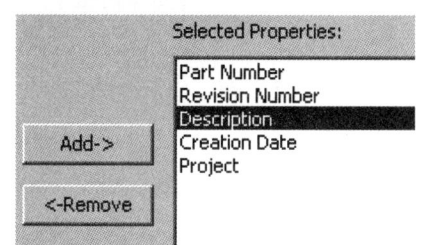

30. Add the Description field and move it into the position below revision number.

    Remove the Status and Stock Number properties.

    Press **Done**.

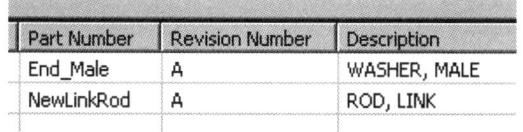

31. The table updates with the selected properties.

Part Number	Revision Number	Description
End_Male	A	WASHER, MALE
NewLinkRod	A	ROD, LINK

    NOTE: *I was not able to get the parts list to update until I selected Bill of Materials and then closed that dialog. Another method is to open the top assembly file and save.*

32. Close the Design Assistant.

2	1	NewLinkRod	A	ROD, LINK
1	2	End_Male	A	WASHER, MALE
ITEM	QTY	PART NUMBER	REV	DESCRIPTION

    Parts List

    The parts list updates.

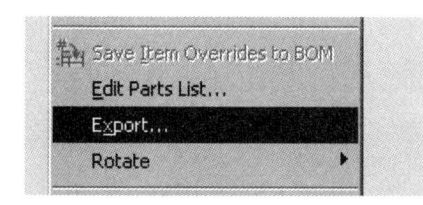

33. Select the **Parts List**.

    Right-click and select **Export**.

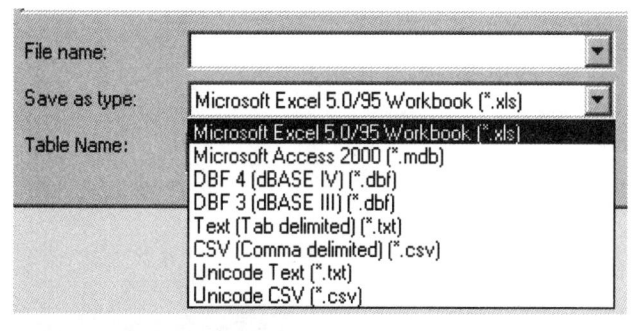

34. You can export your parts list into the following file types: mdb, xls, dbf, txt, or csv.

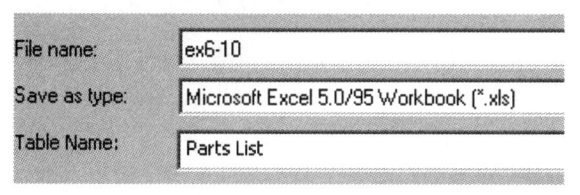

35. Save the file as *ex6-10.xls*.

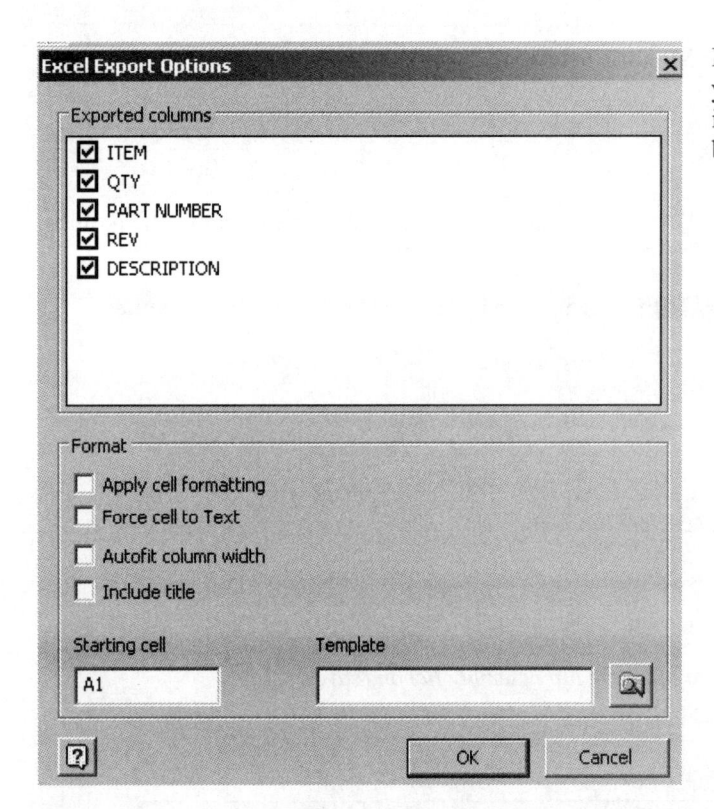

The Power of Design: An Introduction to Autodesk Inventor

If you select the **Options** button, you can select which columns to include and how the columns should be formatted.

36. 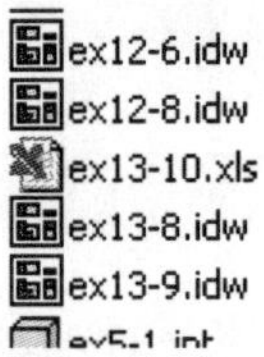 Locate *ex6-10.xls* using Explorer.
    Double click on the file name to launch **Excel**.

37. Your parts list opens as an Excel spreadsheet.

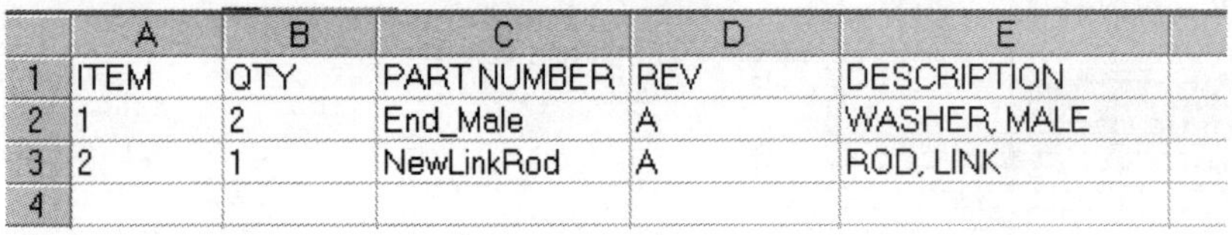

	A	B	C	D	E
1	ITEM	QTY	PART NUMBER	REV	DESCRIPTION
2	1	2	End_Male	A	WASHER, MALE
3	2	1	NewLinkRod	A	ROD, LINK
4					

NOTE: You must have Excel installed for this part of the exercise to work.

Close the Excel spreadsheet without saving.

38. Save the Inventor file as *ex6-10.idw*.

6-34

Exercise 6-11
Adding a Hole Chart

File: Ex6-7.idw
Estimated Time: 15 minutes

1. Open *ex6-7.idw*.

2. Select the **Hole Table-Selection** tool.

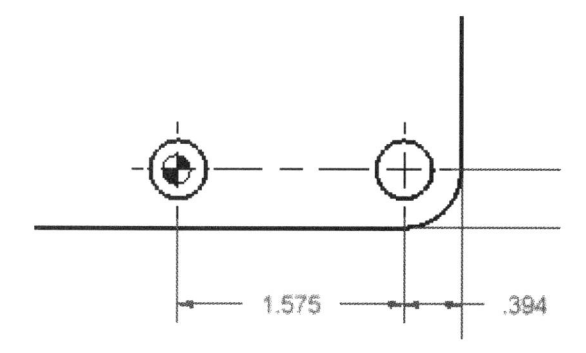

3. Select the center of the lower left hole for the hole table origin, then select all four holes.

   Right-click and select **Create**.

   Pick to place the Hole Chart.

4. The Hole Chart is added and the holes are each labeled.

5. You can control the visibility of the origin and the hole labels.

   Select the **Hole Chart**, right-click, and disable the **Visibility** of the **Origin**.

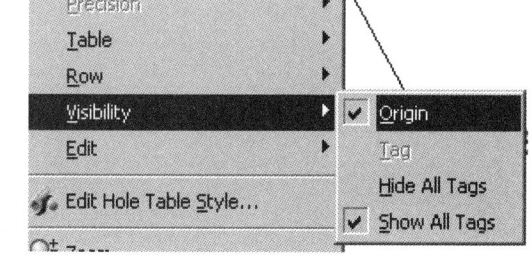

   We no longer see the Origin symbol.

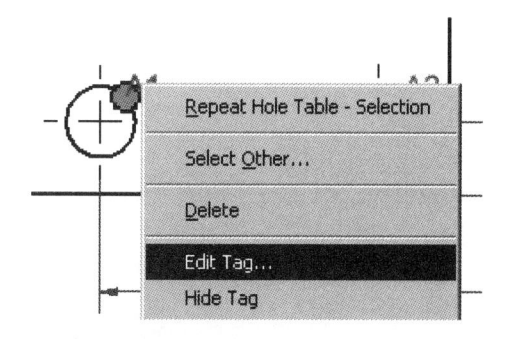

6. Select the **A1** hole label.
   Right-click and select **Edit Tag**.

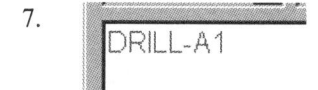

7. Type the word '**DRILL-**' in front of the default label.
   Press **OK**.

   This only changes a single hole label.

6-35

The Power of Design: An Introduction to Autodesk Inventor

8. Select the **Hole Table**.
    Right-click over the **A-2** cell and select **Edit→Edit Tag**.

			DESCRIPTION
DRI			Ø10.00 THRU
			Ø10.00 THRU
A3	1.57	3.15	Ø10.00 THRU
A4	0.00	3.15	Ø10.00 THRU

    Type the word '**DRILL-**' in front of the default label.

    Press **OK**.

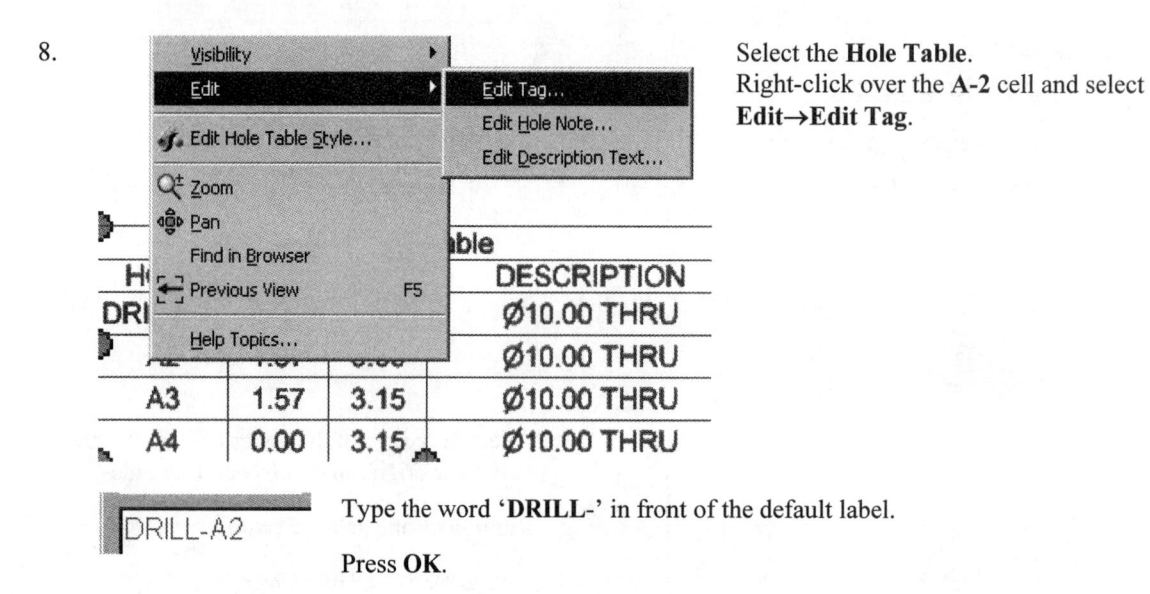

9. Save the file as *ex6-11.idw*.

TIP: You can use the grips on the Hole Tag to reposition or add a leader to the Hole Tag.

6-36

Exercise 6-12
Adding Revision Tags and Tables

File: Ex6-11.idw
Estimated Time: 15 minutes

1. Open *ex6-11.idw*.

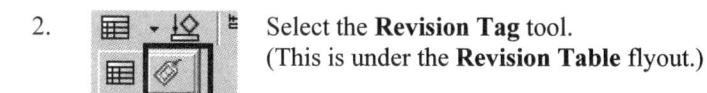

2. Select the **Revision Tag** tool.
   (This is under the **Revision Table** flyout.)

3. Place the tag next to the hole.

   Right-click and select **Continue**.
   Right-click and select **Done**.

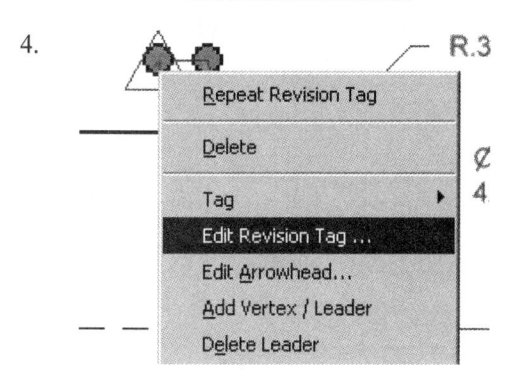

4. Select the tag.
   Right-click and select **Edit Revision Tag**.

5. Change the tag to **E-1**.

   Press **OK**.

6. The tag updates.

7. Select the **Revision Table** tool.

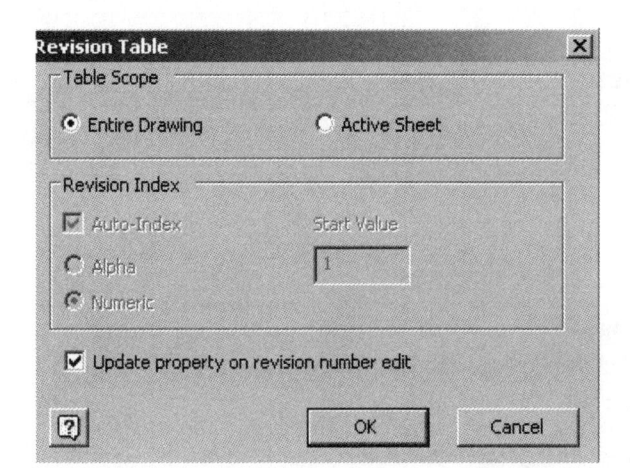

The Power of Design: An Introduction to Autodesk Inventor

Press **OK**.

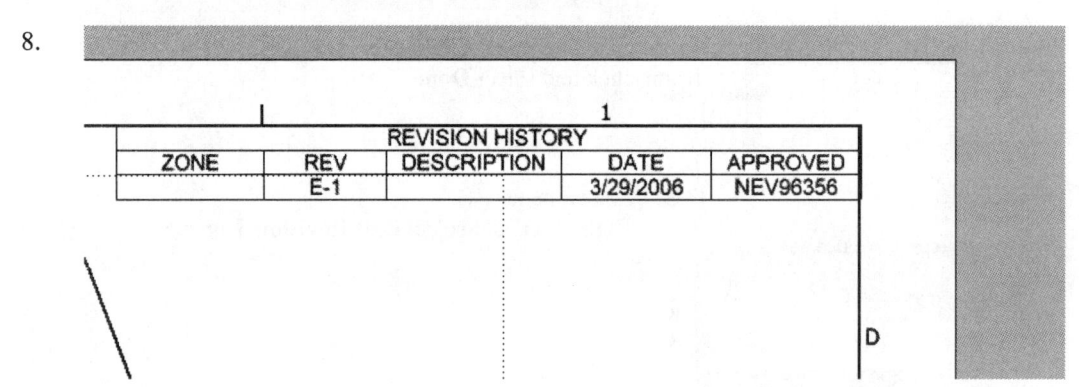

8. Place the **Revision Table** in the upper right corner of the sheet.

9. Select the **Revision Table**.
   Right-click and select **Edit**.

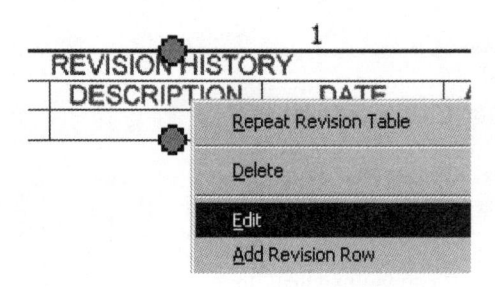

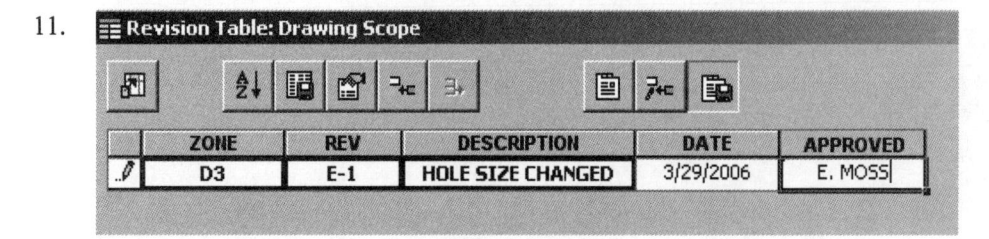

10. Press **OK**.

11. Edit the Description field to read **HOLE SIZE CHANGED**.
    Edit the values in the **Zone**, **Date**, and **Approved** columns.
    The Revision Table updates.
    Exit the dialog.

Drawing Annotation

12. Select the **Revision Table**.

Right-click and select **Edit Revision Table Style**.

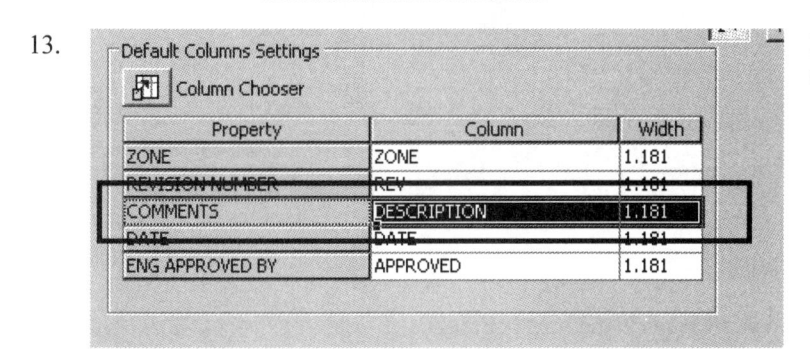

13. Left pick on the **Description** column.

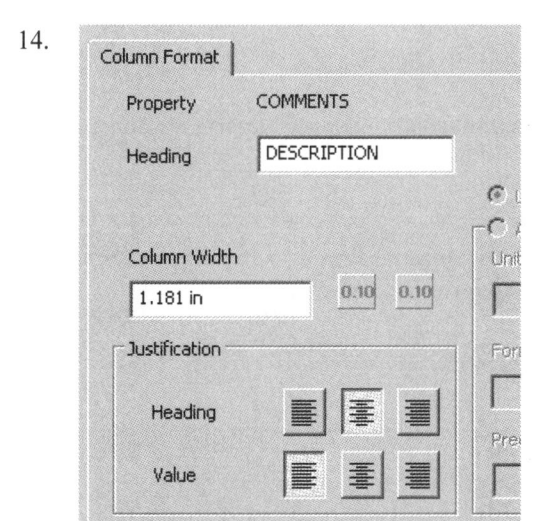

14. Change the **Value** to **Left Justified**. Press **OK** and **Done**.

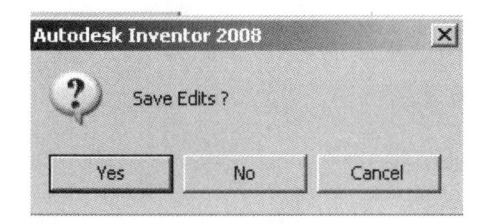

Press **Yes**.

15. Save as *ex6-12.idw*.

6-39

Review Questions

1. You can select and delete only one ordinate dimension created using the Ordinate Dimension Set tool.

    A. True
    B. False

2. When placing ordinate dimensions using the Ordinate Dimension Set tool, you can switch between the two axes – that is, place a dimension along the x-axis, place a dimension along the y-axis, then place a dimension along the x-axis.

    A. True
    B. False

3. You can not hide the Origin Indicator placed with the Ordinate Dimension tool.

    A. True
    B. False

4. Inventor only has one style of Datum Target.

    A. True
    B. False

5. To modify a linear dimension:

    A. Select the dimension, right-click and select 'Edit Dimension'
    B. Select Edit from the Menu.
    C. Double click on the dimension
    D. Right-click in the graphics window and select 'Edit Dimension'.

6. When you place dimensions on elements in a custom title block, you need to hide the dimensions when you insert the title block into a drawing.

    A. True
    B. False

7. The Projected View tool can create an isometric view.

    A. True
    B. False

8. Isometric views are aligned with the base/primary view.

    A. True
    B. False

9. You can use the Design View window dialog to create a base view.

    A. True
    B. False

10. To have model dimensions automatically appear when creating a view:

    A. Enable 'Get Model Dimensions' in the Create View dialog box.
    B. Enable 'Get Model Dimensions' in the Drawing Options dialog.
    C. Enable 'Get Model Dimensions' in the Browser
    D. A & B, but NOT C

11. To place views so they are not aligned, hold down this key as you move and place views:

    A. CONTROL
    B. TAB
    C. SHIFT
    D. ALT

12. The draft view is used to:

    A. Create draft views
    B. Redline a drawing
    C. Create model geometry
    D. All of the above

13. Drawing Resources include all of the items listed below EXCEPT:

    A. Sheet Formats
    B. Borders
    C. Sketch Tools
    D. Title Blocks

14. You can insert the following image type file into a title block:

    A. JPEG
    B. BMP
    C. PCX
    D. GIF

15. You can define property fields for a title block. A Model Properties property fields uses the data stored here:

    A. Under File Properties
    B. Under Model Properties
    C. Under Sheet Properties
    D. Under Design Properties

ANSWERS: 1) A; 2) B; 3) B; 4) B; 5) A; 6) B; 7) A; 8) B: 9) A; 10) D; 11) A; 12) B: 13) C; 14) B; 15) A

NOTES:

Lesson 7
Assembly Tools

Learning Objective

At the conclusion of this lesson, the user will have a good overall understanding of the tools used in constraining and managing assemblies.

The Assembly tool bar is divided into three sections: Component, Assembly Constraints, Component Management, and Component Viewing.

NOTE: The toolbar shown is for Inventor Professional. Some of the tools are not available in Inventor Standard.

Assembly components can be individual parts or subassemblies that behave as a single unit. For example, a single-part base plate and a multi-part air cylinder subassembly are both components when placed in an assembly.

To make sure that they are always available when you open the assembly, add the paths for all components to the project file for the assembly.

The behavior and characteristics of a component depend on its origin.

A Mechanical Desktop part placed as a component in an Autodesk Inventor assembly acts much like any assembly component. You can add assembly constraints, set its visibility, and perform other assembly operations. However, you cannot edit the part in Autodesk Inventor.

Each Mechanical Desktop part is linked to the assembly through a special file called a proxy file. The proxy file contains the linking information so that the assembly component updates when you edit the part in Mechanical Desktop.

Parts or subassemblies created using another CAD system can be inserted as components in the active assembly. You cannot change the size or shape of external components, but you can customize them by adding features.

Adaptive parts can change size and shape to satisfy assembly design requirements. When an adaptive part is constrained to other assembly components, underconstrained geometry in the adaptive part resizes.

When a part is first placed in an assembly, it is not defined as adaptive in the assembly context. You can create fixed-size geometry, and then place the part in an assembly. Select one occurrence in the assembly and designate it as adaptive.

Most assemblies contain a combination of existing components and components (parts and subassemblies) created in the assembly environment.

When you create components in place, you can use geometry from other parts (such as edges and hole centers) in feature sketches. Parts based on existing geometry are sized and positioned in relation to that geometry. Parts created in place have an automatic mate constraint applied between the part XY sketch plane and the part face you sketch on. You can define a part created in place as adaptive so that its size and shape can adjust as assembly requirements change.

Any part in an assembly may have all of its degrees of freedom removed and be fixed in position, relative to the assembly coordinate system. The origin of a grounded part will not move when you place assembly constraints, but a grounded part can still be designated as adaptive. The features on a grounded, adaptive part can change size or shape although its position is fixed.

> **TIP:** If you make extensive changes to any component in an assembly, some assembly constraints may not compute correctly when the Autodesk Inventor assembly file is updated. These constraints must be recreated.

> **TIP:** The first component placed in an assembly is automatically grounded, so that subsequent parts may be placed and constrained in relation to it. If necessary, you can remove the grounded status of a part.

Placing the First Component

The first component placed in an assembly should be a fundamental part or subassembly, such as a frame or base plate, on which the rest of the assembly is built.

The first component in an assembly file sets the orientation of all subsequent parts and subassemblies. The part origin is coincident with the origin of the assembly coordinates and the part is grounded (all degrees of freedom are removed).

If necessary, you can restore degrees of freedom to the grounded part (the base component) and reposition it. Any components you have constrained to it will also move.

Although there is no distinction in an assembly between components, you can think of the first component you place as the base component because it is usually a fundamental component to which others are constrained. If you place a first component and then want to change to a different base component, you can place a new component, specify it as grounded, and then reconstrain any components you placed earlier, including the first component. Right-click on the first component, clear the Grounded check box, and then constrain it to the new base component.

There is no limit to how many components can be grounded, but most assemblies have only one grounded component. Grounded components are appropriate for fixed objects in assemblies because their position is absolute (relative to the assembly coordinate origin) and all degrees of freedom are removed. Grounded components have no dependencies on other components.

Assembly Tools

You can use the Move button on the Assembly toolbar to relocate the grounded component. You can drag the component to its new position. The component is grounded in the new location.

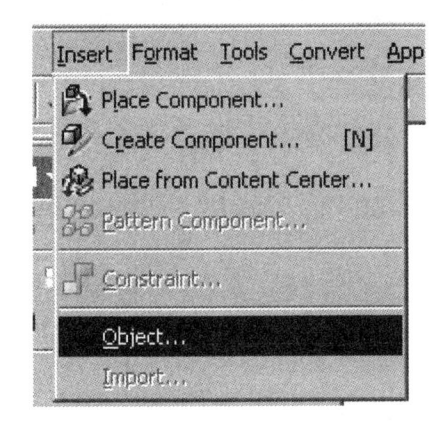

Assemblies can also be managed from the menu. You can place an Existing Component, Create a New Component, Access the Standard Parts Library using Shared Content, Pattern a Component, or Add a Constraint under the Insert menu.

Tips & Tricks

- Double-click on top of the part you wish to edit to activate Part Edit mode in an assembly file.
- A mate constraint is automatically placed between the new sketch and the face or work plane. To omit this constraint, clear the check box in the Create Part In-Place dialog box when you create the part file.
- You can set options on the Adaptive tab of the Options dialog box to control feature termination.
- Place parts and subassemblies in the order in which they would be assembled in manufacturing.
- You can drag and drop a component from Windows Explorer or a Browser window, but the component includes undisplayed default work planes that may offset the part from the cursor. Drag and drop places a single instance, unlike multiple occurrences as described above. Whether you place components through the dialog box or drag and drop, use assembly constraints to position components and remove degrees of freedom.
- Typing a **P** will initiate the Place Component command.

Exercise 7-1
Place Component

File: New Assembly file using Standard (mm)
Estimated Time: 10 minutes

1. Go to **File→New**.

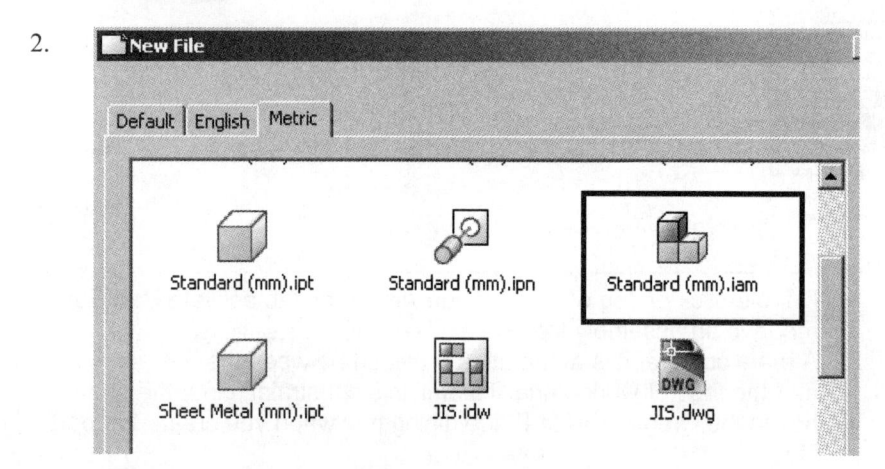

2. Select the **Metric** tab.

   Start a New Assembly file using the *Standard (mm.) iam* template.

3. Select the **Place Component** tool from the Panel Bar.

4. Browse to the *Samples\Models\Assemblies\Scissors\Components* folder under the Inventor installation in Program files.

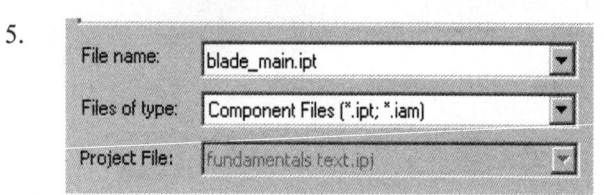

5. Locate *the blade_main.ipt* file.

   Press **Open**.

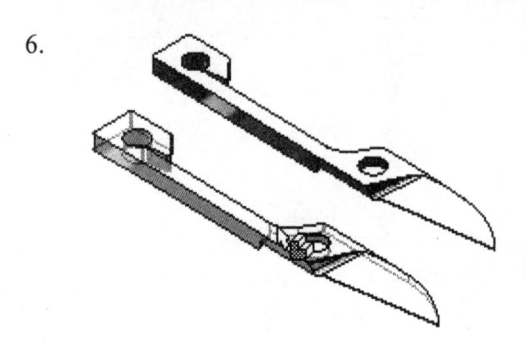

6. Inventor automatically assumes you want to place more than one instance of each selected part.

   Notice how the cursor image changes to indicate that you are in Part Placement mode.

   Left-click the mouse to insert the part into the assembly.

Assembly Tools

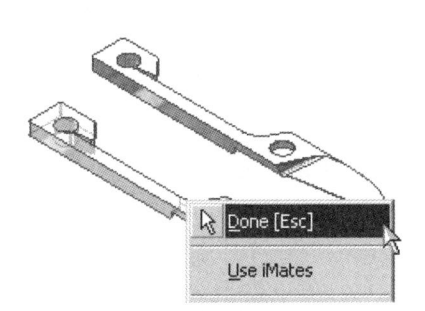

7. After placing one instance of the part, right-click and select **Done**.

8. The first part placed is automatically grounded. This is indicated by the pushpin next to the part name in the Browser.

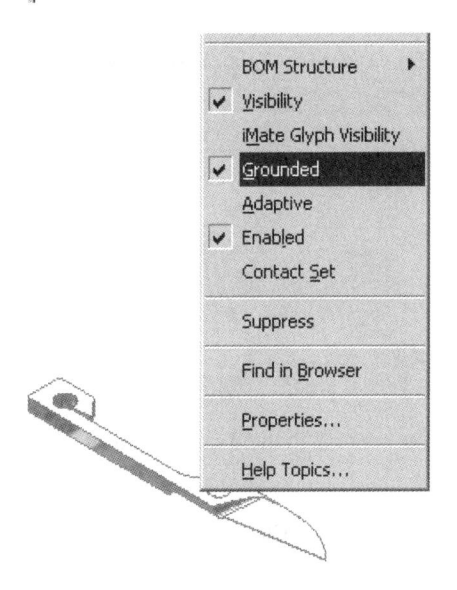

Grounded means the part is fixed in place and cannot be moved.

To remove the ground, you can select the part in the window or in the browser, right-click and disable **Grounded** on the right-click menu.

9. Save the assembly file as *Ex7-1.iam*.

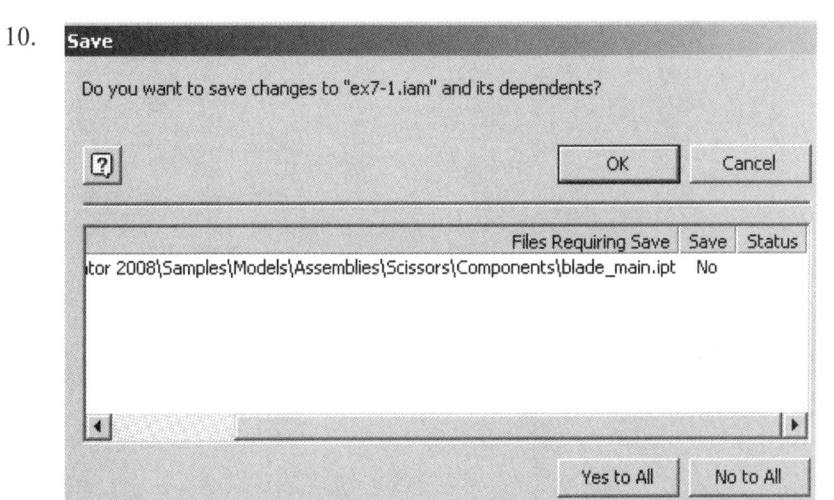

10. Press **OK**.

Note that you will not be saving any changes to the part file.

7-5

Exercise 7-2
Place Component using Windows Explorer

File: New Assembly file using Standard (inches)
Estimated Time: 10 minutes

1. Start a New Assembly file.

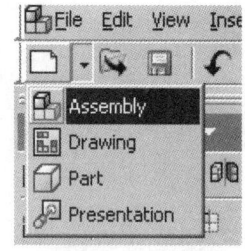

2. Open Windows Explorer.

   Locate the path:
   Program Files\Autodesk\Inventor <Version>\Samples\Models\Assemblies\Tuner\Tuner Components

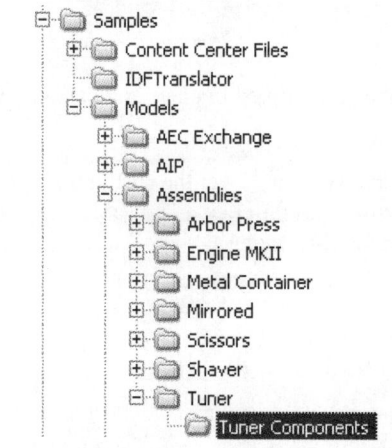

3. Select all the components in the **Tuner Components** folder.

   You can use **Edit→Select All**, Shift-Control, or hold down the Control key and pick to select all the files.

   Note: you can use the Thumbnails option in Explorer to preview the files.

Assembly Tools

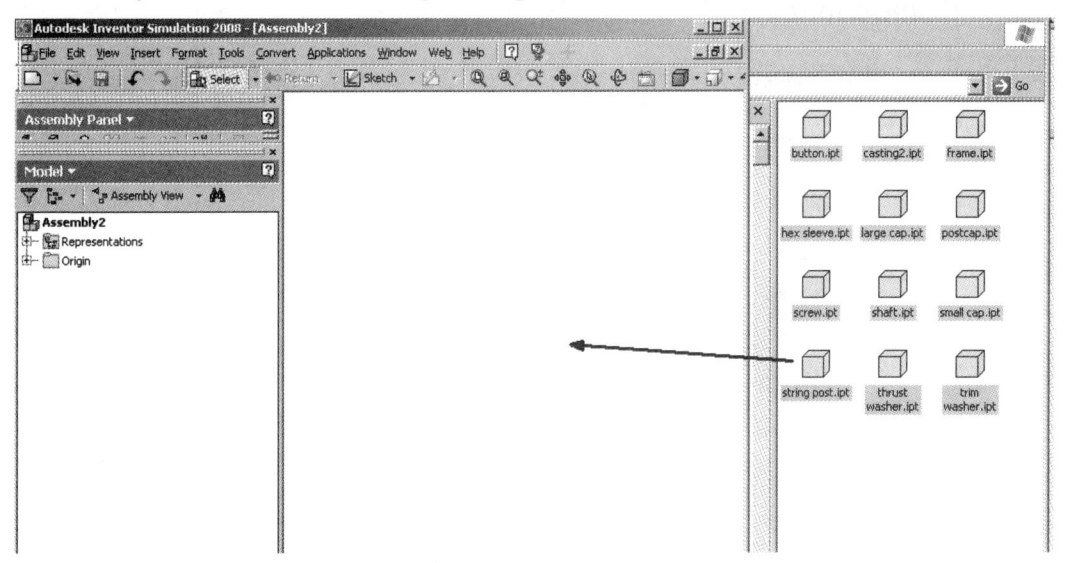

4. Hold down your left mouse button, drag and drop all the files into your assembly file.

5. All the parts are automatically inserted.

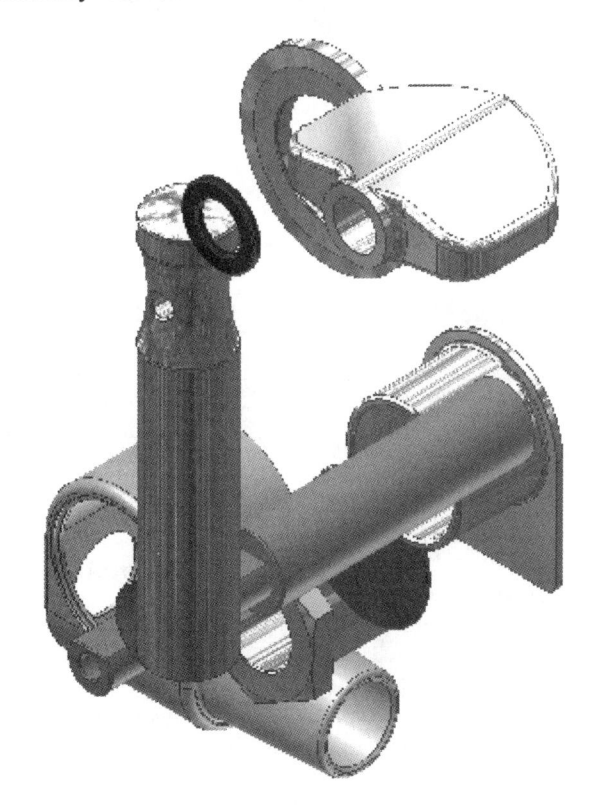

6. Save your file as *Ex7-2.iam*.

7. You may be asked if you want to save all the dependent files along with the assembly.

   Press **OK**.

8. You may be asked if you want to migrate the files from the previous release. Press **OK**.

The Power of Design: An Introduction to Autodesk Inventor

Exercise 7-3
Create Component

File: Ex7-3.iam (*download from publisher's website at www.schroff.com/resources*)

Estimated Time: 20 minutes

1. Open *ex7-3.iam* file.

2. This is the metal container included in the Samples folder. The part file is *ebox.ipt*.

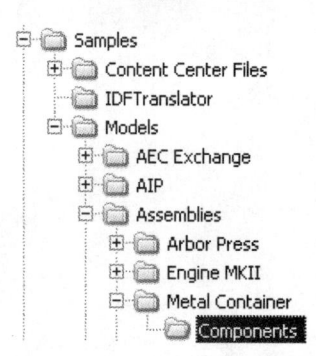

   If you get an error message that Inventor is unable to locate the file, browse to the *Samples\Models\Assemblies\Metal Container\ Components folder* under Inventor.

   Locate the file and press **Open**.

3. Select the **Create Component** tool.

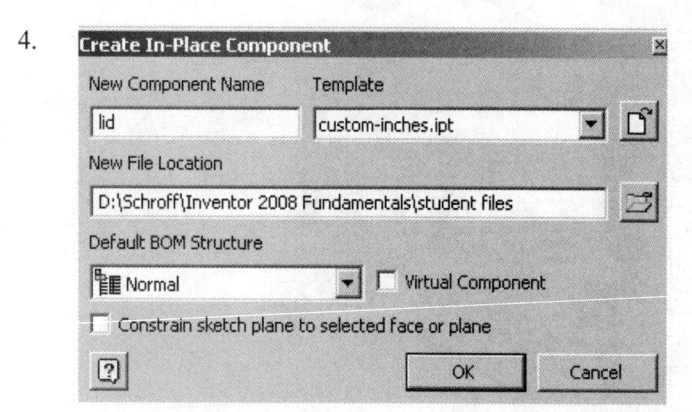

4. Enter *rail* for the new file name.

   Set the template to use *custom-inches.ipt*. It should be available from the drop-down list.

   Disable the **Constrain** sketch plane option.

   Press **OK**.

   Virtual Component is used when you want to create a component, like adhesive or lubricant, that you would like to appear on the bill of materials, but is not an actual component.

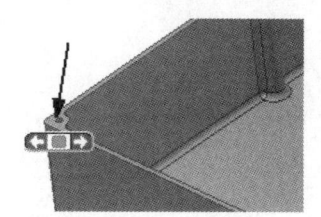

5. Select the top face of the *ebox* as the sketch plane.

6. Select **Project Geometry**.

7-8

Assembly Tools

7. Select the top of the lid face to project all the geometry including the holes.

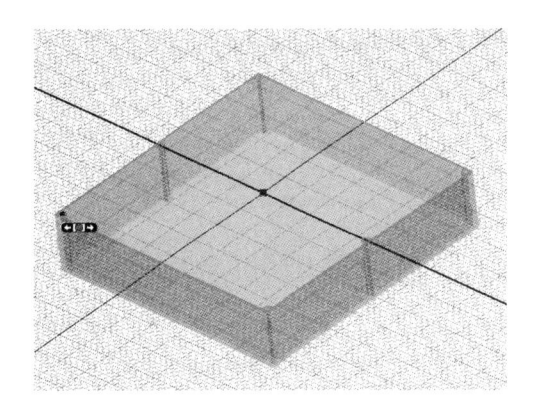

8. Look in the Browser and expand the sketch.
The sketch is shown as Adaptive. This means that it is linked to the geometry of the bottom part.
The Reference shown indicates the projected geometry.

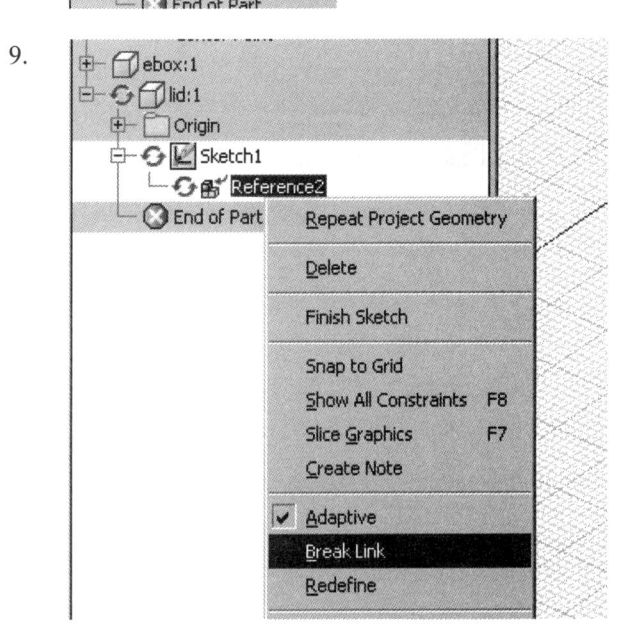

9. Selecting the reference in the browser can break the link between the projected geometry and the new part. Right-click and select **Break Link**.

If you make edits to the source part that affects the projected geometry, it can cause an error with the dependent part. In order to eliminate potential future errors, I will usually break the link and remove the adaptivity.

10. Select the drop-down arrow next to the **Return** button.

Note that there are two Return paths, one is for the Parent (this is the part where the sketch resides) and one to the Top (this is the top assembly).

11. Selecting the Return button brings you up one level back to the part environment.
Select the **Return** button.

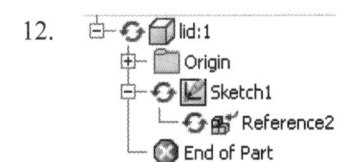

12. In the Browser, you see that the part is highlighted and the remainder of the assembly is grayed out. The Panel Bar indicates that you are in Part Features mode.

13. Select the **Extrude** tool.

7-9

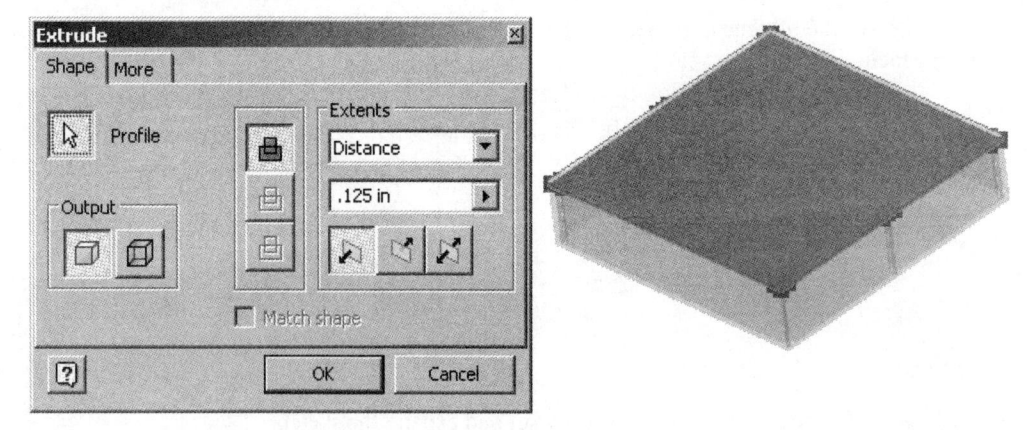

14. Extrude as a solid rectangle. Be sure to select the circles and arcs to be filled in.

    Set the distance to **.125**.

    Press **OK**.

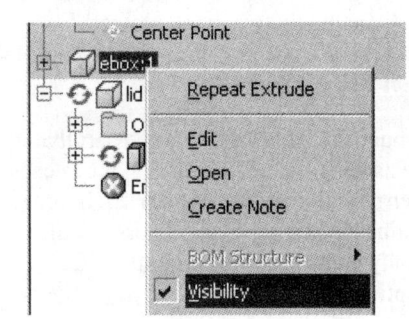

15. Highlight the *ebox*.

    Right-click and disable **Visibility**.

    This will turn off the display of the bottom component. This does not delete the component from the assembly.

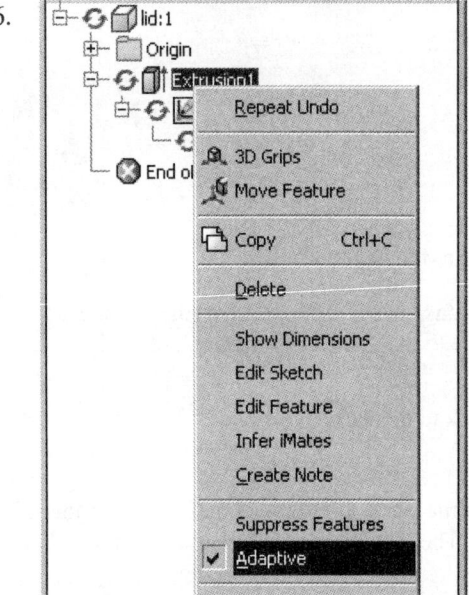

16. Highlight the **Extrusion** for the lid.

    Right-click and disable **Adaptive**.

    Adaptive means that some of the features of the adaptive part are dependent on a source part.

17. The Extrusion no longer is indicated as Adaptive.

    Note that the part is still considered Adaptive.

7-10

Assembly Tools

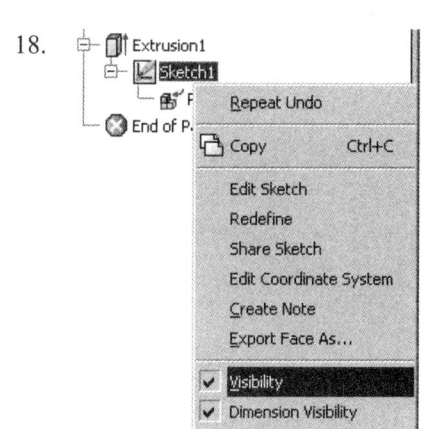

18. Set the **Sketch** under the **Extrusion** with **Visibility** turned on.

19. Select the **Extrude** tool.

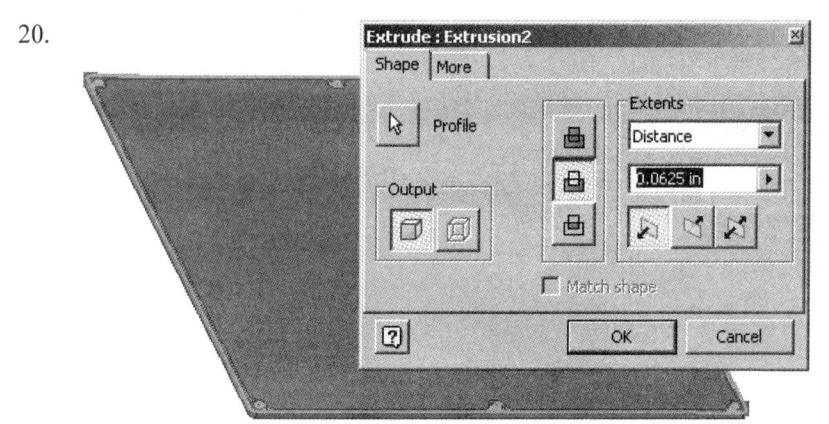

20. Select the inside of the sketch as the profile.

    Set the extrude as a **Cut**. Set the distance to **.0625**. Change the direction so it goes into the part.

    Press **OK**.

21. Select the **Hole** tool.

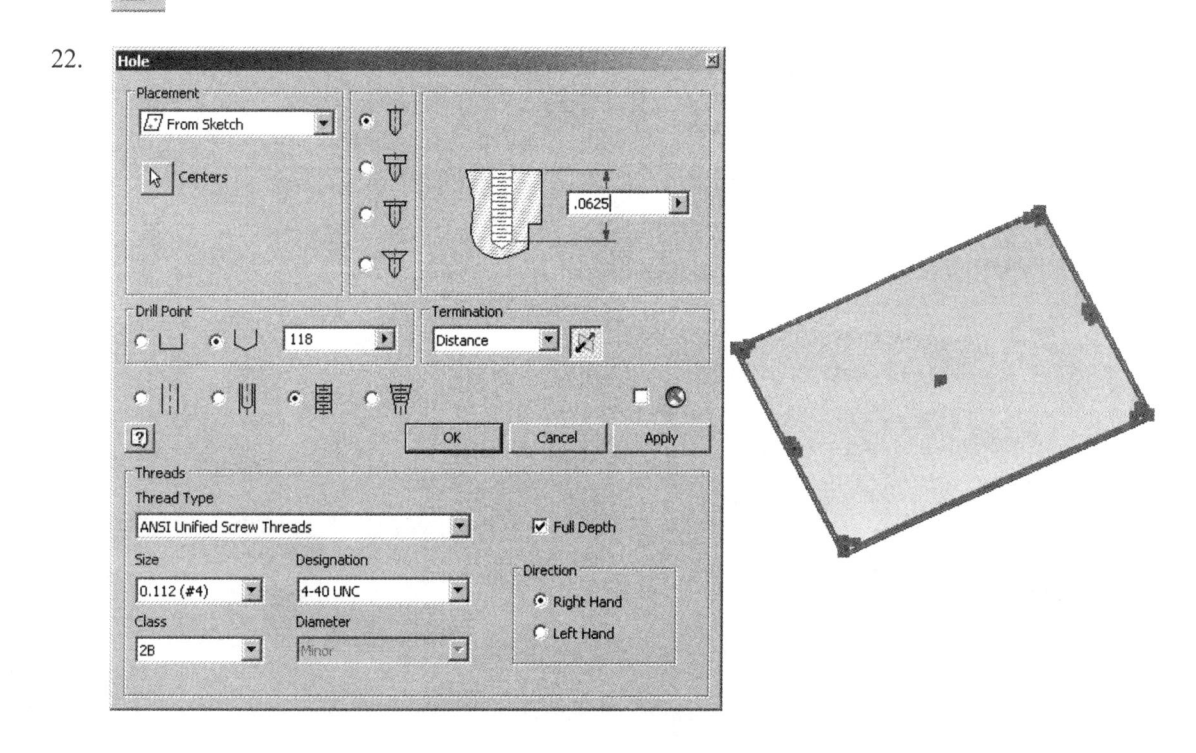

22.

7-11

Set the Placement as **From Sketch**.
(You also could select Concentric placement and select each cylindrical face)

Set the hole distance to **.0625**.
Enable **Tapped**.

Set the hole size to **40-40 UNC**.
Enable **Full Depth**.

Verify that the direction of the hole is going into the part.

Press **OK**.

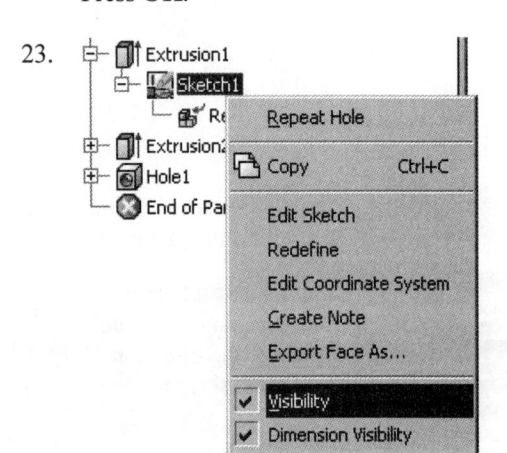

23. Turn off the **Visibility** of the sketch you used to create three different features.

24. Turn on the **Visibility** of the *ebox*.

25. Press **Return** to return to the Assembly environment..

26. Save as *ex7-3a.iam*.

Place Constraint

Assembly constraints determine how components in the assembly fit together. As you apply constraints, you remove degrees of freedom, restricting the ways components can move.

To help you position components correctly, you can preview the effects of a constraint before it is applied. After you select the constraint type, the two components, and set the angle or offset, the components move into the constrained position. You can make adjustments in settings as needed, then apply it.

Assembly constraints eliminate the degrees of freedom for each component. You can use **View→Degrees of Freedom** to turn on symbology to indicate the remaining degrees of freedom for a part. Each assembly constraint removes one or more degrees of freedom.

Constraints are used when creating exploded views and motion simulations, so the types of constraints applied should reflect the way an assembly will actually be built.

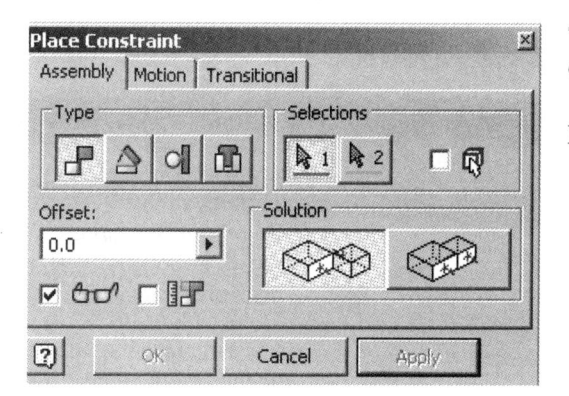

The Place Constraints dialog box creates constraints to control position and animation.

Motion constraints do not affect position constraints.

The Assembly tab has constraints to control position. The four types of assembly constraints are: Mate, Tangent, Angle, and Insert.

- A **mate** constraint positions selected faces normal to one another, with faces coincident or aligns parts adjacent to one another with faces flush. The faces may be offset from one another.
- An **angle** constraint positions linear or planar faces on two components at a specified angle.
- A **tangent** constraint between planes, cylinders, spheres, and cones causes geometry to contact at the point of tangency. Tangency may be inside or outside a curve.
- An **insert** constraint positions cylindrical features with planar faces perpendicular to the cylinder axis.

To create a complex assembly, create several small assemblies and save each one as a separate file. Combine them in larger assemblies, constraining them to other subassemblies and parts as a single unit.

Group parts in subassemblies if you want to use them in more than one assembly. Modify small subassemblies or regroup parts to change assembly configuration.

The Power of Design: An Introduction to Autodesk Inventor

Exercise 7-6
Place INSERT Constraint

File: Ex7-6.iam
Estimated Time: 10 minutes

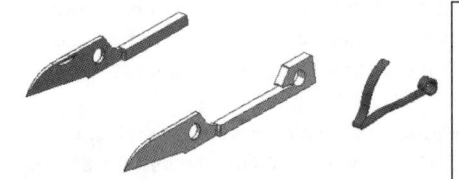

This file can be downloaded from the publisher's website. It consists of the parts available in the *Scissors Components* folder under *Samples\Assemblies*.

1. Open the *ex7-6.iam* file.

2. You can use the **Degrees of Freedom** tool under View in the menu to help you see how your parts are constrained together.

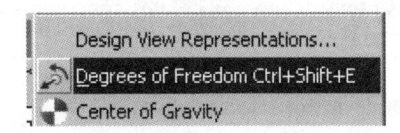

   Go to **View→Degrees of Freedom**.

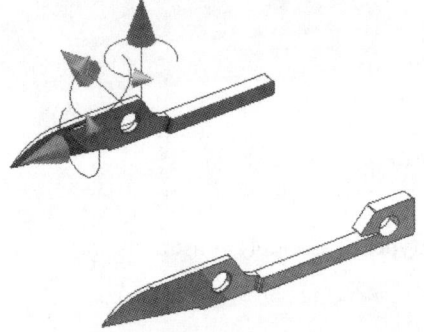

3. Turning on the Degrees of Freedom shows us how the parts are constrained

   The top blade has all degrees of freedom available as we have not added any constraints yet.

   The blade_main does not show any degrees of freedom because it is grounded.

4. Select the **Constraint** tool.

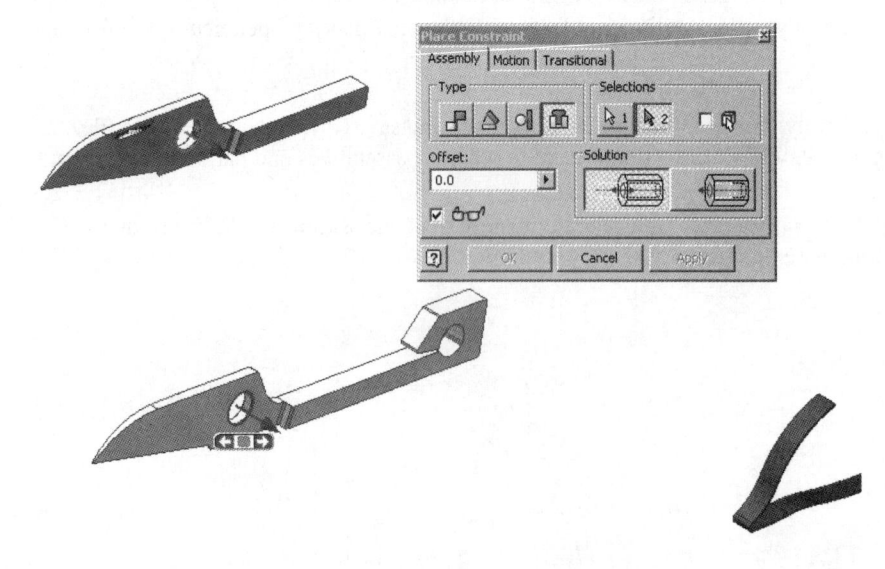

5. Select the **INSERT** type.

   Select the holes in the top blade and select the corresponding hole in the blade_main.

   Press **Apply**.

   *NOTE: You can also right-click and select **Apply**.*

7-14

Assembly Tools

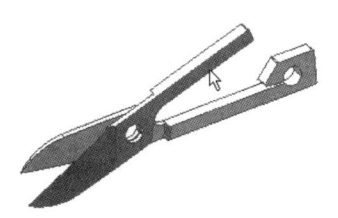

6. Place your mouse cursor on the top blade.

Hold down the left mouse button and move the blade up and down.

7. Go to **View→Degrees of Freedom**.

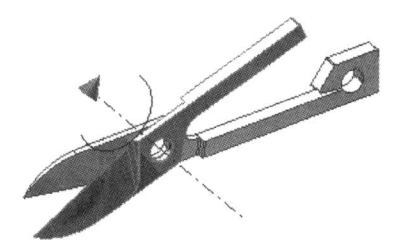

8. Note that the only degree of freedom remaining is rotation.

This is what allows you to move the blade up and down.

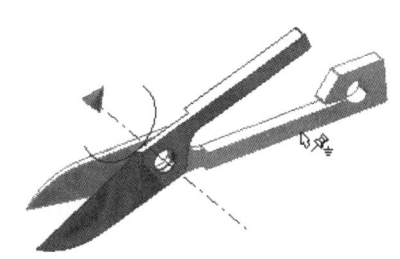

9. *Try to use your mouse cursor to move the blade_main up and down. It is grounded in place so it can not be moved.*

You will see a small push pin next to your cursor to remind you that the part is fixed in place.

10. Press **Ctl+Shift+E** to turn off the DOF display.

11. Select the **Constraint** tool.

12.

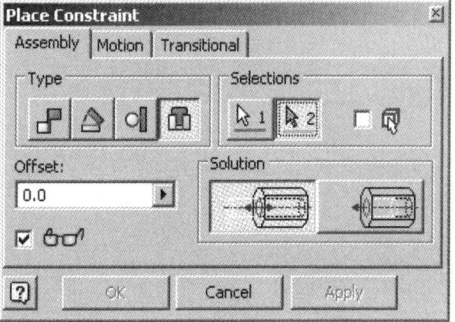

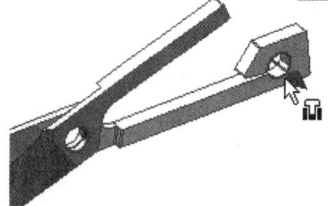

Select the **INSERT** type.

Select the hole in the **blade _main** and select the hole in the spring.

Press **Apply**.

7-15

*NOTE: You can also right-click and select **Apply**.*

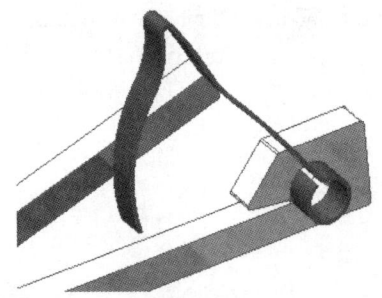

13. The spring is extruding out from the blade and we want it inserted inside of the blade.

14. Locate the **Insert** constraint for the scissor_spring in the Browser.

    Right-click and select **Edit**.

15. Select the Aligned button for the constraint.

    Press **OK**.

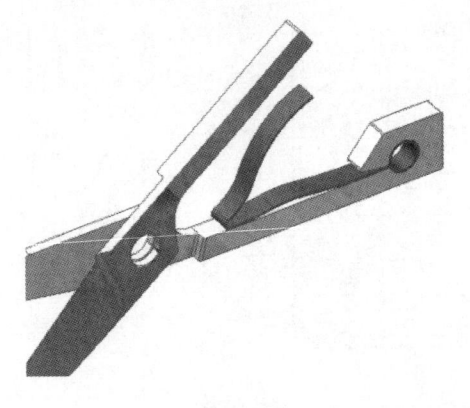

16. The spring is now inserted correctly.

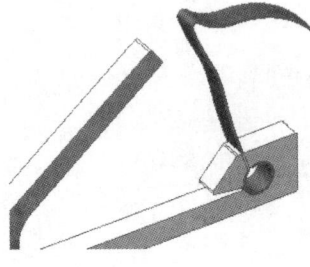

17. Use your cursor to reposition the spring and top blade if it is not in the correct position..

18. Save the file as *ex7-7.iam*.

Exercise 7-7
Place TANGENT Constraint

File: Ex7-7.iam
Estimated Time: 5 minutes

1. Open the *Ex7-7.iam* file or continue working in the open file.

2. Select the **Constraint** tool.

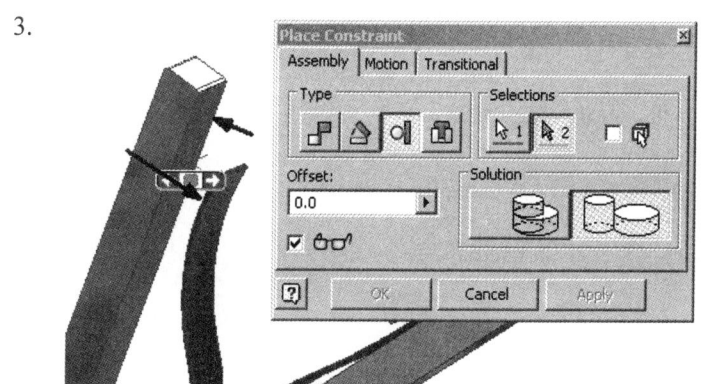

3. Select the **TANGENT** constraint.

   Set the Outside button.

   Select the top of the curved portion of the spring where it will make contact with the top blade.

   Select the bottom face of the top blade where it will contact the spring. Press **Apply**.

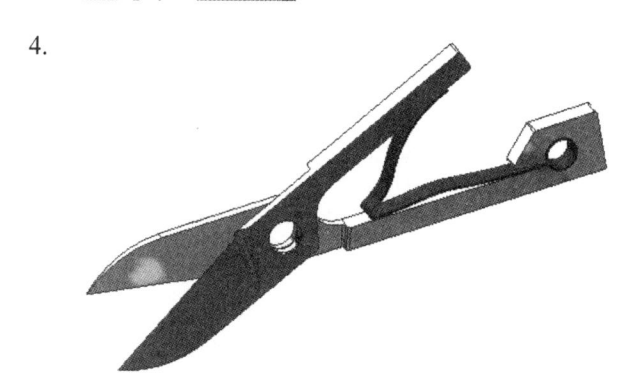

4. The scissors adjusts position.

   Use your mouse cursor to move the blade and see how the spring adjusts.

5. Save as *ex7-8.iam*.

7-17

The Power of Design: An Introduction to Autodesk Inventor

Exercise 7-8
Place MATE Constraint

File: Ex7-8.iam
Estimated Time: 10 minutes

1. Open the *ex7-8.iam* file.

2. Select the **Move** tool.

    This tool allows you to move parts in an assembly without deleting any constraints. It basically suspends the assigned constraints to allow you to further manipulate the assembly.

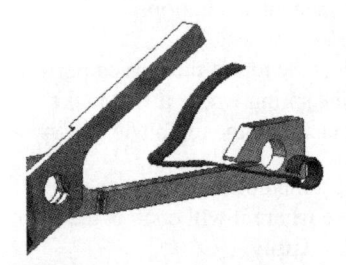

    Move the spring out of position.

3. Select the **Constraint** tool.

4. Select the face indicated of the main blade and the small flat face near the cylindrical end of the spring.

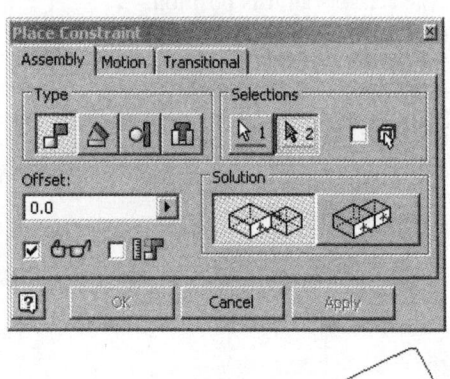

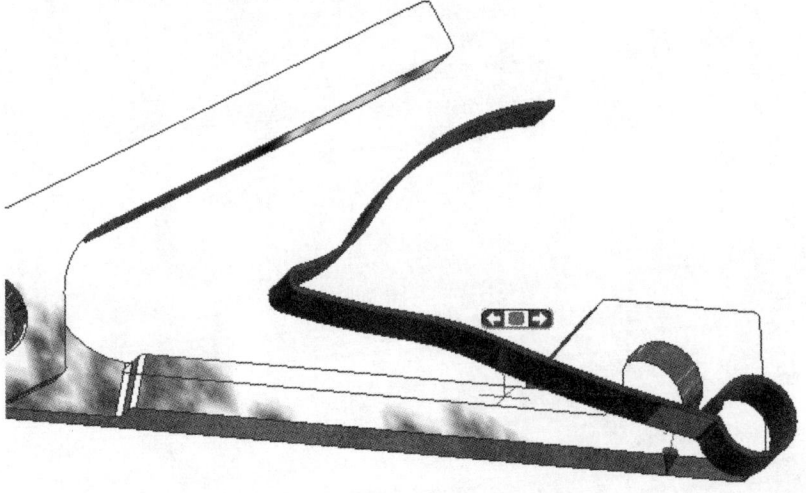

7-18

Assembly Tools

5. Use a **MATE** constraint with an offset of **0.00** to place the bracket on top of the rail.

6. Press **Apply**.

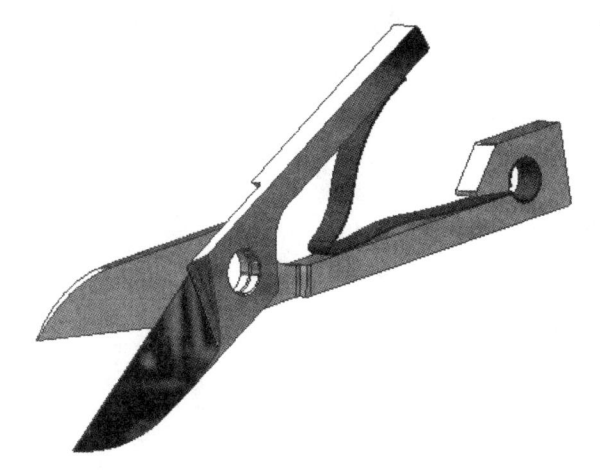

7. The spring pops into place.

8. Save the file as *ex7-9.iam*.

Tips & Tricks

➤ Select a component to be replaced either in the Browser or by picking a part in the drawing window.
➤ From Quarter section and Three-quarter section views, you can right-click and select the opposite view.
➤ It is possible to turn off component visibility, but have the component still be enabled. This may be useful for quickly removing a needed component from view. Enabled components are fully loaded in an assembly file, while only the graphic portion of components not enabled is loaded. The assembly calculates faster because the data structure of not-enabled components is not present, but its graphics are useful for a frame of reference.
➤ The adaptive status of an occurrence controls all occurrences in the assembly. When an adaptive part resizes, all occurrences of the part in other assemblies also resize.

Review Questions

1. The first component placed in an assembly is automatically _____.

   A. constrained
   B. placed on Plane XY
   C. grounded
   D. adaptive

2. A grounded part can not be made adaptive.

   A. True
   B. False

3. In a large assembly, you can ground _____ component(s).

   A. Only one
   B. Eight
   C. All
   D. Six

4. Use 'Place Component' to:

   A. Insert an existing part file into an assembly
   B. Create a new part in-place in an assembly
   C. Insert a Part from another assembly
   D. Move a component into position

5. You can drag and drop one or more part files from WINDOWS Explorer into an Inventor Assembly file.

   A. True
   B. False

6. When you create an in-place component in an assembly file, you must do all the items listed below EXCEPT:

   A. Specify a file name
   B. Specify a file location
   C. Specify a template
   D. Specify a material

7. ![Ex18-6.ipt]
   The symbol shown in front of the part name indicates that the part is:

   A. Grounded
   B. Adaptive
   C. Recycled
   D. Rotated

ANSWERS: 1) C; 2) B; 3) C; 4) A; 5) A; 6) D; 7) B

The Power of Design: An Introduction to Autodesk Inventor

Lesson 8
Presentations

Learning Objectives

The user will learn how to:

- Create a Presentation File
- Create an Exploded View
- Create an Animation

Presentation views are used to:

- Create exploded views that can be used for assembly instructions
- Create an animation to show how components interact
- Turn off the visibility of components in a large assembly to make it easier to identify components

Each presentation file can contain as many animations, exploded views, stylized views, section views, etc. that you need for an assembly. Once you set up the views, you can then insert them into your drawing layout.

NOTE: Presentation files can only link to an assembly file.

You can create exploded views either by manually moving components around or automatically expanding the distance between the parts.

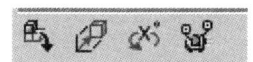

The Presentation Panel has four tools:
- Create View
- Tweak Components
- Precise View Rotation
- Animate

Tips & Tricks

➤ You can tweak more than one component at a time. Hold down the Control key to select multiple components.
➤ New trails are generated using the tweak values. If the original trail has been moved, the trails may not match. Changing a trail does not alter the path of a tweak.
➤ Presentation views do not recognize assembly constraints for any purpose other than creating the first automatic explosion. You can manually tweak (move) a component along any axis or rotate it.
➤ The first view that you add associates the presentation file to an assembly. You can add as many presentation views as needed. All measurements in the presentation file assume the same units as the selected assembly. To create several presentation views from the same design view, add the first view, and then use Copy and Paste to create additional copies. You can then modify the copies independently.

The Power of Design: An Introduction to Autodesk Inventor

Exercise 8-1:
Creating an Exploded View

File: New Presentation using Standard.ipn
Estimated Time: 15 minutes

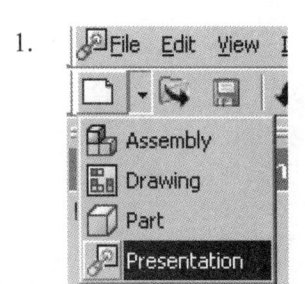

1. Select **Presentation** from the file drop down.

2. Select the **Create View** tool.

   The **Create View** tool prompts the user to select an assembly file. The assembly file is then linked to the presentation file. If you update the assembly, any changes will automatically be reflected in the presentation file.

3. Select the **Browse** button to select an assembly file.

4. Browse to the *Scissors* folder under Inventor.

5. Open the *scissors.iam* file.

8-2

Presentations

6. Select the **Manual** method.

Press **OK**.

7. Your assembly appears in the graphics window.

8. Highlight the **Explosion** in the Browser.
Right-click and select **Copy**.

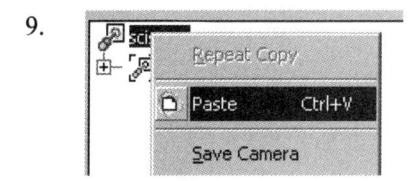

9. Highlight the top assembly.
Right-click and select **Paste**.

10. We have two exploded views available in the presentation file.

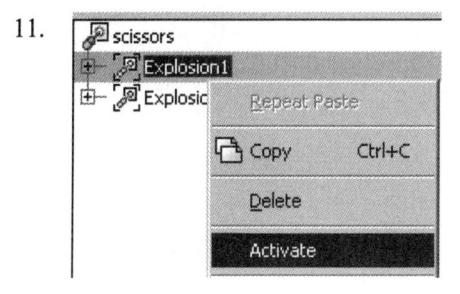

11. Select the first exploded view.
Right-click and select **Activate**.

8-3

The Power of Design: An Introduction to Autodesk Inventor

12. The first explosion will be used to create an exploded view. The second explosion will be used to create an animation showing how the assembly works.

13. Select the **Tweak Components** tool.

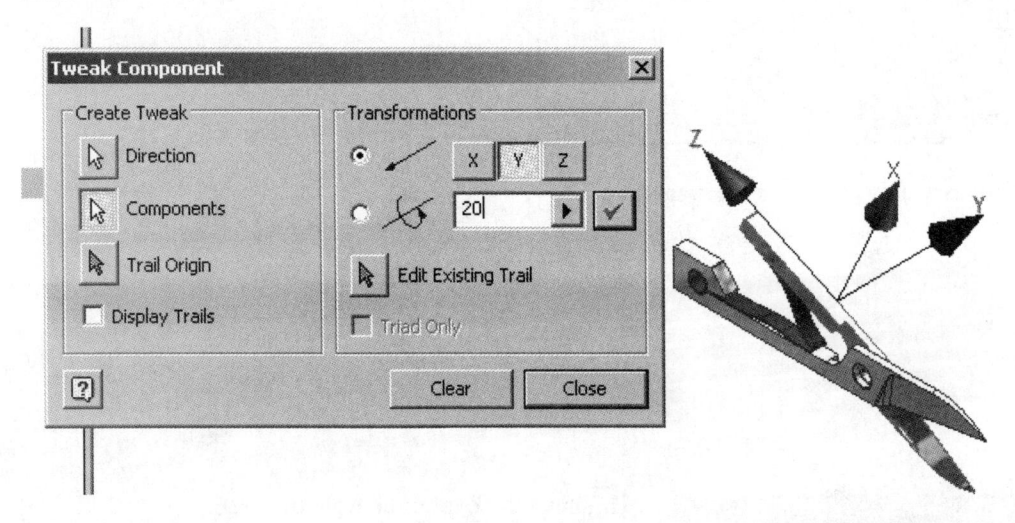

Select the top blade; you can see how the UCS is orientated. You can determine which axis you should select.

14. Select the **Y** axis.
    Select the top blade.
    Set the **Distance** to **20**.
    Select the **green check** button to apply the tweak.

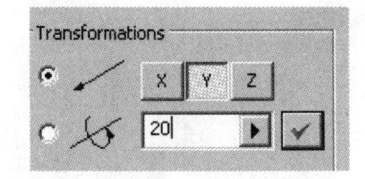

15. Press **Clear** to reset the Tweak dialog.
    This will allow you to select the next component if you wish to continue adding tweaks.

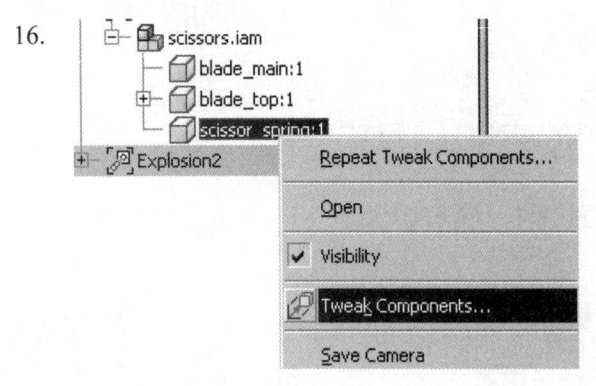

16. Highlight the **scissors_spring** in the Browser.

    Right-click and select **Tweak Components**.

8-4

Presentations

17.

Move the spring over.
To do this, pick an axis and simply drag the bolt into the desired position.

Press the check mark to apply the tweak.

Press **Close**.

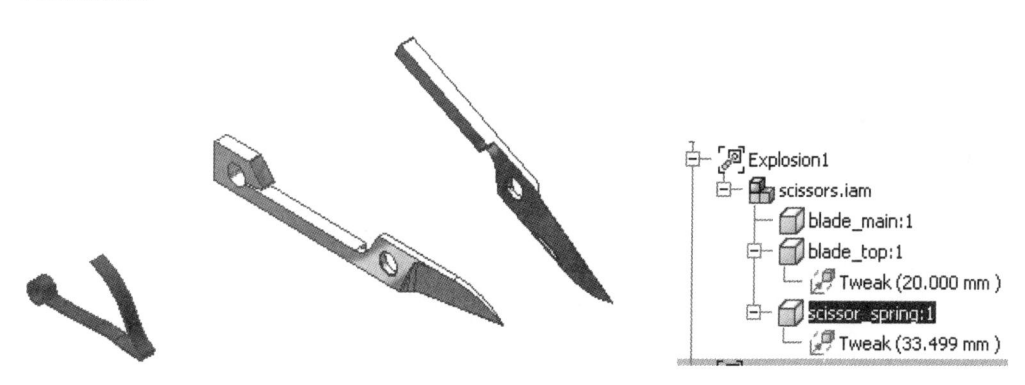

The tweaks are listed in the Browser.

Your tweak values may be different depending on how you applied the tweak.

If you highlight a tweak in the Browser, the tweak value will appear in an edit field. You can then modify the value if needed to adjust the tweak.

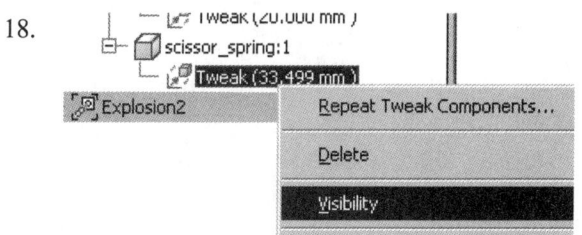

18.

Highlight one of the tweaks in the Browser.

Right-click and select **Visibility**.

8-5

19. The trail is now visible.

20. Highlight the trail in the display window.

    Right-click and select **Hide Trails**.

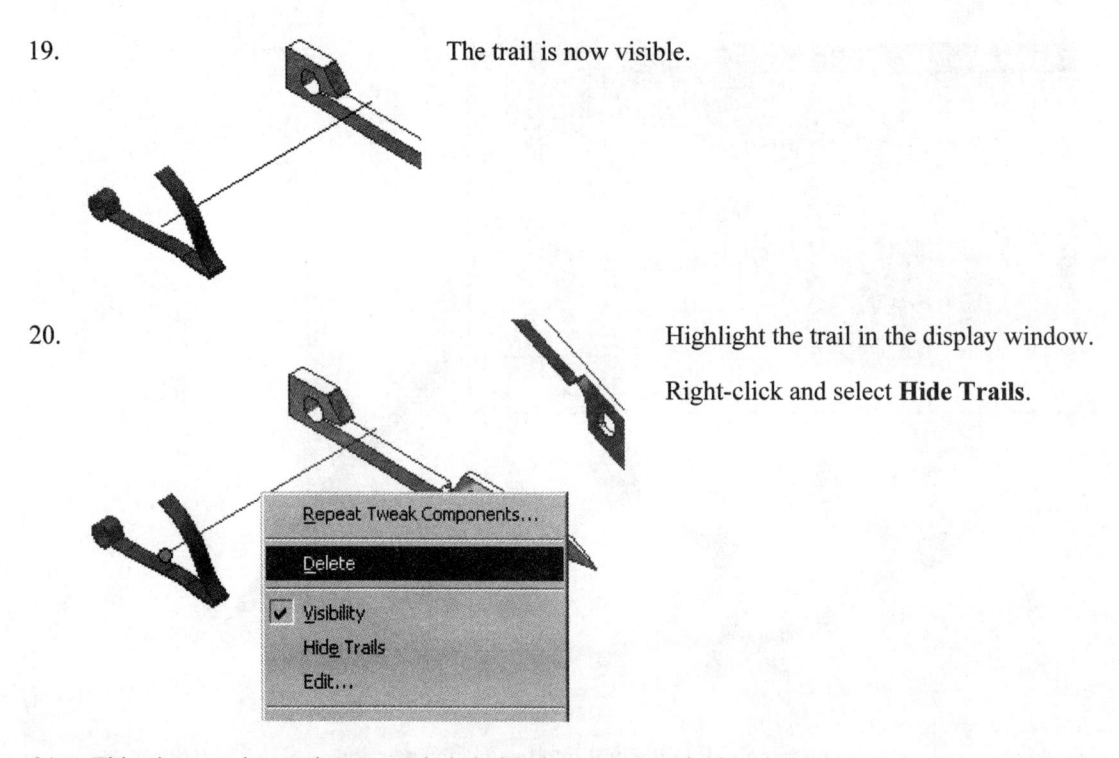

21. This view can be used as an exploded view in our assembly drawing.

    Save the file as *ex8-1.ipn*.

Presentations

Exercise 8-2:
Creating an Animation

File: Ex8-1.ipn
Estimated Time: 30 minutes

1. Select **Presentation** from the file drop down.

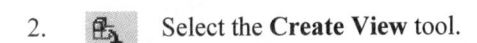

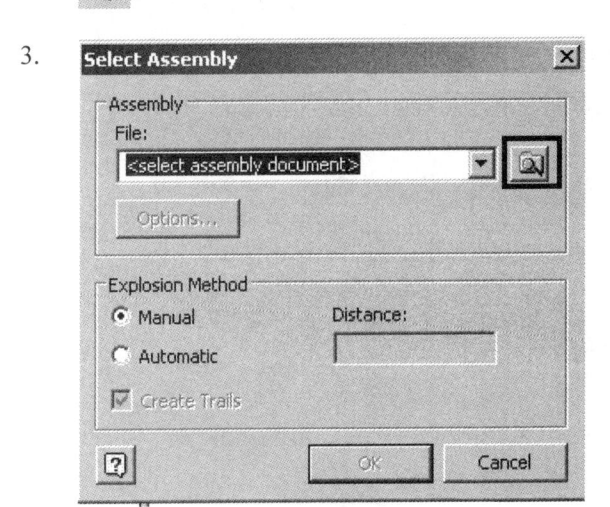

2. Select the **Create View** tool.

3. Select the **Browse** button to select an assembly file.

4. Browse to the *Arbor Press* folder under Inventor.

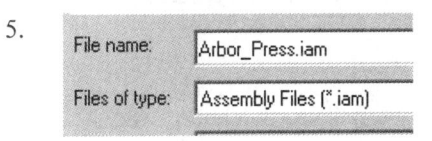

5. Open the *Arbor Press.iam* file.

8-7

6.

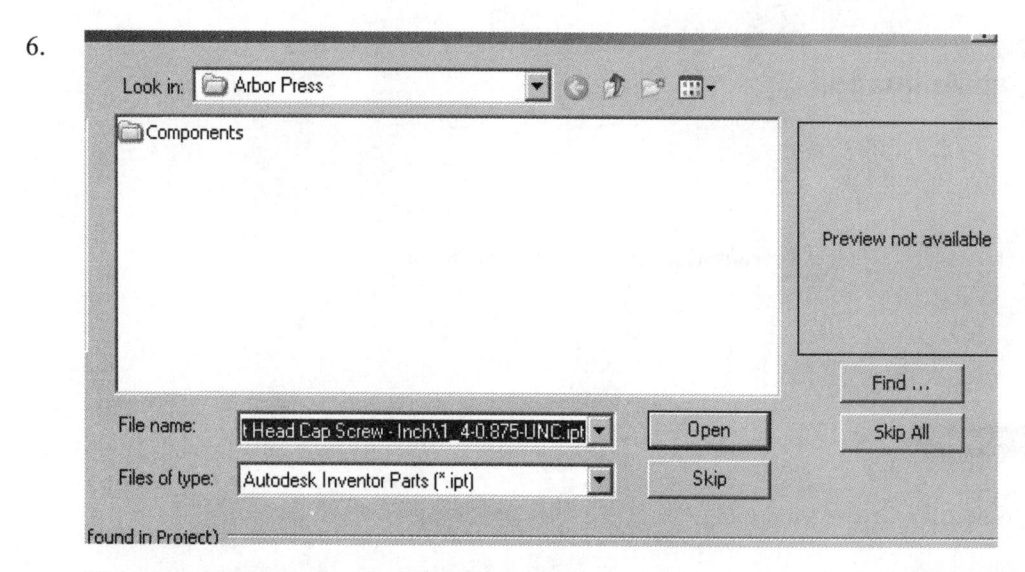

If you are prompted to resolve a link, press **Skip All**.

Inventor will search for any hardware components that were created using Content Center. If Content Center is not installed or not installed correctly, you will not be able to resolve the missing links.

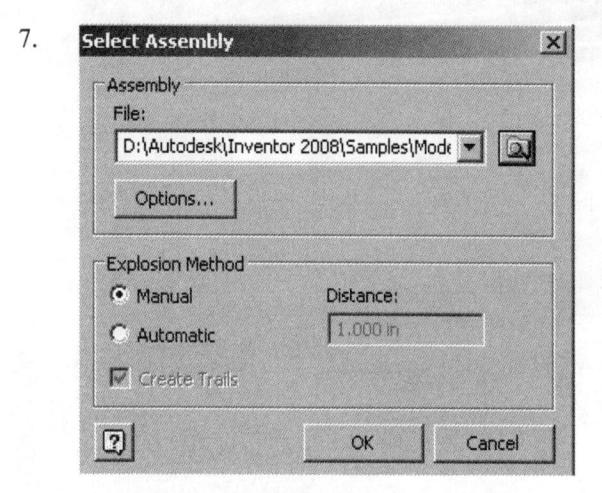

7. Select the **Manual** method.

   Press **OK**.

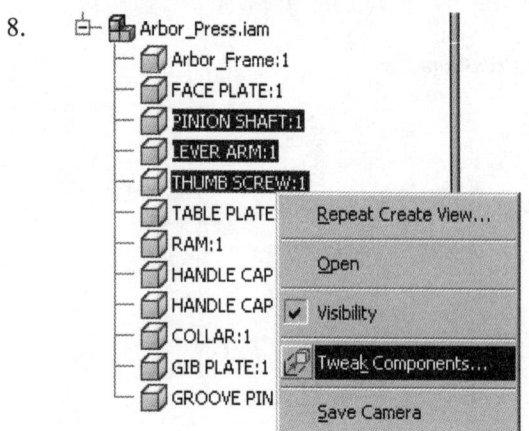

8. Highlight the PINION SHAFT, the LEVER ARM, and the THUMB SCREW in the Browser.

   Right-click and select **Tweak Components**.

9.

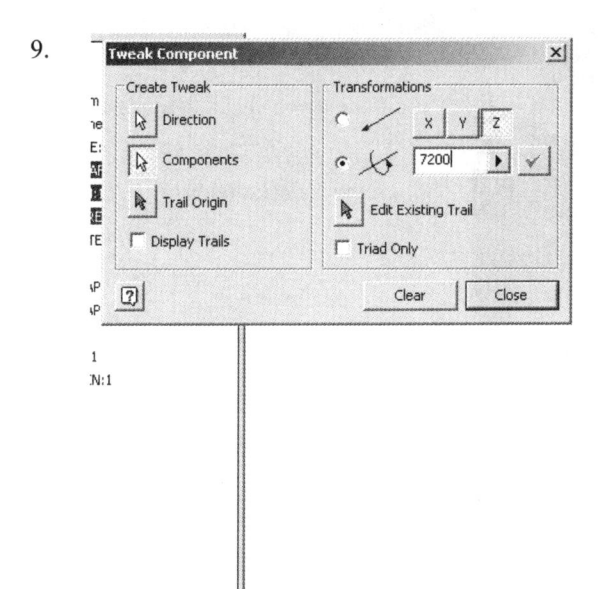

Pick the center of the circular face on the pinion shaft to locate the triad.
Select the **Z** axis.
Enable **Rotation**.
Set the Value to **7200**.
Press the green check to apply the tweak.

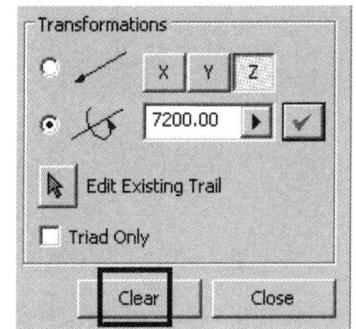

10. Press the **Clear** button to add another tweak.

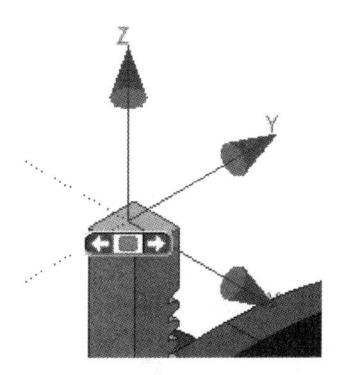

11. Select the top of the RAM.

Then select the ram so it is highlighted in the Browser.

12. Move the RAM down -1.5.

Press the green check to apply the tweak.

Close the Tweak dialog.

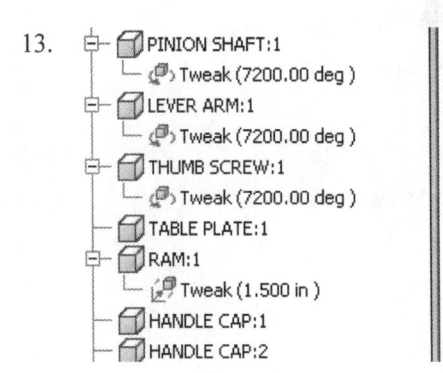

13. Verify the tweaks in the browser.

If there are any errors, delete or edit the tweaks.

- PINION SHAFT:1
  - Tweak (7200.00 deg)
- LEVER ARM:1
  - Tweak (7200.00 deg)
- THUMB SCREW:1
  - Tweak (7200.00 deg)
- TABLE PLATE:1
- RAM:1
  - Tweak (1.500 in)
- HANDLE CAP:1
- HANDLE CAP:2

14. Select the **Animate** tool.

15. Select the **More** button.

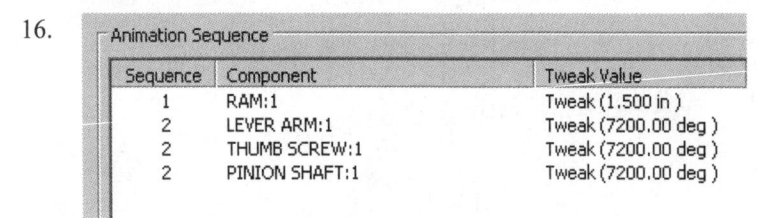

16. The components and their tweaks are displayed.

Note that some components are moved in step 1 and some are moved in step 2.

Sequence	Component	Tweak Value
1	RAM:1	Tweak (1.500 in)
2	LEVER ARM:1	Tweak (7200.00 deg)
2	THUMB SCREW:1	Tweak (7200.00 deg)
2	PINION SHAFT:1	Tweak (7200.00 deg)

17. Press the **Play** button.

18. We forgot to tweak the handle caps.

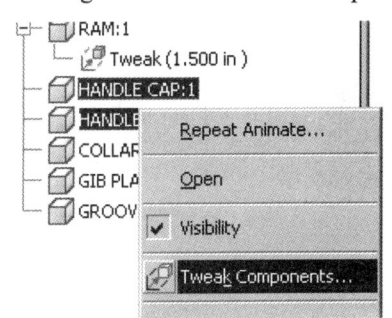

    Highlight the two HANDLE CAPs.

    Right-click and select **Tweak Components**.

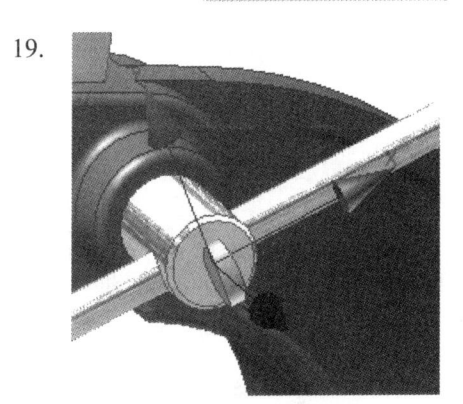

19. Select the center of the pinion shaft to place the UCS direction indicator.

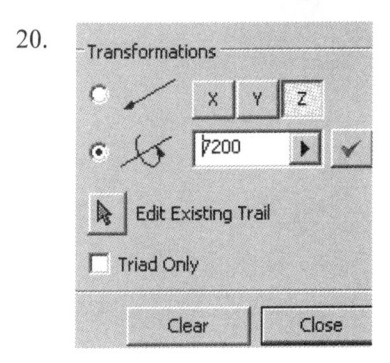

20. Select the **Z** axis.
    Enable **Rotation**.
    Set the **Value** to **7200**.
    Press the green check to apply the tweak.

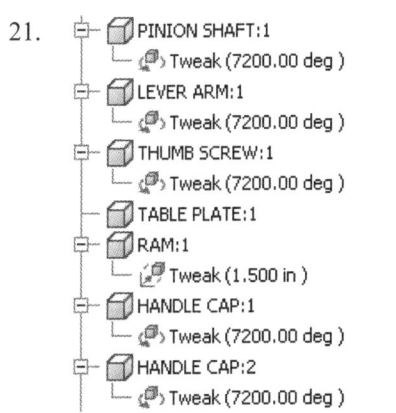

21. Verify the tweaks in the Browser.

    If there are any errors, delete or edit the tweaks.

22. Select the **Animate** button.

23. Select the **More** button.

The Power of Design: An Introduction to Autodesk Inventor

24.

Sequence	Component	Tweak Value
1	HANDLE CAP:2	Tweak (7200.00 deg)
1	HANDLE CAP:1	Tweak (7200.00 deg)
2	RAM:1	Tweak (1.500 in)
3	LEVER ARM:1	Tweak (7200.00 deg)
3	THUMB SCREW:1	Tweak (7200.00 deg)
3	PINION SHAFT:1	Tweak (7200.00 deg)

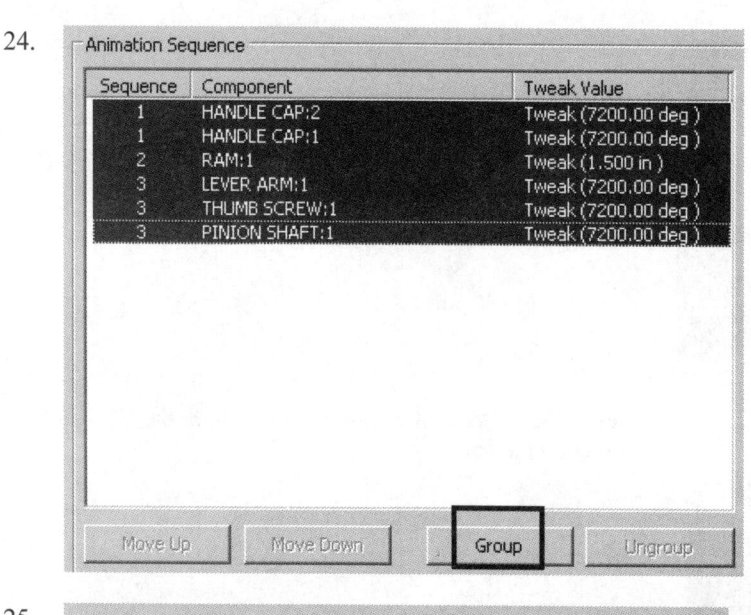

Highlight all the components.

Press the **Group** button.

This will set it so that all the components move together.

25.

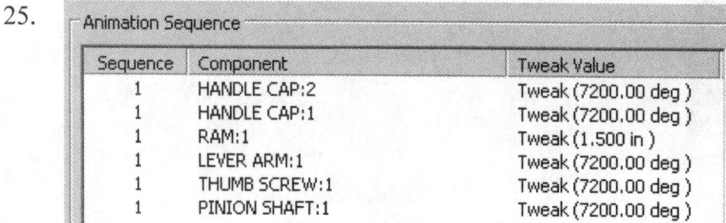

Sequence	Component	Tweak Value
1	HANDLE CAP:2	Tweak (7200.00 deg)
1	HANDLE CAP:1	Tweak (7200.00 deg)
1	RAM:1	Tweak (1.500 in)
1	LEVER ARM:1	Tweak (7200.00 deg)
1	THUMB SCREW:1	Tweak (7200.00 deg)
1	PINION SHAFT:1	Tweak (7200.00 deg)

Note that the sequence for the components is now the same number.

26.  Press **Apply**.

27. Press the **Play** button to see how the assembly moves.

28.  Use the **Reset** button to re-initialize your animation.
Close the dialog.

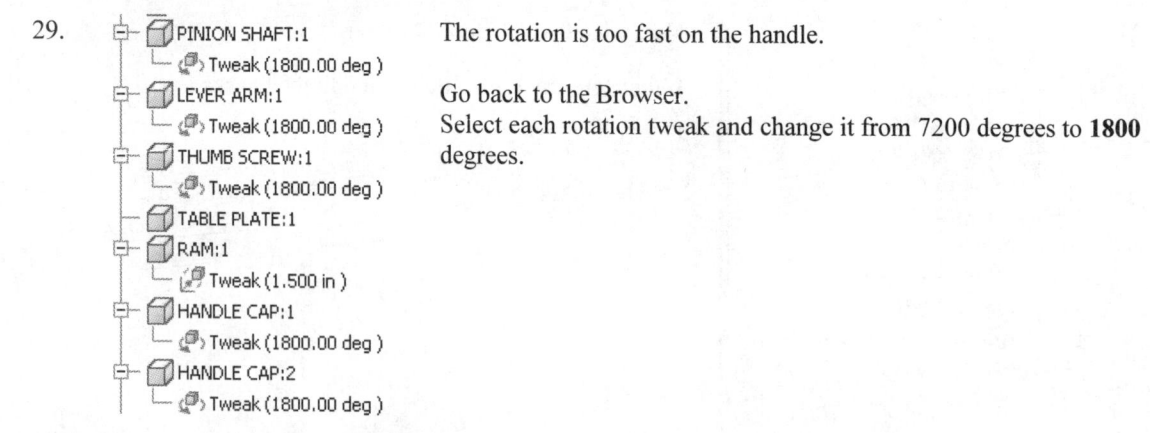

29.
- PINION SHAFT:1
  - Tweak (1800.00 deg)
- LEVER ARM:1
  - Tweak (1800.00 deg)
- THUMB SCREW:1
  - Tweak (1800.00 deg)
- TABLE PLATE:1
- RAM:1
  - Tweak (1.500 in)
- HANDLE CAP:1
  - Tweak (1800.00 deg)
- HANDLE CAP:2
  - Tweak (1800.00 deg)

The rotation is too fast on the handle.

Go back to the Browser.
Select each rotation tweak and change it from 7200 degrees to **1800** degrees.

30.  Select the **Animate** button.

8-12

31. Set the Interval to **100**.

32. Press the **Play** button to see how the assembly moves.

33. Save the file as *ex8-2.ipn*.

TIP: Turn off the visibility of the 3D Indicator before you create your animation to make your animation look more professional. This is controlled under **Tools→Application Options** in General tab.

The Power of Design: An Introduction to Autodesk Inventor

Exercise 8-3:
Recording an Animation

File: Ex8-2.ipn
Estimated Time: 30 minutes

1. Open *ex8-2.ipn*.

2. Select the **Animate** tool.

3. Set the **Reset** button to make sure the assembly is set at interval 0.

4. Enable **Minimize dialog during recording** to hide the animation dialog during the video.

5. Select the **Record** button.

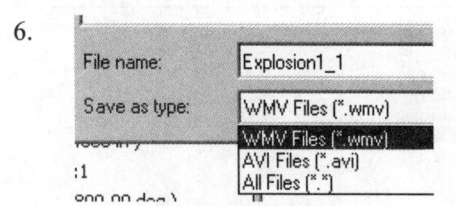

6. Animation files can be saved as *avi* or *wmv* file formats.

7. Name your animation *press.avi*.
   Press **Save**.

8. Set the **Compressor** to **Microsoft Video 1**.

   This ensures that the avi you create playable using Microsoft software.

   Press **OK**.

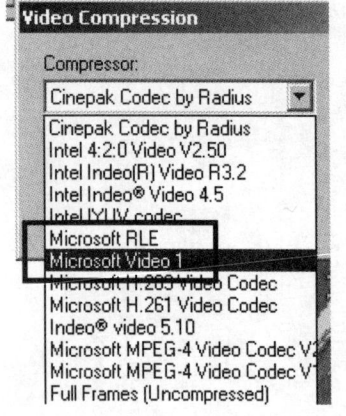

9. Press the **Play** button.

10. When the bolt reaches the end of the travel, select the **Record** button again to stop the recording.

11. Locate the **press.avi** file using Windows Explorer.

8-14

Presentations

12. Double-click on the file to play the animation.

13. Save as *ex8-3.ipn*.

Exercise 8-4:
Changing Views in an Animation

File: Ex8-3.ipn
Estimated Time: 30 minutes

1. Open *ex8-3.ipn*.

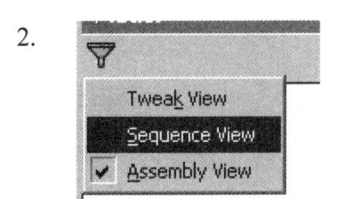

2. Under the **Filter** icon drop-down, we can select three different view styles for our Browser.

   Select **Sequence View**.

3. Under the **Explosion**, you should see one **Task** in the Browser and one **Sequence**.

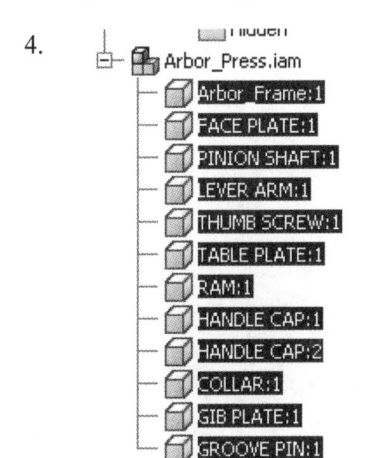

4. Highlight all the components.

8-15

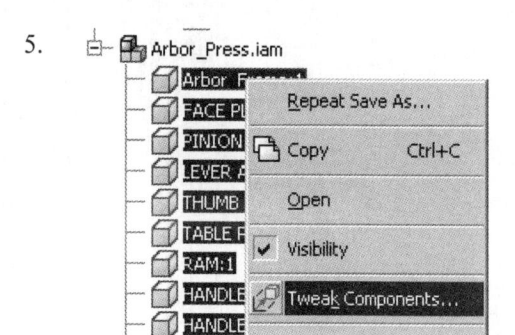

5. Right-click and select **Tweak Components**.

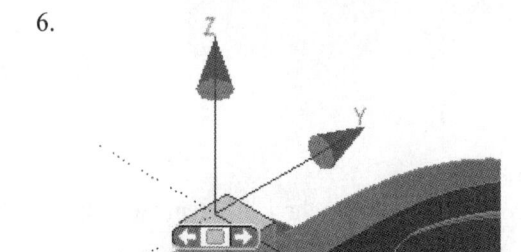

6. Place the directional UCS on top of the RAM.

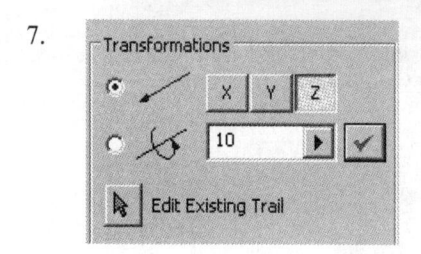

7. Add a tweak in the **Z** direction of **10** units.

   Press the green check button to accept.

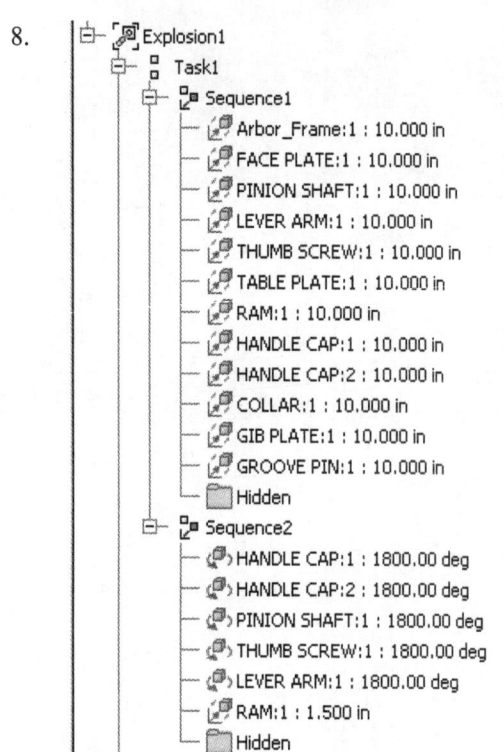

8. We now have two sequences... one where the entire assembly moves, and one with the motion added in the previous lessons.

8-16

Presentations

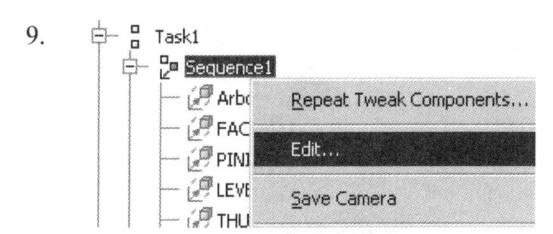

9. Highlight **Sequence 1** in the Browser.
Right-click and select **Edit**.

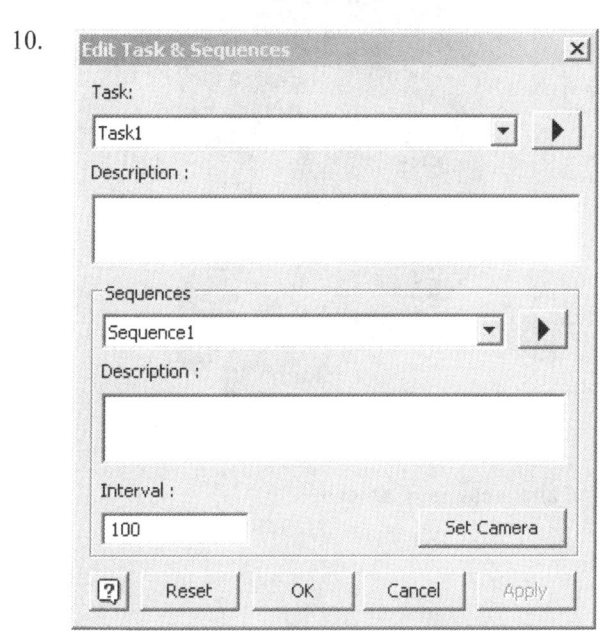

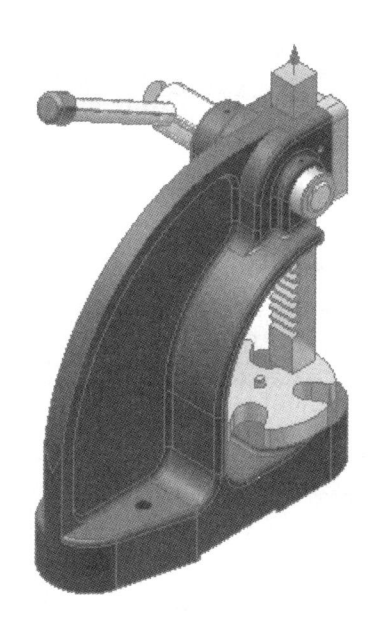

10.

Rotate the view so you see the back view of the assembly.
Press **Set Camera**.
Press **Apply**.

11. Select **Sequence2** from the **Sequences** drop-down.

8-17

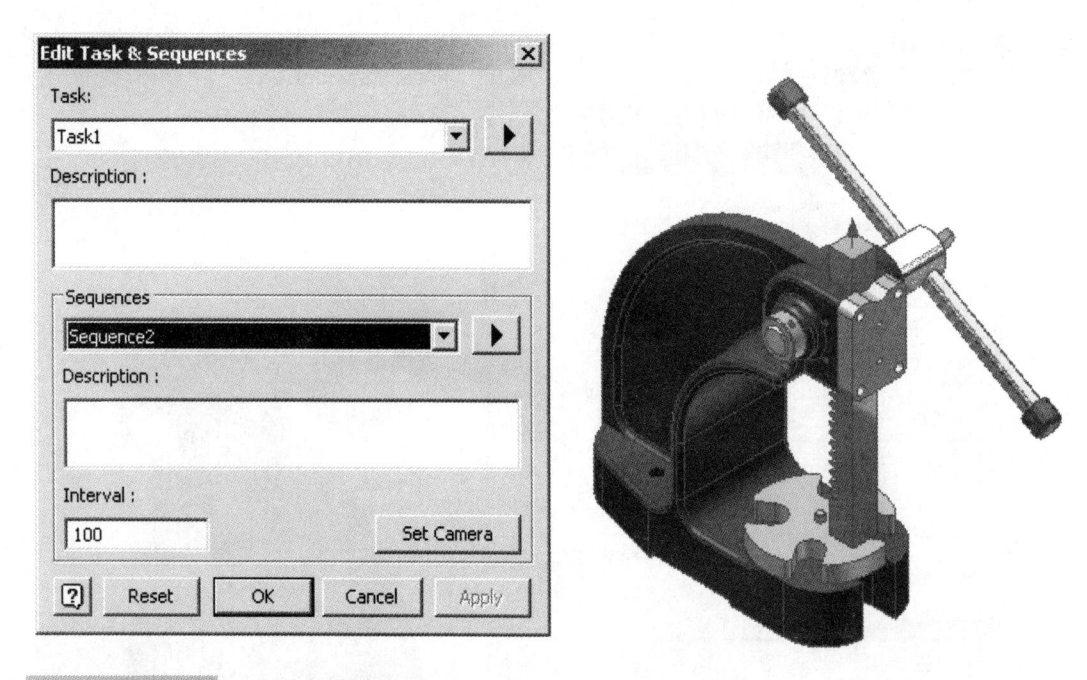

12.

Turn the view back around and press **Set Camera**.

Press **Apply**.

13. Select the **Animate** tool.

14. Press **Play Forward**.

You see that your camera views now change with each sequence.

15. Save as *ex8-4.ipn*.

Presentations

Review Questions

1. Identify the tool shown.

    A. Move Component
    B. Rotate Component
    C. Add a Tweak
    D. Animate

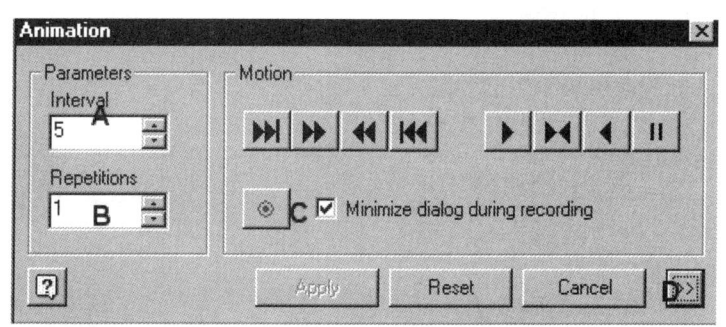

2. Select here to speed up or slow down an animation.

3. Select here to Record an Animation

4. Select here to see all the tweaks added.

5. Identify the tool shown.

    A. Rotate component
    B. Animate Presentation
    C. Precise View Rotation
    D. Move component

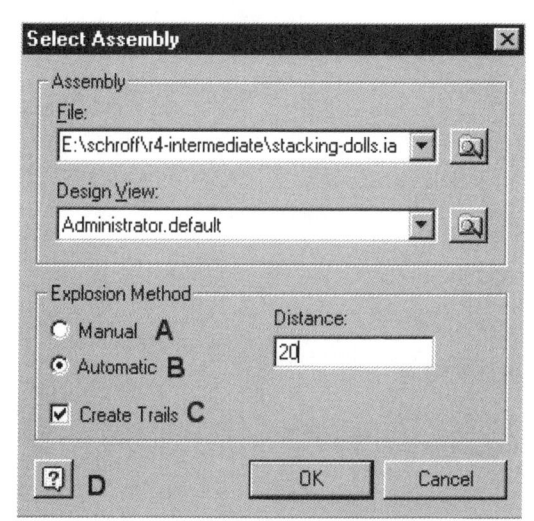

6. Select here to bring up HELP on Presentations.

7. Select here to automatically place Trails.

8-19

The Power of Design: An Introduction to Autodesk Inventor

8. Name the toolbar shown:

   A. Animation
   B. Scene
   C. Assembly
   D. Presentation

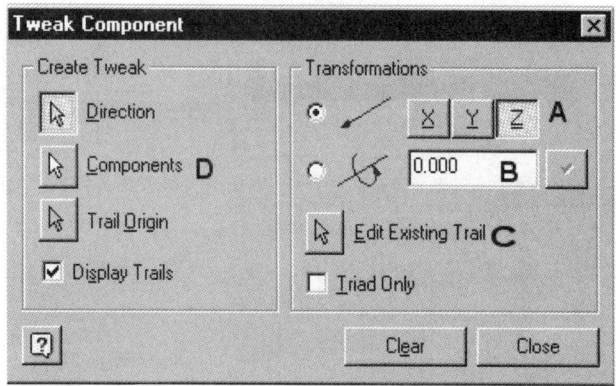

9. To add a rotating motion to a component, select here.

10. To select the axis the component will move along for a tweak, select here.

11. Presentation files are linked to Part files.

    A. True
    B. False

12. To speed up an animation, increase the number of Intervals.

    A. True
    B. False

13. When creating an Animation, there is no need to Reset between Plays.

    A. True
    B. False

14. The tool shown is:

    A. Animate
    B. Mickey Mouse
    C. Movie
    D. Camera

ANSWERS: C; 2) A; 3) C; 4) D; 5) C; 6) D; 7) C; 8) D; 9) B; 10) A; 11) B; 12) B; 13) B; 14) A

About the Author

Elise Moss has worked for the past twenty years as a mechanical designer in Silicon Valley, primarily creating sheet metal designs. She has written articles for Autodesk's Toplines magazine and AUGI's PaperSpace. She is President of Moss Designs, creating custom applications and designs for corporate clients. She has taught at DeAnza College, Evergreen Valley College and at San Francisco State University's CEL. She holds a BSME from San Jose State University.

She has been married more than twenty-five years to Ari Stassart, a computer scientist. They have three sons. Benjamin is an electrical engineer at a Fortune 500 firm in Silicon Valley. Daniel is a project manager for a local construction firm. Isaiah is still a student.

She can be contacted via email at elise_moss@mossdesigns.com

More information about the author and her work can be found on her website at www.mossdesigns.com.

More books from SDC Publications by Elise Moss:

Autodesk AutoCAD 2008: Fundamentals

Table of Contents

1. The AutoCAD Environment
2. Zooming and Panning
3. Drawing Lines
4. Toolbars and Profiles
5. Draw and Modify Commands
6. Dimensions
7. Object Properties
8. Drafting Settings and Object Snaps
9. Titleblocks and Templates
10. Text Tools
11. Viewports and Layouts

551 Pages
ISBN: 978-1-58503-344-7

Description

This is a basic training course for new AutoCAD users. It is geared for students and professionals with little or no prior experience using AutoCAD. The course will cover fundamental skills necessary for effectively using AutoCAD and will provide a strong foundation for advancement.

This text goes beyond just introducing new users to AutoCAD, it also fills in many important gaps. The major gap is applying the knowledge of AutoCAD to mechanical drafting. Knowing how to draw a line in AutoCAD is not the same as understanding which line type is required when creating technical drawings. This text provides the necessary information on how to use AutoCAD as a tool to work as a mechanical drafter or designer.

More Information

For a complete Table of Contents and to download a sample chapter please visit our website at www.schroff.com.

More books from SDC Publications by Elise Moss:

AutoCAD Architecture 2008 Fundamentals

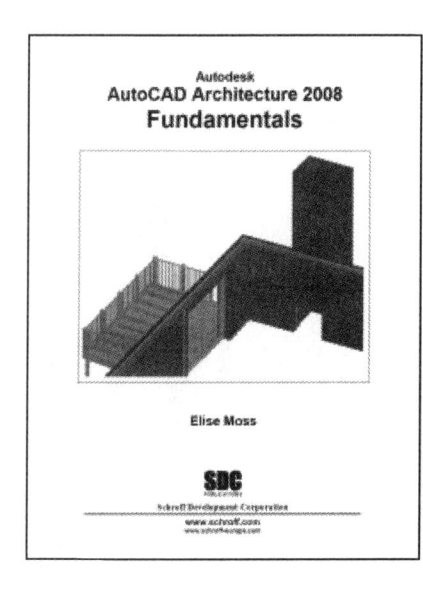

Table of Contents

1. Desktop Features
2. Site Plans
3. Floor Plans
4. Space Planning
5. Roofs
6. Structural Members
7. Layouts
8. Dimensioning
9. Schedules
10. Creating a Video

299 Pages
ISBN: 978-1-58503-358-4

Description

For users familiar with Architectural Desktop, the 2008 release of AutoCAD Architecture marks a serious advance in the software's features. This textbook is meant for beginning users who want to gain a familiarity with the tools and interface of AutoCAD Architecture before they start exploring on their own. By the end of the text, users should feel comfortable enough to create a standard model, and even know how to customize the interface for their own use. Knowledge of basic AutoCAD and its commands is helpful, but not required.

This Fundamentals text introduces the beginning student or professional to Autodesk's AutoCAD Architecture software. The text covers the Layer Manager, Design Center, Structural Members, Doors, Windows, and Walls. Step by Step lessons take the reader from creation of a site plan, floor plan, space planning, elevations, all the way through to creating a video of the finished building - a standard three bedroom, two bathroom residence. The reader is provided with in-depth coverage of toolbars, dialog boxes and commands.

More Information

For a complete Table of Contents and to download a sample chapter please visit our website at www.schroff.com.

NOTES:

NOTES:

NOTES:

NOTES:

NOTES:

NOTES:

NOTES: